nothing more recent 11/94

Diffuse Reflectance Spectroscopy in Environmental Problem-Solving

Author:

R.W. Frei
Head of Analytical Research
Sandoz, Ltd.
Basle, Switzerland

Co-author:

J.D. MacNeil
Research Station
Canada Department of Agriculture
Summerland, B.C.

published by:

A division of The Chemical Rubber Co. 18901 Cranwood Parkway, Cleveland, Ohio 44128

International Standard Book Number 0-87819-022-8

Library of Congress Card Number 73-78163

THE EDITORS

Roland W. Frei, Ph.D., is the head of analytical research for Sandoz, Ltd., Basle, Switzerland.

Dr. Frei obtained his Ph.D. in Analytical Chemistry from the University of Hawaii, Honolulu in 1964.

Prior to his present appointment, Dr. Frei worked as a chemical engineer for Union Carbide in Geneva. He has served as an instructor in physical science and chemistry in Pago Pago, American Samoa; Tokyo, Japan; Halifax, Nova Scotia; and Lausanne, Switzerland.

Dr. Frei has filled appointments as Coordinator of the Association of Atlantic Universities' Science Branch (1966-68); member of the Executive of the Canadian Institute of Chemistry's Analytical Chemistry Division (1969); editor-in-chief of *Environmental Analytical Chemistry* (1970); vice-chairman of Analytical Division of Chemical Institute of Canada (1972); member of "Commission for Water Quality" of International Assoc. of Hydrological Scientists. He has also served as a consultant to various corporations.

Dr. Frei is a full member of Sigma Xi and Phi Kappa Phi Honor Societies and of the American Chemical Society, The Swiss Chemical Society, and the Canadian Institute of Chemistry.

He has published over 100 papers including three chapters on reflectance spectroscopy in edited volumes.

James Daniel MacNeil, M.Sc., Ph.D., is a research scientist in environmental chemistry for the Canada Department of Agriculture at Research Station, Summerland, British Columbia.

Dr. MacNeil obtained his B.Sc. (1966) and his M.Sc. (1969) from St. Francis Xavier University, Antigonish, Nova Scotia. He received his Ph.D. (1972) from Dalhousie University, Nova Scotia where Dr. R. W. Frei was his research supervisor.

He has worked as an analyst with the Nova Scotia Research Foundation.

Dr. MacNeil has published 8 papers and has 6 currently in press. His subjects include thin-layer chromatography of pesticides and indolic compounds with quantitative analysis by reflectance or fluorescence techniques and high-speed liquid chromatography of ergot alkaloids.

PREFACE

Diffuse reflectance spectroscopy is rapidly gaining in acceptance as a technique not only for surface studies in physical chemistry but also in inorganic and organic chemistry and, especially, in analytical chemistry. It is particularly suited for the quantitative analysis of compounds separated on thin-layer chromatograms and for experiments involving materials adsorbed on surfaces. The former is the more recent application of the technique and has led to the development of instruments designed especially for this purpose.

The main subject divisions of this book include the theory of diffuse reflectance spectroscopy; measurement and standardization of diffuse reflectance; instrumentation; application to color measurement and physical, inorganic, and organic chemistry; and applications in chromatographic analysis.

While the use of reflectance spectroscopy dates from the 1920's, it has only been in the last decade that its analytical potential has been developed. Interestingly, much of the early research involved industrial uses where measurement of color was required. The development and acceptance of thin-layer chromatography has opened up new areas of analysis for the application of this technique.

It is not the purpose of this book to delve deeply into the theoretical aspects of reflectance spectroscopy, as this has already been done in several previous books. Insofar as it is possible, this book is an up-to-date guide to instruments and techniques intended primarily for the chemical analyst, though it is hoped that it may contain information of interest to other scientists. The potential for the application of this technique is great and the authors feel confident that the coming decade will see many interesting developments in this type of spectroscopic study, particularly in the field of analysis.

TABLE OF CONTENTS

Chapter 1

INTRODUCTION

Transmission spectroscopy is the technique that takes advantage of the transparent nature of samples to measure their spectral properties. With this method it is possible to measure the specific absorption of substances dissolved in a suitable solvent by transmitting monochromatic light through the resulting solution. This approach becomes difficult when applied to turbid or colloidal systems where substantial amounts of energy are lost due to light-scattering processes. Transmission measurements become completely unsuitable or even impossible to determine when absorption spectra of substances embedded or adsorbed on solid surfaces are to be measured. It is for problems of this nature that a solution can be provided by the technique of reflectance spectroscopy. Instead of measuring the light flux that is transmitted through a medium as a corresponding percent transmittance on the instrument meter, one measures the amount of light reflected from the sample surface with the aid of a suitably modified transmission spectrophotometer. The data are reported as percent reflectance (%R) read on the transmittance scale of commercial instruments and correspond to $R = I/Io$ (where Io is the intensity of incident radiation and I is the intensity of radiation reflected from the medium). Compared to transmission spectroscopy, reflectance spectroscopy is still very much in its infancy. The technique is, however, not as new as one might expect, particularly if one is speaking about diffuse reflectance spectroscopy, which deals with the measurement of light reflected in a diffuse manner (in contrast to specular or direct reflectance). The paper, paint, dye, textile, printing, and ceramics industries, for example, have long made use of diffuse reflectance methods for measurement of color in routine quality-control processes.

Initial attempts to use diffuse reflectance spectroscopy as a potentially quantitative technique date back to about 1920 and resulted in the design by Taylor[1] and Benford[2] of the first relatively sophisticated filter-type reflectometers. A few years later Hardy[3] developed a recording spectrophotometer-type reflectometer. These instruments had one common design feature, viz., they all used an integrating sphere (or "Ulbrichtkugel"[4]), which permitted the collection of diffusely reflected light and, in addition, either the exclusion of the generally undesirable specular component or alternatively, its inclusion to give total reflectance.

It was not until 1961[5] that specular reflectance techniques, also known as "Internal Reflectance" or "Attenuated Total Reflectance" (ATR), emerged as an analytical tool of great potential. Specular reflectance methods will not be discussed in this book and the various terms used, such as reflectance spectroscopy, spectral reflectance, or reflectance techniques will refer only to diffuse reflectance spectroscopy unless otherwise specified.

As indicated earlier, a number of industries were among the first users of diffuse reflectance spectroscopy, and the first book on the subject of color measurements appeared in 1936.[6] The most authoritative text has been written by Judd and Wyszecki;[7] it contains many references to this technique as used in the industries already mentioned. Recently, Billmeyer and Saltzman[8] published an excellent treatise on the principles of color measurement. This text gives a good introduction to the field, as well as an up-to-date treatment of the recent developments in color-measurement theory, instrumentation, and techniques. In analytical chemistry, however, there has been until recently little interest in the application of diffuse reflectance spectroscopy to the solution of qualitative or quantitative analytical problems. The situation now is rapidly changing due to the ready availability of diffuse reflectance attachments for practically all commercial spectrophotometers. This change, in turn, has been reflected in a rapid growth of the literature in the field. The most prominent contribution to the popularity of diffuse reflectance techniques in the field of chemistry, particularly for investigations of adsorption phenomena and the study of kinetics and equilibria on surfaces, must be credited to Kortüm. His work recently has culminated in a book[9] that is also available in an English translation. The first comprehensive treatment on the subject of reflectance spectroscopy was written by Wendlandt and Hecht in 1966.[10] Both these texts include a discussion of specular reflectance techniques. Also of interest are the published proceedings of the 1967 American Chemical Society Symposium on Reflectance Spectroscopy.[11]

Reflectance spectroscopy has received a continuing increase in attention during the past decade. It is predicted that this trend will continue during the 1970's and, in particular, that diffuse reflectance spectroscopy will become an increasingly useful tool in the field of chromatography.[12]

REFERENCES

1. **Taylor, A. H.,** *J. Opt. Soc. Am.,* 4, 9 (1919).
2. **Benford, F.,** *Gen. Elec. Rev.,* 23, 72 (1920).
3. **Hardy, A. C.,** *J. Opt. Soc. Am.,* 18, 96 (1929).
4. **Ulbricht, T.,** *Elektrotech. Z.,* 21, 595 (1900).
5. **Fahrenfort, J.,** *Spectrochim. Acta,* 17, 698 (1961).
6. **Hardy, A. C.,** *Handbook of Colorimetry,* M.I.T. Press, Cambridge, Mass., 1936.
7. **Judd, D. B. and Wyszecki, G.,** *Color in Business, Science and Industry,* 2nd ed., John Wiley & Sons, New York, 1963.
8. **Billmeyer, F. W., Jr. and Saltzman, M.,** *Principles of Color Technology,* John Wiley & Sons, New York, 1966.
9. **Kortüm, G.,** *Reflexionsspektroskopie,* Springer-Verlag, Berlin, 1969.
10. **Wendlandt, W. W. and Hecht, H. G.,** *Reflectance Spectroscopy,* John Wiley & Sons, New York, 1966.
11. **Wendlandt, W. W.,** Ed., *Modern Aspects of Reflectance Spectroscopy,* Plenum Press, New York, 1968.
12. **Frei, R. W.,** in *Progress in Thin-Layer Chromatography and Related Methods,* Vol. II, Niederwieser, A. and Pataki, G., Eds., Ann Arbor Science Publishers, Ann Arbor, Mich., 1971.

Chapter 2

THEORY

1. The Kubelka-Munk Theory

The most generally accepted theory concerning diffuse reflectance and the transparency of light-scattering and light-absorbing layers has been developed by Kubelka and Munk.[1,2] It may be said that this theory has acquired an importance in the field of reflectance spectroscopy comparable to the importance of the Bouger-Beer law in transmittance spectroscopy. Development of the theory for an infinitely thick opaque layer yields the Kubelka-Munk equation, which may be written as

$$\frac{(1 - R'_\infty)^2}{2R'_\infty} = \frac{k}{s} \qquad (2.01)$$

where R'_∞ is the absolute reflectance of the layer, k is its molar absorption coefficient, and s is the scattering coefficient. A derivation of the equation is presented in the Appendix.

Instead of determining R'_∞, however, it is customary to work with the more convenient relative diffuse reflectance, R_∞, which is measured against a standard such as MgO or $BaSO_4$. In these cases it is assumed that the k values for the standards are zero and that their absolute reflectance is one. However, since the absolute reflectance of the standards exhibiting the highest R'_∞ values never exceeds 0.98 to 0.99, one is actually dealing in such instances with the relationship

$$\frac{R'_\infty \text{ sample}}{R'_\infty \text{ standard}} \equiv R_\infty \qquad (2.02)$$

and it is essential to specify the standard used. If this expression is introduced into the initial equation it will assume the form

$$F(R_\infty) \equiv \frac{(1-R_\infty)^2}{2R_\infty} = \frac{k}{s}, \qquad (2.03)$$

which indicates that a linear relationship should be observed between $F(R_\infty)$ and the absorption coefficient k, provided s remains constant. S is rendered independent of wavelength by using scattering particles whose size is large in relation to the wavelength used.

When the reflectance of a sample diluted with a non- or low-absorbing powder is measured against the pure powder, the absorption coefficient k may be replaced by the product $2.30\epsilon C$, where ϵ is the extinction coefficient and C is the molar concentration.[3] The Kubelka-Munk Equation 2.03 can then be written in the form

$$F(R_\infty) \equiv (1-R_\infty)^2/2R_\infty = C/k' \qquad (2.04)$$

where k' is a constant equal to $s/303\epsilon$. Since $F(R_\infty)$ is proportional to the molar concentration under constant experimental conditions, the Kubelka-Munk relationship is analogous to the Beer-Lambert law of absorption spectrophotometry. At high enough dilutions, the regular reflection from the sample approximates that from the standard and is thus canceled out in any comparison measurement.

A straight-line relationship between $F(R_\infty)$ and C is only observed, however, when dealing with weakly absorbing substances and only when the grain size of the powders used is relatively small (ideally around 1 μ in diameter).[4-6] Furthermore, any significant departure from the state of infinite thickness of the adsorbent layer assumed in the derivation of the Kubelka-Munk equation results in background interference, which in turn is responsible for nonideal diffuse reflectance. When either adsorbents with a large grain size or large concentrations of the absorbing species are used, plots of $F(R_\infty)$ vs. concentration deviate from straight lines in that there is a decrease in slope at higher concentrations.

In his explanation of this phenomenon, Kortüm[4-7] postulates that the reflected radiation is the result of both regular and diffuse reflectance. The first can be described as a mirror reflection, whereas the second occurs when impinging radiation is partly absorbed and partly scattered by a system so that it is reflected in a diffuse manner, i.e., with no defined angle of emergence. Regular reflectance for cases involving normal incidence is described by the Fresnel equation

$$R_{reg} \equiv \frac{I_{refl}}{I_O} = \frac{(n-1)^2 + n^2k^2}{(n+1)^2 + n^2k^2} \qquad (2.05)$$

where k is the absorption coefficient and n is the refractive index. Diffuse reflectance is described by the Kubelka-Munk function given earlier. Since regular reflectance is superimposed on diffuse reflectance, a distortion of the diffuse reflectance spectrum results; this distortion is responsible for the anomalous relationship observed between $F(R_\infty)$ and k at high concentrations of the absorbing species. It is, therefore, essential to eliminate as far as possible the interference caused by regular reflectance, R_{reg}. This can be accomplished by selecting appropriate experimental conditions. Especially effective is the use of powders having a small grain size and the dilution of the light-absorbing species with suitable diluents.

Most equations derived by other investigators[7-10] have proved to be special cases or adaptations of the Kubelka-Munk equation. Kortüm and Vogel[7] have summarized the theory and the derivation of the Kubelka-Munk function for special cases involving infinitely thick opaque layers which, in the case of fine powders, would be those having a depth of about 1 mm. Judd and Wyszecki[10] have compiled many of the different forms of the Kubelka-Munk function and have pointed out some of their specific uses and applications.

2. Other Theories

The phenomenon of radiative transfer for diffuse incident radiation and isotropic scatter can be treated more rigorously than the Kubelka-Munk equations, particularly in cases where the refractive index of scattering particles is unity. Techniques for such treatment have been given by various authors.[11-13] Two of the models proposed for the explanation of reflectance phenomena deserve special mention. One is the reflectance model proposed by Johnson,[14] in which the powdered sample is approximated by a stack of parallel plates with thickness d and characterized by a constant index of refraction and surface reflectivity. All the energy radiated onto the plate is assumed to be absorbed or reflected, so that the reflected light would be the sum of the rays passing through n particles ($n = 0$ to ∞) by means of refraction or reflection at the particle-air interfaces. Phenomena such as multiple reflection and losses due to scattering processes are accounted for by a semiempirical approach; an adjustable parameter is introduced in the expression for the sum of the reflected rays. The shortcomings of the Johnson model, however, were particularly apparent with strongly absorbing samples or with surfaces with large particle size, due to an inadequate correction for the scattering and multiple reflection losses. In 1963, Melamed[15] proposed a model that explicitly includes multiple reflection. In this model the sum of transmitted, reflected, and scattered rays was considered in an assembly of particles of uniform size but arbitrary shape. Companion[16] discussed and compared the two models and the Kubelka-Munk theory on the basis of experimental data. From her work it can be concluded that the diffuse reflectance models proposed by Johnson and Melamed do not yield absolute absorption coefficients for powder bulk samples, but they do serve to explain the reflectance process. Neglect of variations in the index of refraction and surface reflectivity under highly absorbing conditions can contribute to this fact. Improvement of these models requires additional elaborate mathematical treatment, which does not seem justified for analytical purposes.

The author feels that the reflectance function obtained on the basis of the Kubelka-Munk theory is no more in error than the models discussed above, and the deviations encountered by the assumption of different scattering phase functions[17] are just as serious. Therefore, the use of the Kubelka-Munk theory for general analytical applications seems as appropriate now as it was in the early stages of development of diffuse reflectance spectroscopy.

3. Relevance of the Kubelka-Munk Theory to Chromatographic Systems

Since chromatography is becoming an increasingly interesting field for the application of diffuse reflectance spectroscopic techniques it seems appropriate to examine the validity of the Kubelka-Munk theory in relation to systems of this nature.

Commercially available TLC-grade adsorbents with particle sizes between 5 and 40 μ have been reported suitable for use as diluents in both visible and UV reflectance spectroscopic work.[18-20] With these powders, simple mixing of the sample in a small mortar produces a sufficiently homogeneous mixture of reproducible texture. The resulting system usually conforms to the Kubelka-Munk equation over a concentration range that has analytical use.[18-20]

Although Kortüm reported that a layer at least

1 mm thick of a powder with a particle diameter of about 1 μ is needed to achieve infinite layer thickness and hence adherence to the Kubelka-Munk function, it was found[18] that the portion of visible light that can penetrate a compressed layer of silica gel of 0.4-mm thickness is negligibly small for analytical purposes. Cells with white, gray, or black backing paper were used to investigate background interference; the reflectance values for silica gel samples packed in each of these cells in 0.4-mm-thick layers differed by no more than 1%. Similar experiments were carried out with Whatman No. 1 filter paper.[18] It was found that five layers were required to eliminate completely any background interference. This finding agrees well with data obtained by other workers.[21] Jork[19] has investigated the relationship between the layer thickness of the adsorbent and the amount of light transmitted through the layer. The UV region of the spectrum was included in this study (see Figure 2.01). With a silica gel layer of 300 μ thickness, for example, he found a transmission of 2.2% at 500 nm and of 1% at 300 nm. The amount of transmitted light ($\lambda > 300$ nm) at the more realistic plate-layer thickness of 160 μ is, however, already quite appreciable, i.e., > 5%. The use of white backing paper inserted behind the plate as a reflecting background has, therefore, been recommended if spectra are to be recorded directly from the plate and in the visible region of the spectrum.[22] This results in an increase of method sensitivity and a reduction of background noise. The light that would otherwise be absorbed by the dark background is now reflected back through the layer, which results in enhancement of the signal. The Kubelka-Munk theory can, however, hardly be expected to be valid with such an experimental setup, and approximations would have to be used with the original function.

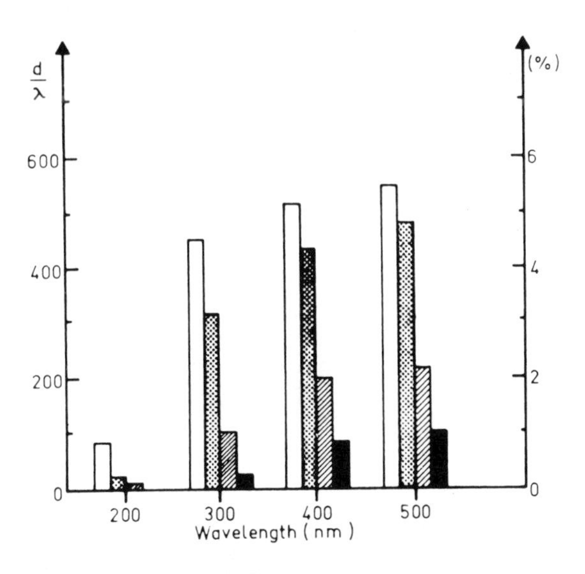

FIGURE 2.01 Determination of the degree of transmission (as a function of wavelength) with Silica Gel G layers of different thicknesses.[19] □ 130 μ; ▨ 160 μ; ▥ 300 μ; ■ 700 μ. (From Jork, H., *Z. Anal. Chem.*, 221, 17 (1966). With permission of Springer-Verlag.)

At first glance, the percentage of light transmitted in the UV region (see Figure 2.01: 1% with a 0.3-mm layer at 300 nm) may be surprisingly small since the radiation is more energetic at this wavelength. The reason for this phenomenon could be explained satisfactorily on the basis of the high self-absorption of UV light on the substrate. At a wavelength below 300 nm this self-absorption becomes appreciable for silica gel and cellulose, for example. An additional factor is the UV absorption of the glass support of the chromatogram. (UV-absorbing plastic materials are also used for commercially manufactured sheets.) All these factors often result in Kubelka-Munk calibration plots, which do not go through zero and deviate from linearity even at relatively low adsorbate concentrations. In these circumstances one tends to agree with Klaus[22,23] that the use of the Kubelka-Munk function is not justified and that simpler empirical functions resulting in linear calibration curves for a particular system and over a specific concentration range would be equally suitable.

Goldman and Goodall[24] have verified experimentally the validity of the Kubelka-Munk theory of radiative transfer in connection with a TLC system. They have derived a semiempirical relationship

$$0.434\ KX = 2\exp[-2A_0]\ (A + 0.4\ A^2), \qquad (2.06)$$

which was developed from a suitable form of the Kubelka-Munk equation for transmittance (see Appendix, Equation 1). In Equation 2.06, A is the absorbance of the light-absorbing species on the layer relative to the background absorbance A_0. 0.434 KX is proportional to the light-absorbing component per unit area (absorptivity x mg x cm^2); $[-2A_0]$ accounts for the background absorbance and $(A + 0.4\ A^2)$, for the curvature response.

The approach of Goldman and Goodall[24] is

actually based on two separable effects, "hyperchromaticity" and "curvature of response," and has been verified experimentally with a band of Sudan III dye on silica gel layers coated on a microscope slide. Figure 2.02 shows the familiar experimental curvature of the calibration curve if absorbance A (obtained by transmitting light through the layer) is plotted vs. a suitable form of the concentration. A reasonably linear curve is obtained by plotting the above-mentioned function f(A) in relation to KX. The same workers[25] also proposed a method for the direct analysis of chromatographic spots with computer evaluation of the measurements based on the same theory.

Certain rules have to be observed if the Kubelka-Munk equation is to be used under meaningful conditions, no matter whether the spots are investigated in the transmission or reflectance mode. One condition (sufficient monochromaticity of the light source) is easily fulfilled with most monochromator instruments. With filter instruments, the choice of interference filters that enable the isolation of a narrow slice from the spectrum (< 20 nm) is very important.

Another requirement is that the absorption be adequately uniform over the illuminated area. This is difficult to achieve in a chromatographic zone, since the distribution of the substance over the spot area is rather irregular. This is particularly so when, during the separation step, diffusion processes cause tailing and deformation of the chromatographic spots. The method described by Frodyma et al.,[26] which involves the removal of the spot followed by mixing the substance and adsorbent before measurement, results in reasonably homogeneous systems. For direct scanning of spots on the layer, the use of a narrow slit or, even better, a "flying spot,"[25] which irradiates only a small portion of the chromatographic zone at one particular time, is recommended. The small element of area illuminated will have a uniform gradient of absorbance, at least at a first approximation. The numerous values obtained by this method for one single spot can then be integrated by means of a suitable recorder to give the typical chromatographic band.

In an experiment carried out with dinitrophenylhydrazone on microscope slides coated with silica gel, Goldman and Goodall[25] used the "flying spot" technique (Figure 2.03) for scanning the chromatographic spot zone in the transmission mode. (The arguments for using this technique would be valid for the spectral reflectance mode as well.) In order to maintain a constant angular distribution of the incident light and a constant angular orientation of the light-collecting device (which is another condition necessary for maintaining the validity of the Kubelka-Munk theory), the chromatogram rather than the light spot was moved. (A typical device for scanning spots by the "flying spot" method is described in Chapter IV, Section 4b.) The spot was scanned in a sawtooth

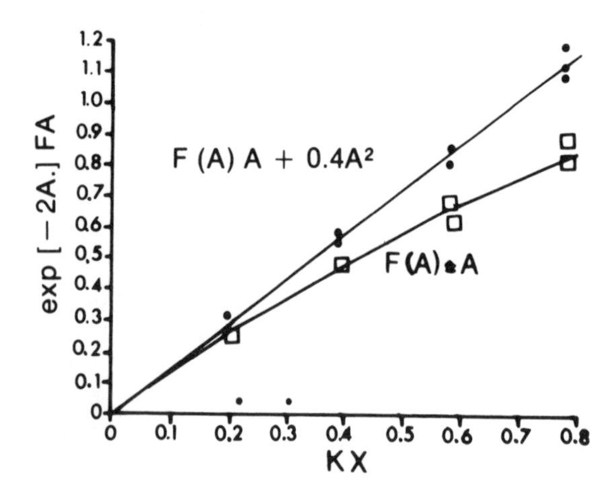

FIGURE 2.02 Experimental curvature of response as absorbance (□), and linearity when plotted in the theoretical expression (O).[24] (From Goldman, J. and Goodall, R. R., *J. Chromatogr.*, 32, 24 (1968). With permission of Am. Elsevier.)

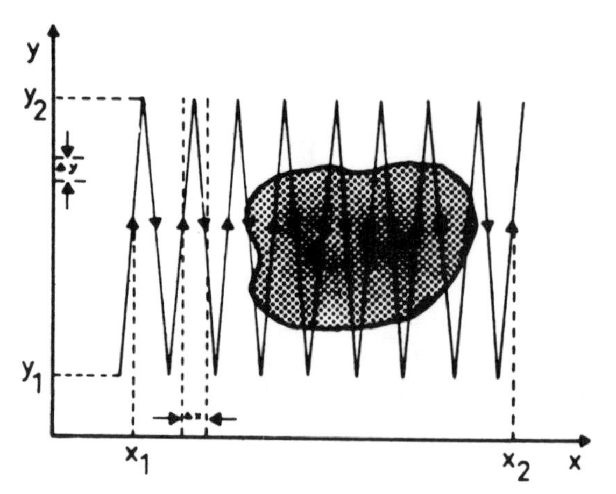

FIGURE 2.03 Path of the light spot (→) over a chromatographic zone (dotted area).[25] (From Goldman, J. and Goodall, R. R., *J. Chromatogr.*, 40, 345 (1969). With permission of Am. Elsevier.)

motion as depicted in Figure 2.03. In this particular case the spacings for Δx and Δy were of the order of 0.05 cm, with a total scan of 1.5 cm in both the x and y direction. Since this results in about 900 expressions of the form $2\exp[-2A_o](A + 0.4 A^2)$, and since each of these had to be calculated before addition, the operation was carried out by a computer. An overall standard deviation of 2.7% was reported[25] with this method for 10 analyses on separate plates after development of the substance over a distance of 5 cm. This is the reproducibility one can expect for data gained from spots developed on the same plate and evaluated by standard reflectance or transmittance techniques.[27] The mathematical function proposed by Goldman and Goodall (Equation 2.06) is only valid for the visible region, since in the UV region the absorbance of the adsorbent as well as the support have to be taken into account. Such additional absorption effects would be considerably less of a problem if the reflected rather than the transmitted light were to be measured. Another problem that arises for transmission work in the visible region of the spectrum is the reflectance from the glass support, which may constitute up to 5% of the total light energy.

Nevertheless, the results discussed above are valuable in that they present a first attempt to verify the validity of the Kubelka-Munk theory in conjunction with chromatography and to provide a less empirical approach to the problem of direct evaluation of chromatograms.

4. Factors Influencing Reflectance Spectra

It is possible to use reflectance spectra for identification purposes if an appropriate set of experimental conditions is selected. This technique is particularly useful for the in situ identification of substances after their resolution on paper and thin-layer chromatograms. Such an approach is possible not only for substances that are colored or have a characteristic UV spectrum, but also for substances that have to be reacted with a suitable spray before reflectance spectroscopic investigation.[27] In order to interpret reflectance spectra properly and to obtain the best use from this method, it is useful to know something about the parameters that influence these spectra. Therefore, a brief discussion of the major influencing factors is given in this section.

a. Specular Reflectance

It has been pointed out that reflected radiation consists of two distinct components:[4-7] regular (specular) and diffuse reflectance. The two components are superimposed on one another, resulting in the distortion of diffuse reflectance, which generally shows much less structure than the transmission spectrum. It is essential, therefore, to eliminate as far as possible the interference caused by regular reflectance. This can be accomplished by dilution of the light-absorbing species with a non- or low-absorbing powder. Figure 2.04 provides an example of the effect of dilution on the reflectance spectra of anthraquinone.[3] Dried sodium chloride is used as a diluting agent and reflectance measurement is made against pure sodium chloride. It is evident, with increasing dilution, that the two main absorption peaks at 29,600 and 39,000 cm^{-1} become sharper. At low enough concentration, the spectrum becomes independent of the degree of dilution and is similar to the spectrum of anthraquinone dissolved in diluted alcoholic solution. It is worth noticing that the spectrum of pure anthraquinone, as a

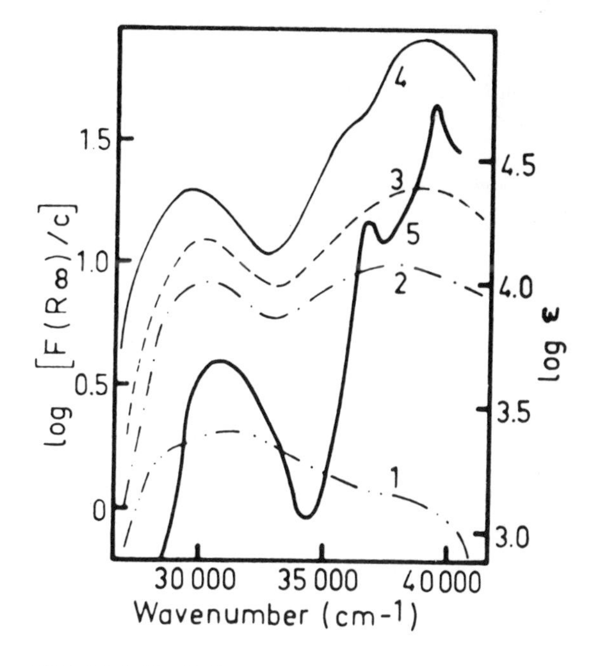

FIGURE 2.04 Reflectance spectra of anthraquinone. 1, undiluted; 2, adsorbed on NaCl (mol fraction $x = 1.26 \times 10^{-2}$); 3, adsorbed on NaCl ($x = 5.0 \times 10^{-3}$); 4, adsorbed on NaCl ($x = 1.9 \times 10^{-4}$); 5, absorption spectrum of a dilute alcoholic solution of anthraquinone.[3] (From Kortüm, G., Braun W., and Herzog, G., *Angew. Chem. Int. Ed. Engl.*, 2, 333 (1963). With permission of Verlag Chemie GMBH.)

consequence of the superimposition of regular reflection, bears almost no resemblance to the spectra of the anthraquinone diluted with sodium chloride or its alcoholic solution.

For highly light-absorbing material, the dilution technique has the added advantage of permitting the material to be studied in the optimum reflectance range (20 to 65%) for analysis. Other studies on the effect of superimposition of these two modes of reflectance have been carried out by Kortüm and Vogel.[5] They were able to separate diffuse from direct reflection by measuring the reflectance spectra of powders between crossed polarization foils.

b. Adsorbent

Similar to the solvent effect known in absorption spectrophotometry, a relationship exists between the reflectance spectrum of a sample and its environment or the type of material used as a diluting agent. For a sample that is sufficiently diluted (this generally can be achieved with mole fractions between 10^{-3} and 10^{-5}) so that no complete monomolecular layer of the material under investigation is formed on the adsorbent, the spectrum obtained is that of the adsorbed single molecule and differs somewhat with the use of different adsorbents. As indicated in Figure 7.06, which contrasts the reflectance spectra of 25×10^{-3} mg of eosine B adsorbed on filter paper and alumina and silica gel, the spectra obtained are influenced by the nature of the adsorbent employed.

When comparing reflectance spectra with spectra obtained in transmittance measurement, a broadening is observed, and the absorption peaks are displaced. Vibrational structure, if present at all, is usually strongly suppressed.[3,28,29] Curves 4 and 5 of Figure 2.04 provide an example of the contrast observed between reflectance and transmittance spectra. In cases where strong interaction occurs between adsorbent and the material under investigation (chemisorption), greater changes appear in the reflectance spectra in comparison to the spectra of free molecules.[6,30-32] (See also Figure 6.05.)

c. Particle Size

Zeitlin and Niimoto[33] found that the particle size of the adsorbent or diluent can affect reflectance spectra and that the adsorption bands tend to broaden as the particle size increases. For weakly or strongly light-absorbing material that has been diluted with non- or low-absorbing powder, the observed extinction of a sample decreases with decreasing particle size. This is in accordance with the observation that weakly absorbing material, such as $CuSO_4 . 5H_2O$, appears lighter as the material becomes more finely divided.[34] This phenomenon is attributed to diffuse scattering becoming more efficient with smaller particles, so that the radiation does not penetrate as deeply. The need for standardization of particle size, particularly for measurements of a more precise nature, is, therefore, obvious. Sifting or grinding procedures can be used.

d. Moisture

The presence of moisture tends to increase the extinction of a sample diluted with non- or low-absorbing powder. This is in accordance with the general observation that samples appear darker when moist. Kortüm et al.[3] ascribed this phenomenon to the dependence of the scattering coefficient on the ratio of refractive indices of the powder and on the surrounding medium. The displacement of air by water, as the surrounding medium in the moist sample, reduces this ratio and thus the scattering coefficient. As can be observed from the Kubelka-Munk equation, $F(R_\infty) = k/s$, a reduction of the value of the scattering coefficient will result in a greater extinction.

Sometimes the presence of moisture results in the development of a vibrational structure in the spectrum of a compound. For example, hardly any vibrational structure is shown in the reflectance spectrum of anthracene adsorbed on absolutely dry sodium chloride. When the sample is exposed to moisture, however, the adsorbed anthracene molecules evidently are displaced from the adsorbent by the water molecules and the vibrational structure appears on the spectrum.[3]

e. Presentation of Spectra

Reflectance spectra may be displayed in a number of ways. Converting the Kubelka-Munk equation (Equation 2.03) into a logarithmic form results in

$$\log F(R_\infty) = \log \epsilon + \log 2.303\ C/s. \qquad (2.07)$$

Since the scattering coefficient is practically independent of wavelength, a plot of $\log F(R_\infty)$ as a function of the wavelength or wave number should be equivalent to the real absorption spectrum obtained by transmission measurement, except for a displacement by –log 2.303 C/s in the direction of the ordinate. This type of plot is known as the "typical color curve" of the sample, and facilitates the comparison of spectra in a qualitative analysis. The comparison of spectra of samples of different concentrations can be achieved by simply displacing the ordinate up or down and establishing the identity or nonconformity.

More frequently, however, reflectance spectra are presented in the forms of log $(1/R_\infty)$ or % R_∞ vs. wavelength (or wave number). This corresponds to absorbance or percent transmittance curves in regular transmission spectroscopy. The quantity log $(1/R_\infty)$ is often referred to as the "apparent absorbance."

5. Optimum Analytical Concentration Range

For systems exhibiting no deviation from the Kubelka-Munk equation, the optimum conditions for accuracy can be deduced by computing the relative error dC/C. In terms of Kubelka-Munk Equation 2.04, the error in C is

$$dC = k'(R^2_\infty - 1)dR_\infty/2R^2_\infty \tag{2.08}$$

and the relative error in C is

$$dC/C = (R_\infty + 1)dR_\infty/(R_\infty - 1)R_\infty. \tag{2.09}$$

Assuming a reading error in reflectance amounting to 1%, that is $dR_\infty = 0.01$,

$$(dC/C)100 = (R_\infty + 1)/(R_\infty - 1)R_\infty = \text{percent error in C}. \tag{2.10}$$

To determine the value of R_∞ that will minimize the relative error in C, d(percent error in C)/dR_∞ is equated to zero. The positive solution of the resulting equation

$$R^2_\infty + 2R_\infty - 1 = 0 \tag{2.11}$$

indicates that the minimum relative error in C occurs at a reflectance value of 0.414, corresponding to a reflectance reading of 41.4% R. This is presented graphically in Figure 2.05, where the percent of error, computed by Equation 2.10, is plotted as a function of the percent of reflectance. The minimum in the resulting curve corresponds to the 41.4% R value obtained by the solution of Equation 2.11.

Reflectance spectrophotometric methods of analysis usually involve the use of a calibration curve prepared by measuring the reflectance of samples containing known amounts of the substance of interest. Conformity to the Kubelka-Munk equation is indicated if the data, when plotted in the form $F(R_\infty)$ vs. C, follow a straight line. From the standpoint of use in analysis, however, it is immaterial whether the system in question conforms to the Kubelka-Munk equation. Provided some sort of near-linear relationship is found to exist between reflectance and concentration, it is more important, if the method is to be used for analysis, to select a suitable concentration range for the analysis and to evaluate its accuracy. To illustrate how this might be done, two systems were selected as models. These were rhodamine B, which absorbs in the visible region, and aspirin, which absorbs in the UV, both adsorbed on silica gel. The experimental data of the first system are presented in Figure 2.06 as curve 1, and of the second as curve 3. When these data are plotted in the form $F(R_\infty)$ vs. C, (Figure 2.07), curve 1 results for rhodamine B and curve 3 for aspirin. Although both curves are linear over a considerable portion of the concentration ranges investigated, neither conforms to the Kubelka-

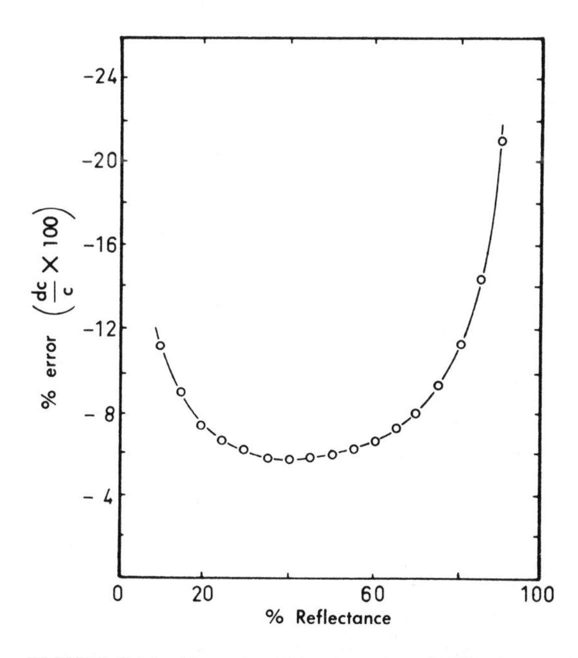

FIGURE 2.05 Percentage error, computed with the use of Equation 2.10, as a function of percentage reflectance.[20] (From Lieu, V. T. and Frodyma, M. M., *Talanta*, 13, 1319 (1962). With permission of Pergamon Press.)

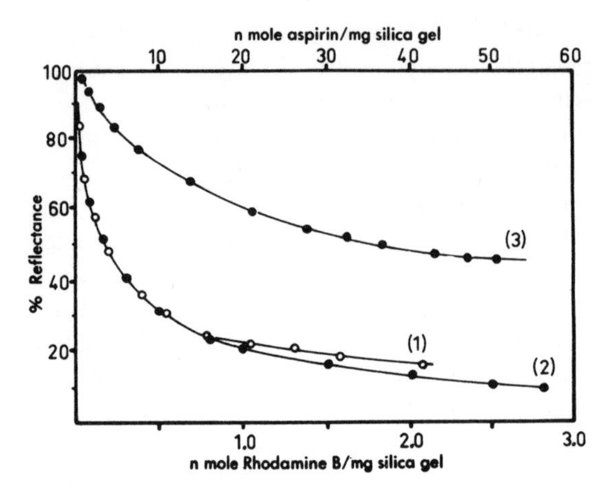

FIGURE 2.06 Percentage of reflectance as a function of concentration. (1) Experimental and (2) theoretical values of Rhodamine B at 545 nmole. (3) Experimental values of aspirin at 302 nmole.[20] (From Lieu, V. T. and Frodyma, M. M., *Talanta,* 13, 1319 (1962). With permission of Pergamon Press.)

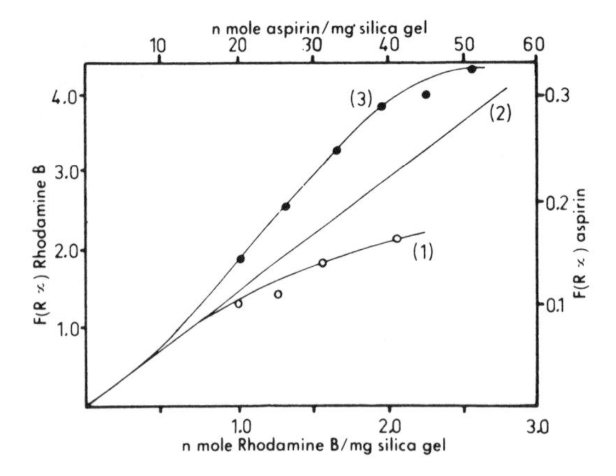

FIGURE 2.07 Kubelka-Munk values as a function of concentration. (1) Experimental and (2) theoretical values of Rhodamine B at 543 nmole. (3) Experimental values of aspirin at 302 nmole.[20] (From Lieu, V. T. and Frodyma, M. M., *Talanta,* 13, 1319 (1962). With permission of Pergamon Press.)

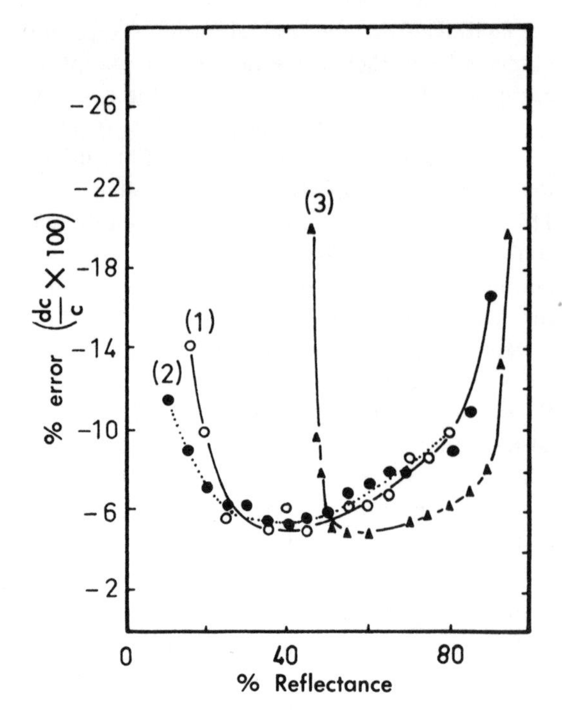

FIGURE 2.08 Percent of error arising from a reading error of 1% R (estimated graphically) as a function of percent reflectance. (1) Experimental and (2) hypothetical Rhodamine B values. (3) Experimental aspirin values.[20] (From Lieu, V. T. and Frodyma, M. M., *Talanta,* 13, 1319 (1962). With permission of Pergamon Press.)

Munk equation. This becomes more evident when the pair of curves obtained with rhodamine B (curves 1 of Figures 2.06 and 2.07) is contrasted with the pair of hypothetical curves that would have been obtained had the system behaved ideally (curves 2 of Figures 2.06 and 2.07). The departure from linearity observed with both rhodamine B and aspirin at higher concentrations may be ascribed to the approaching saturation of the adsorbent surface by the first monomolecular layer of the adsorbed species.[3] Variations in the optical properties of the reference standard were at least partly responsible for the nonlinearity that was also observed at lower concentrations in the case of aspirin. When the relative error in concentration arising from a 1% error in R is plotted against percent of reflectance (Figure 2.08), there is close agreement between the experimental curve of Figure 2.06 and the hypothetical values curve for the rhodamine B system. The scatter in curve 2 arises from the method of calculation. There is less agreement with the experimental curves for rhodamine B and aspirin because of the departure from the Kubelka–Munk equation. Nevertheless, the curves do show the range of reflectance that gives minimum error.

Another approach to the selection of the optimum range for reflectance analysis is one suggested by Ringbom,[35] and later by Ayres,[36] who evaluated relative error and defined the suitable range for absorption analysis by plotting the absorbance (1–transmittance) against the logarithm of the concentration. When the reflectance data obtained with rhodamine B and aspirin and the hypothetical data computed for rhodamine B are plotted as percent of reflectance vs. the logarithm of concentration (Figure 2.09), the advantages of this method become manifest. The optimum range for analysis corresponds to that portion of each curve exhibiting the greatest slope. Since a considerable portion of each of the curves shown in Figure 2.09 is also fairly linear in this region, it is apparent that reasonable accuracy can be expected over a wide concentration range. The maximum accuracy can be estimated from the equation

$$\frac{dC}{C}100 = 2.303d(\log C)100 = 2.303\frac{d(\log C)}{dR}100dR. \quad (2.12)$$

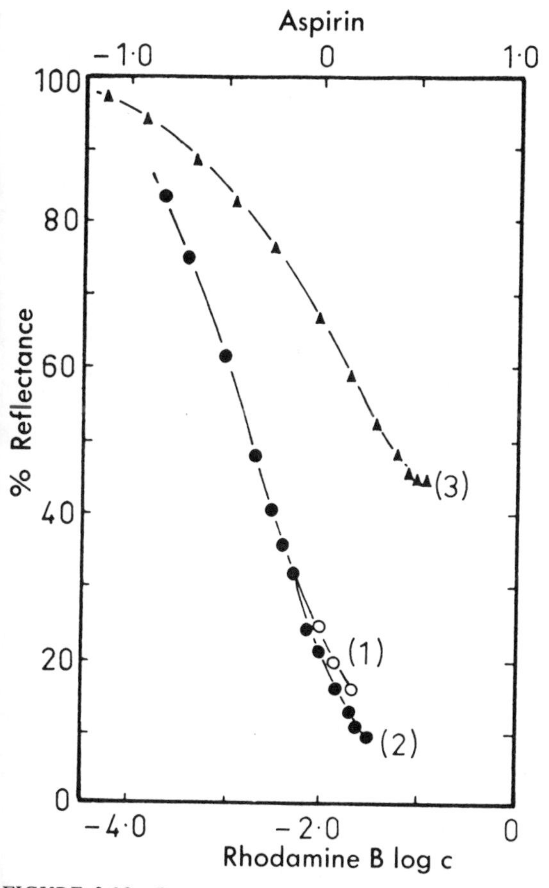

FIGURE 2.09 Percent reflectance as a function of the logarithm of concentration. (1) Experimental and (2) hypothetical Rhodamine B values. (3) Experimental aspirin values.[20] (From Lieu, V. T. and Frodyma, M. M., *Talanta,* 13, 1319 (1962). With permission of Pergamon Press.)

Assuming a constant reading error of 1% R, i.e., dR = 0.01,

$$\text{percent error} = 2.303\frac{d(\log C)}{dR}. \quad (2.13)$$

For the three systems under consideration the optimum range for analysis can be found by a consideration of the curves in Figure 2.09, and the percent of error resulting from a reading error of 1% R can be computed with the use of Equation 2.13 and the slope of the appropriate curve. When this is done, the data obtained (presented in Table 2.01) are found to be in accordance with those resulting from the application of the graphic method. Values for the hypothetical rhodamine B system also agree with the figures obtained for an ideal system and computed with the use of Equation 2.13. It would seem, therefore, that the minimum error in reflectance spectrophotometric analysis is about 6%/1% R reading error, regardless of whether the system conforms to the Kubelka-Munk equation. This value obviously can be decreased by reducing the reading error. Although a reading error amounting to 0.5% R should not be too difficult to attain, it would be unrealistic to expect a precision better than 0.1 to 0.2% R, and, therefore, a minimum error smaller than 1 to 2%. Similarly, regardless of whether a system conforms to the Kubelka-Munk equation, the optimum range for analysis can be found after plotting the reflectance data according to either of the two procedures described. The Ringbom method[35] has the advantage not only of making available the optimum range and maximum accuracy but also of providing a plot usable as a calibration curve. Kortüm[37] has reported similar studies on an error analysis of systems, measured by diffuse reflectance spectroscopy.

6. Differential Measurement of Reflectance

From Figure 2.08 it can be seen that the accuracy of spectral reflectance techniques leaves much to be desired when samples of too high or too low a concentration are used. With powder samples of high concentration, dilution with a suitable powder diluent often will be convenient for bringing the concentration down into the optimum concentration range. Many samples, such as textiles or paint surfaces, cannot be diluted. The problem becomes even greater if one deals with samples of very low concentration, which is often the case in chromatographic systems where submicrogram

TABLE 2.01

Optimum Range of Reflectance and Error of Analysis[20]

Curve number (see Figure 2.09)	System	Optimum range of reflectance (% R)	% Error of analysis per 1% reading error
1	Rhodamine B (experimental)	25-65	6.0
2	Rhodamine B (hypothetical)	20-65	6.0
3	Aspirin (experimental)	55-85	6.0

(From Lieu, V. T. and Frodyma, M. M., *Talanta,* 13, 1319 (1962). With permission of Pergamon Press.)

quantities are to be analyzed. In many cases, however, it is possible to reduce the error of the reflectance measurements considerably by modifying the approach in which the measurements are carried out. These techniques, which are differential in nature, can be quite analogous to differential transmission spectrophotometric methods.[38-40] Differential reflectance spectroscopic methods were first discussed by Lermond and Rogers,[41] particularly in conjunction with highly absorbing powder and textile samples (see the discussion in Chapter 5, Section 3g). The scale expansion effect that can be achieved with differential methods is demonstrated clearly in Figure 5.17. Basically, the differential measuring techniques can be classified into high-reflectance methods applicable to samples of low concentration and low-reflectance methods that can be used with samples that contain high amounts of light-absorbing substance. Both modes of operation have been studied in detail by Lieu et al.[42,43] and investigated as to their usefulness in chromatographic work.

a. High-reflectance Method

In the standard procedure of reflectance measurement, two preliminary adjustments are required in order to use the full reflectance scale. The first of these is performed with a shutter in the beam, so that no radiation strikes the detector; electrical compensation for the dark current is then made to set the instrument to read zero reflectance. The second adjustment is made when the beam is reflected from a non- or low-absorbing standard and impinges on the detector; here, the radiation intensity is varied until it reads full scale, 100% R. For differential high-reflectance measurements, the instrument is set to read zero when the photocell is exposed to light reflected from a reference sample that is somewhat more concentrated than the sample under investigation. As in absorption spectroscopy, it is assumed that the instrument reading is a linear function of the light reflected from the sample. The basic equation relating reflectance R to light power measured is

$$R = I_X/I_O, \tag{2.14}$$

where I_X and I_O are the light intensities reflected by sample and nonabsorbing reference standard, respectively. In the differential method, where zero reflectance is set to correspond to that of a standard rather more concentrated than the test sample, $R_O = I_s/I_O$, so that in effect the scale is expanded (cf. Reilley and Crawford[40]) and the differential reflectance R_d is given by

$$R_d = (I_X - I_s)\,(I_O - I_s). \tag{2.15}$$

When the sample is diluted with a non- or low-absorbing powder and its reflected light is measured relative to the light reflected by the pure diluent powder, the Kubelka-Munk Equation 2.04 can be rewritten in the form

$$C = k'(1 - R_{XO})^2/2R_{XO}, \tag{2.16}$$

where C is the molar concentration, R_{XO} is the reflectance of the sample relative to a non- or low-absorbing standard, and k' is a constant.

A similar expression for differential spectrophotometry can be obtained from Equation 2.15 by converting light intensities into the corresponding reflectances. Division by I_O gives

$$R_d = (R_{xO} - R_{sO})(1 - R_{sO}), \qquad (2.17)$$

where R_{sO} is the reflectance of the differential standard. Rearranging and combining Equations 2.16 and 2.17 gives

$$C = k'[(1 - R_{sO})(1 - R_d)]^2/[2R_d(1 - R_{sO}) + R_{sO}]. \qquad (2.18)$$

If the concentration of the differential standard is so great that the incident light is completely absorbed, Equation 2.18 reduces to Equation 2.16.

Equation 2.18 indicates that slopes of plots obtained for systems conforming to the Kubelka-Munk equation should be independent of the concentration of the differential standard. To illustrate this, the nickel-dimethylglyoxime complex, adsorbed on cellulose, was taken as a model. The results, with varying concentrations of the complex used as differential standards, are shown in Figure 2.10. When these results are graphed in the form $[(1 - R_{sO})(1 - R_d)]^2/[R_d(1 - R_{sO}) + R_{sO}]$ vs. concentration of the complex, the plot in Figure 2.11 results. Data obtained when the zero adjustment was made with the photocell in darkness are included for comparative purposes; they correspond to those obtained in conventional reflectance spectrophotometry. These results demonstrate the close conformity of the nickel-dimethylglyoxime-cellulose system to the Kubelka-Munk equation, and so it is not surprising that the plots of the data obtained by means of the high-reflectance method take the form of straight lines which can be superimposed.

Regardless of whether a system conforms to the Kubelka-Munk equation, however, the optimum concentration range for the high-reflectance spectrophotometric method can be deduced by computing the relative error, dC/C, caused by reading error. Taking the derivative of Equation 2.18 with respect to R_d, the relative error in C is

$$\frac{dC}{C} = \frac{k'(1 - R_{sO})^2 (1 - R_d)(R_{sO}R_d - R_d - R_{sO} - 1)dR_d}{2C[R_d(1 - R_{sO}) + R_{sO}]^2}. \qquad (2.19)$$

If the reading error is assumed to amount to one reflectance unit, i.e., $dR_d = 0.01$, it is a simple

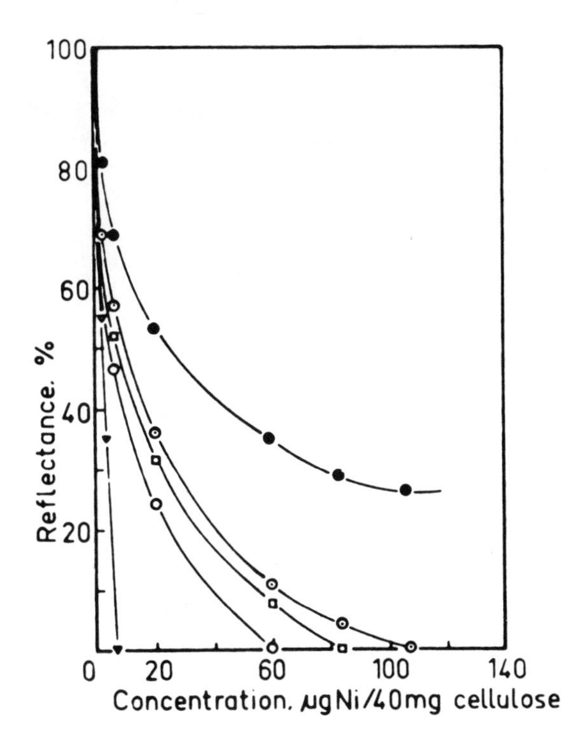

FIGURE 2.10 Reflectance at 540 nm of nickel-dimethylglyoximate adsorbed on cellulose, as a function of concentration. Differential standard concentrations (µg Ni/40 mg cellulose):-●-photocell in darkness;-⊙-107;-□-83; —○— 59; —▽— 5.9.[42] (From Lieu, V. T., Zaye, D. F., and Frodyma, M. M., *Talanta,* 16, 1289 (1969). With permission of Pergamon Press.)

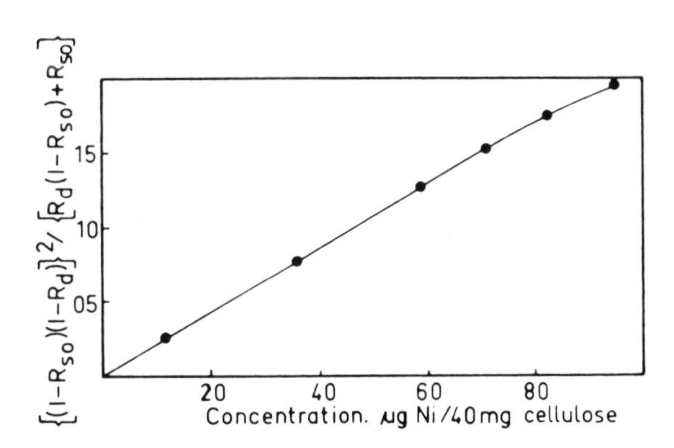

FIGURE 2.11 $[(1 - R_{sO})(1 - R_d)]^2/[R_d(1 - R_{sO}) + R_{sO}]$ vs. concentration of the nickel-dimethylglyoxime complex. Each point represents the mean value obtained for a given concentration measured relative to a series of differential standards having different concentrations. The range is indicated by the size of the spots.[42] (From Lieu, V. T., Zaye, D. F., and Frodyma, M. M., *Talanta,* 16, 1289 (1969). With permission of Pergamon Press.)

matter to compute the relative error (in percent) in the calculated concentration. The inverse of the slope at concentration C, viz., k′/2, may be obtained from the Kubelka-Munk plot in Figure 2.11, while R_{sO} may be obtained by determining the reflectance of the differential standard conventionally. When this was done over a range of concentrations for nickel-dimethylglyoximate adsorbed on cellulose, the plots shown in Figure 2.12 resulted. Similar families of curves were obtained when the error in the concentration was estimated by the graphical method described in the previous section.[20]

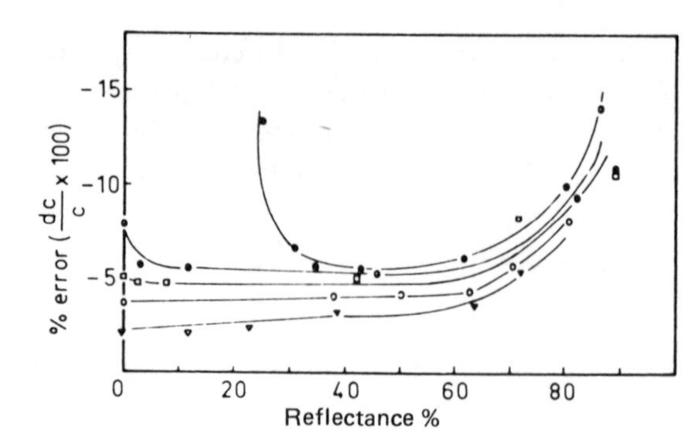

FIGURE 2.12 Relative error arising from an error of 1% R, computed with the use of Equation 2.19, as a function of reflectance for nickel-dimethylglyoximate adsorbed on cellulose. Concentration of differential standard, μg Ni/40 mg cellulose:-●-photocell in darkness; -⊙-107;-□-83;-○-59; -▽-5.9.[42] (From Lieu, V. T., Zaye, D. F., and Frodyma, M. M., *Talanta,* 16, 1289 (1969). With permission of Pergamon Press.)

The optimal reflectance and concentration ranges for analyses involving the nickel-dimethylglyoximate system, as well as the analysis error to be expected as a result of an error amounting to 1% R, are indicated in Figure 2.12 and summarized in Table 2.02. It shows some of the advantages afforded by the high-reflectance method as contrasted with the conventional procedure for measuring the reflectance of a test sample. For example, the optimum reflectance and concentration ranges for analysis are both extended to lower values with the use of the high-reflectance method, the extension increasing as the concentration of the differential standard decreases, and approaching a value of 0% R when the concentration of the standard falls in the range 2.1 to 2.7 μg/mg of cellulose for nickel. At the same time, a decrease in the concentration of the differential standard is accompanied by an extension of the optimum concentration range to lower concentrations. This limit is lowered from the value 0.12 μg/mg of cellulose, obtained by means of conventional reflectance spectroscopy, to 0.025 μg/mg when a 0.15 μg/mg differential standard is used.

Of equal importance is the fact that at relatively high concentrations within the optimum concentration range, the relative error in the analysis arising from an error of 1% R decreases with a decrease in the concentration of the differential standard. It is thus possible, by a proper choice of the differential standard, to increase the accuracy of the analysis appreciably. For example, the error in the analysis by conventional reflectance spectroscopy of a sample con-

TABLE 2.02

Optimum Reflectance and Concentration Ranges for Analysis[42]

System	Concentration of differential standard (μg/40 mg of cellulose)	Optimum range: Reflectance (% R)	Optimum range: Concentration (μg/mg of cellulose)	% Error per 1%: lower limit	% Error per 1%: mid-range	% Error per 1%: upper limit
Nickel-dimethylglyoxime	5.9	0–70	1.0–5.9	-2		-5
	59	0–70	1.3–59	-3		-6
	71	0–70	1.5–71	-4		-6
	83	0–70	2.3–83	-5	-4	-6
	107	5–70	3.0–76	-6	-5	-7
	Photocell in darkness	30–70	5.0–76	-6		-7

(From Lieu, V. T., Zaye, D. F., and Frodyma, M. M., *Talanta,* 16, 1289 (1969). With permission of Pergamon Press.)

sisting of 5.0 μg of nickel adsorbed on 40 mg of cellulose is ~-7%.[20] The error can be reduced to ~-2% by the simple expedient of using 0.15 μg Ni/mg of cellulose as the differential standard. Therefore, it seems that maximum accuracy would be attained by working in the upper portion of the optimum concentration range, with a differential standard of higher concentration than the test sample.

It has been assumed throughout that dR_d is independent of R_d and of the differential standard employed. The validity of this assumption was tested experimentally by preparing a concentration series of nickel-dimethylglyoxime-cellulose samples with reflectance extended over the entire reflectance scale, and then determining the reflectance of each of its members in turn, relative to differential standards containing 10, 25, 35, and 70 μg of nickel. Several measurements were made for each combination of test and standard samples. The standard derivation obtained for every set of replicate measurements fell randomly in the range 0.1 to 0.3% R.

From these data it seems that the application of the high-reflectance technique makes it possible to analyze samples with an error as low as 2%, and in no case with an error greater than 7%, for a reading error of 1% R. This represents a substantial increase in accuracy over that afforded by the conventional method of measuring reflectance, for which the minimum error is ~7% under optimal conditions.[20]

b. Low-reflectance Method

In the low-reflectance method, a standard somewhat more diluted than the sample is employed for establishing full-scale setting, while the zero adjustment is accomplished in the ordinary fashion. By setting the full-scale setting with such a standard, one effectively increases the size of the reflectance scale, and, accordingly, greater accuracy is possible in the reflectance readings and in the determination of concentrations.

The Kubelka-Munk Equation 2.04 can be written twice in the form

$$\frac{(1 - R_{xO})^2}{2R_{xO}} = \frac{C_x}{k'} \tag{2.20}$$

$$\frac{(1 - R_{sO})^2}{2R_{sO}} = \frac{C_s}{k'} , \tag{2.21}$$

where R_{xO} is the reflectance of the sample relative to a non- or low-absorbing standard; R_{sO} is the reflectance of a differential standard somewhat more diluted than the sample relative to a non- or low-absorbing standard and C_x and C_s are the concentration of the sample and the differential standard, respectively. The following expression is obtained when Equation 2.21 is subtracted from Equation 2.20:

$$\frac{(1 - R_{xO})^2}{2R_{xO}} - \frac{(1 - R_{sO})^2}{2R_{sO}} = \frac{1}{k'}(C_x - C_s). \tag{2.22}$$

By substitution and rearrangement,[43] Equation 2.22 can be expressed

$$C_x = \frac{k'(1 - \sigma R_{xs})^2}{2\sigma R_{xs}} k'h, \tag{2.23}$$

where R_{xs} is the reflectance of the sample relative to the differential standard, σ and h are constants and equal, respectively, to R_{sO} and $\frac{(1 - \sigma)^2}{2\sigma} - \frac{C_s}{k'}$. Since k' and h are constant, Equation 2.23 shows that, as long as the Kubelka-Munk equation is followed, a plot of $(1 - \tau R_{xs})^2/2\tau R_{xs}$ is a straight line.

Labinowich,[44] in an effort to effect an increase in accuracy, determined copper concentrated on a chromatographic column by such a procedure. Plots of % R vs. concentrations were prepared, using copper samples of different concentrations as differential standards. The percent relative errors in concentration of copper arising from a 1% R reading error were estimated graphically[20] for both the ordinary and low-reflectance methods from these plots. The data obtained are presented in Figure 2.13. It is apparent that the use of the low-reflectance method not only extends the optimum reflectance (or concentration) ranges for analysis to higher reflectance values, but also results in a reduction of the relative error in the analysis, arising from an error of 1% R. The greatest reduction in error (or gain in accuracy) occurs with differential standards of high concentration; thus, this technique is most useful for the analysis of samples of high concentrations, particularly samples that cannot be diluted conveniently.[44]

7. Multicomponent Systems

The validity of the theoretical principles discussed was further investigated with respect to

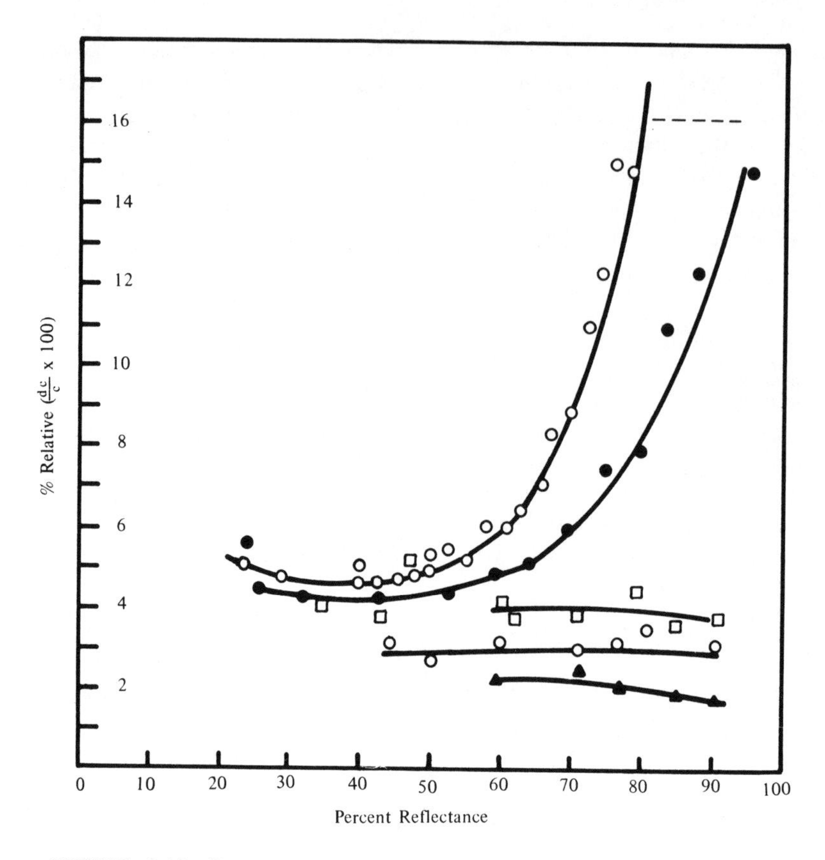

FIGURE 2.13 Percent relative error arising from a reading error of 1% (estimated graphically) as a function of % reflectance. Copper(I)-neocuproine complex adsorbed on magnesium silicate.[44]

multicomponent systems,[45] of interest in cases of incomplete chromatographic separation.

For a powder mixture containing n light-absorbing components whose reflectance functions are additive, the Kubelka-Munk function $F(R_\infty)$ can be adapted for simultaneous analysis. The function of the total reflectance $R_{\infty T}$ of the mixture at some wavelength i may be represented as the sum of all individual reflectance functions:

$$F(R_{\infty T})i = \sum_{j=i}^{n} \tau_{ij} C_j, \tag{2.24}$$

where j refers to components and τ is the slope of the Kubelka-Munk plot of $R(R_\infty)$ vs. C.

Equation 2.24 can be written in a more explicit manner by writing as many equations as there are components in the mixture.

$$F(R_{\infty T})_1 = \tau_{11} C_1 + \tau_{12} C_2 + \ldots \tau_{1n} C_n \tag{2.25}$$

$$F(R_{\infty T})_2 = \tau_{21} C_2 + \tau_{22} C_2 + \ldots \tau_{2n} C_n \tag{2.26}$$

These equations are, of course, valid only for a concentration range within that prescribed by the Kubelka-Munk law. At higher concentrations, interference due to saturation of the first monomolecular adsorption layer[3] results in marked deviations from linearity, and in extreme cases gives a calibration curve asymptotic to the horizontal axis. The useful concentration range can be extended, however, by use of the semi-empirical relationship

$$(1 - R_\infty)^2 / 2R_\infty = k \log C \tag{2.27}$$

suggested earlier,[18] which results in an almost fivefold extension of the range. Other functions that tend to give linear calibration curves under suitable conditions (log R_∞ vs. C^2, R_∞ vs. $3\sqrt{C}$ or $\sqrt{C/(1 - C)}$) have been proposed by Lermond and Rogers[41] and could be applied for simultaneous analysis as well.

The dye pairs Orange G-Crystal Violet and

Fuchsin-Brilliant Green were chosen for study. Figures 2.14 and 2.15 show the individual spectra of the dyes and the spectra of the mixtures of the two pairs; the theoretically computed total spectra of the pairs (by addition of the individual reflectance functions) are essentially identical with the measured spectra. The absorption maxima for all four dyes do not vary significantly throughout the entire concentration range studied. Figure 2.16 shows adsorption characteristics and interferences of Fuchsin and Brilliant Green; these results show the absence of interferences due to molecular dissociation, association, and abnormal adsorbent-adsorbate interactions, and that the dye systems are satisfactory for further investigation. The absorption maxima determined from Figures 2.14 to 2.16 (626 nm for Brilliant Green, 600 nm for Crystal Violet, 545 nm for Fuchsin, and 490 nm for Orange G) were used for quantitative analysis. Calibration plots for the Fuchsin-Brilliant Green pair are shown in Figure 2.17. Within the concentration ranges studied, the Kubelka-Munk relationship was valid for all four dyes, and the τ values for the slopes of the Kubelka-Munk curves

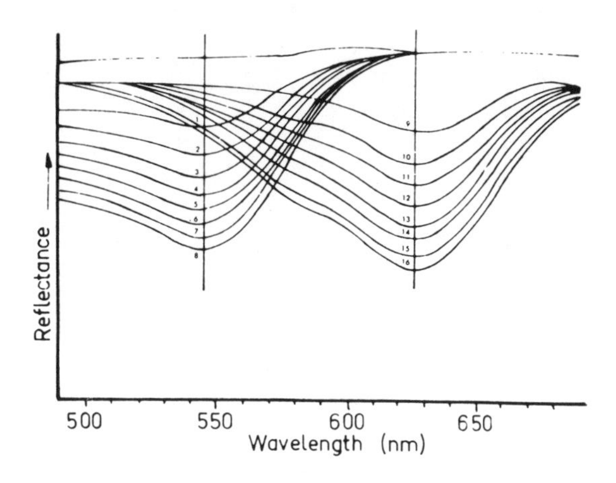

FIGURE 2.16 Reflectance spectra of dilution series of Fuchsin (1-8) and Brilliant Green (9-16) adsorbed on Silica Gel G, recorded against a barium sulfate reflectance standard. The zero line is for pure Silica Gel G. The concentration of dye in μg per sample is as follows. Fuchsin: (1) 0.3, (2) 0.6, (3) 0.9, (4) 1.2, (5) 1.5, (6) 1.8, (7) 2.4, (8) 3.0. Brilliant Green: (9) 0.3, (10) 0.6, (11) 0.9, (12) 1.2, (13) 1.5, (14) 1.8, (15) 2.4, (16) 3.0.[45] (From Frei, R. W., Ryan, D. E., and Lieu, V. T., *Can. J. Chem.*, 44, 1945 (1966). With permission of National Research Council of Canada.)

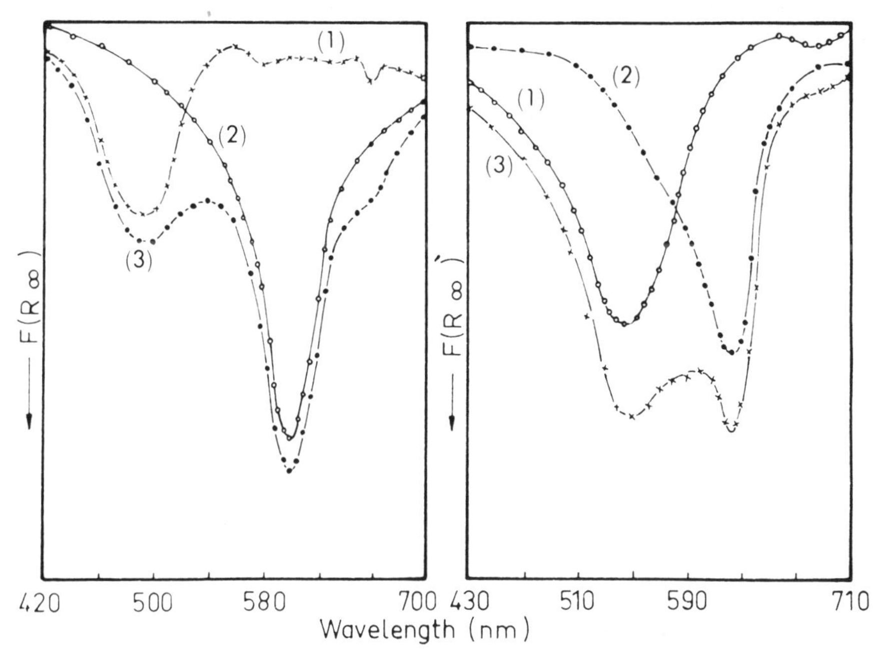

FIGURE 2.14 Reflectance spectra of (1) Orange G; (2) Crystal violet; (3) a mixture of the two.[45] (From Frei, R. W., Ryan, D. E., and Lieu, V. T., *Can. J. Chem.*, 44, 1945 (1966). With permission of National Research Council of Canada.)

FIGURE 2.15 Reflectance spectra of (1) Fuchsin; (2) Brilliant Green; (3) a mixture of the two.[45] (From Frei, R. W., Ryan, D. E., and Lieu, V. T., *Can. J. Chem.*, 44, 1945 (1966). With permission of National Research Council of Canada.)

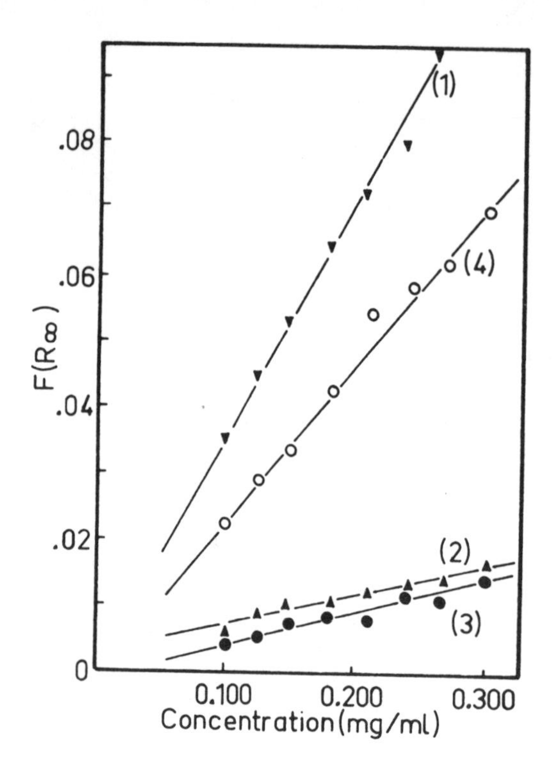

FIGURE 2.17 Kubelka-Munk plots for (1) Brilliant Green at 626 nm; (2) Fuchsin at 626 nm; (3) Brilliant Green at 545 nm; (4) Fuchsin at 545 nm.[45] (From Frei, R. W., Ryan, D. E., and Lieu, V. T., *Can. J. Chem.*, 44, 1945 (1966). With permission of National Research Council of Canada.)

required for Equations 2.25 and 2.26 were readily obtained. To test the reliability of the method, sample mixtures were prepared and analyzed as unknowns by students at the junior and senior university level. To avoid fluctuations due to temperature, humidity changes, and bleaching effects, the dilution series were processed simultaneously with the unknowns. Six readings were taken for each calibration curve and four samples of each dye pair were prepared. The students' results are shown below.

Precision (average standard deviation for four sets of four samples) = ± 2.4%[45]
Accuracy (average deviation of four samples from true value)

Dye pair 1		Dye pair 2	
Orange G	Crystal Violet	Fuchsin	Brilliant Green
1.9%	3.3%	2.1%	3.1%

As expected, the precision is somewhat lower than for single component analysis; the factors limiting precision for such simple systems were found to be the packing reproducibility and the homogeneity of the samples.

Several other workers have reported on the multicomponent analysis of samples by diffuse reflectance spectroscopy. Kortüm and Herzog[46] studied mixtures of rutile and anatase by measuring their spectral reflectance in the UV region (see Chapter 6, Section 3). Doyle and Forbes[47] reported on the simultaneous determination of compounds such as lead monoxide and zinc oxide on substrates such as silicic acid and aluminum oxide (see Chapter 6, Section 3). The application of simultaneous methods to reflectance problems in the color industry, and particularly the investigation of pigment mixtures, was discussed in detail by Duncan[48,49] (see Chapter 5, Section 2). Everhard et al.[50] described similar applications in the pharmaceutical field, e.g., for the study of tablet coatings (see Chapter 5, Section 3f).

REFERENCES

1. **Kubelka, P. and Munk, F.,** *Z. Tech. Phys.,* 12, 593 (1931).
2. **Kubelka, P.,** *J. Opt. Soc. Am.,* 38, 448 (1948).
3. **Kortüm, G., Braun, W., and Herzog, G.,** *Angew. Chem. Int. Ed. Engl.,* 2, 333 (1963).
4. **Kortüm, G.,** *Spectrochim. Acta Suppl.,* 534, (1957).
5. **Kortüm, G. and Vogel, J.,** *Z. Phys. Chem. (Frankfurt),* 18, 230 (1958).
6. **Kortüm, G.,** *Trans. Faraday Soc.,* 58, 1624 (1962).
7. **Kortüm, G. and Vogel, J.,** *Z. Phys. Chem. (Frankfurt),* 18, 110 (1958).
8. **Schreyer, G.,** *Z. Phys. Chem. (Frankfurt),* 18, 123 (1958).
9. **Van den Akker, J. A.,** *Tappi,* 32, 498 (1949).
10. **Judd, D. B. and Wyszecki, G.,** *Color in Business, Science and Industry,* 2nd ed., John Wiley & Sons, New York, 1963.

11. Chandrasekhar, S., *Radiative Transfer,* Clarendon Press, Oxford, 1950.
12. Giovanelli, R. G., in *Progress in Optics,* Vol. II, Wolf, E., Ed., North-Holland Publishing Co. Amsterdam, 1963, 111.
13. Kortüm, G., *Reflexionsspektroskopie,* Springer-Verlag, Berlin, 1969, 160.
14. Johnson, P. D., *J. Opt. Soc. Am.*, 42, 978 (1952).
15. Melamed, N. T., *J. Appl. Phys.*, 34, 560 (1963).
16. Companion, A. L., in *Developments in Applied Spectroscopy,* Vol. 4. Davis, E. N., Ed., Plenum Press, New York, 1965, 221.
17. Hecht, H. G., in *Modern Aspects of Reflectance Spectroscopy,* Wendlandt, W. W., Ed., Plenum Press, New York, 1968.
18. Frei, R. W. and Frodyma, M. M., *Anal. Chim. Acta,* 32, 501 (1965).
19. Jork, H. Z., *Anal. Chem.*, 221, 17 (1966).
20. Lieu, V. T. and Frodyma, M. M., *Talanta,* 13, 1319 (1962).
21. Ingle, R. B. and Minshall, E., *J. Chromatogr.*, 8, 369 (1962).
22. Klaus, R., *J. Chromatogr.*, 16, 311 (1964).
23. Klaus, R., *Pharm. Ztg. Ver. Apotheker-Ztg.*, 112, 480 (1967).
24. Goldman, J. and Goodall, R. R., *J. Chromatogr.*, 32, 24 (1968).
25. Goldman, J. and Goodall, R. R., *J. Chromatogr.*, 40, 345 (1969).
26. Frodyma, M. M., Frei, R. W., and Williams, D. J., *J. Chromatogr.*, 13, 61 (1964).
27. Frei, R. W., in *Progress in Thin-Layer Chromatography and Related Methods,* Vol. II, Niederwieser, A. and Pataki, G., Eds., Ann Arbor Science Publishers, Ann Arbor, Mich., 1971, 1.
28. Kortüm, G. and Schreyer, G., *Angew. Chem.*, 67, 694 (1955).
29. Kortüm, G. and Oelkrug, D., *Z. Phys. Chem. (Frankfurt),* 34, 58 (1962).
30. Kortüm, G., Vogel, J., and Braun, W., *Angew. Chem.*, 70, 651 (1958).
31. Kortüm, G. and Vogel, J., *Chem. Ber.*, 93, 706 (1960).
32. Kortüm, G. and Braun, W., *Liebigs Ann. Chem.*, 632, 104 (1960).
33. Zeitlin, H. and Niimoto, A., *Anal. Chem.*, 31, 1167 (1959).
34. Kortüm, G. and Schöttler, H., *Z. Elektrochem.*, 57, 353 (1953).
35. Ringbom, A. Z., *Anal. Chem.*, 715, 332 (1939).
36. Ayres, G. H., *Anal. Chem.*, 21, 652 (1949).
37. Kortüm, G., *Reflexionsspektroskopie,* Springer-Verlag, Berlin, 1969, 258.
38. Bastian, R., *Anal. Chem.*, 21, 972 (1949).
39. Hiskey, C. F., *Anal. Chem.*, 21, 1440 (1949).
40. Reilley, C. N. and Crawford, C. M., *Anal. Chem.*, 27, 716 (1955).
41. Lermond, C. A. and Rogers, L. B., *Anal. Chem.*, 27, 340 (1955).
42. Lieu, V. T., Zaye, D. F., and Frodyma, M. M., *Talanta,* 16, 1289 (1969).
43. Lieu, V. T. and Frodyma, M. M., Unpublished data.
44. Labinowich, E. P., M.Sc. Thesis, University of Hawaii, Honolulu, Hawaii, 1966.
45. Frei, R. W., Ryan, D. E., and Lieu, V. T., *Can. J. Chem.*, 44, 1945 (1966).
46. Kortüm, G. and Herzog, G., *Z. Anal. Chem.*, 190, 239 (1962).
47. Doyle, W. P. and Forbes, F., *Anal. Chim. Acta,* 33, 108 (1965).
48. Duncan, D. R., *J. Oil Colour Chem. Assoc.*, 32, 296 (1949).
49. Duncan, D. R., *J. Oil Colour Chem. Assoc.*, 45, 300 (1962).
50. Everhard, M. E., Dickcius, D. A., and Goodhart, F. W., *J. Pharm. Sci.*, 53, 173 (1964).

Chapter 3

MEASUREMENT AND STANDARDIZATION OF DIFFUSE REFLECTANCE

1. The Integrating Sphere

Of all the means that have been devised for the collection of diffuse radiation, the integrating sphere is both the best known and the best understood theoretically. For a number of analytical applications, such as scanning the chromatograms with a small slit or a "flying spot" device, the total integrated light energy reflected by the sample can be obtained by mathematical integration of the reflectance value of the individual small areas under examination. This value would normally be proportional to the area of a recorded chromatographic peak.[1] Where larger sample areas are investigated in one single measurement, it is advisable to use an integrating sphere.

The earliest descriptions of such an optical sphere were given by Sumpner[2] and by Ulbricht,[3] hence the name Ulbrichtkugel (Ulbricht sphere). It consists of a sphere-shaped enclosure whose inner walls are coated with a highly reflecting material that can reflect the diffuse radiation with as close to 100% efficiency as possible in the wavelength range of interest to the experimenter. For an ideal sphere the intensity due to reflected light at any point in the sphere should be independent of spatial distribution or location of a particular light source and should be a direct measure of the total flux.

There are various modes of application as well as a large number of experimental arrangements for the integrating spheres used in diffuse reflectance spectroscopy. Two of the more important basic designs are depicted in Figure 3.01. They involve the directional irradiation of monochromatic light onto a sample surface. Figure 3.01(a) shows the principle of the substitution method that is common to reflectance attachments designed for single-beam spectrophotometers. With this method the radiation enters the sphere and impinges on the sample surface. A photometer mounted externally at any point on the sphere measures the radiation intensity at the sphere wall. The sample is replaced by a standard, or vice versa, and the measurement is repeated.

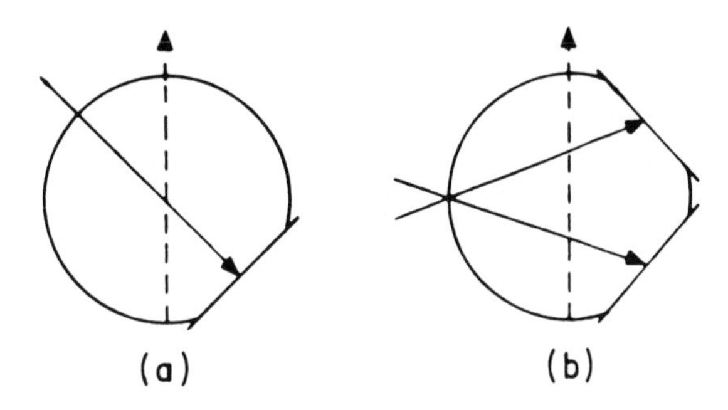

FIGURE 3.01 Schematic representation of the use of integrating spheres in reflectance spectroscopy. (a) substitution method, (b) comparison method.

The diffuse reflectance of a sample is thus measured relative to a suitable standard.

The comparison method, whose working principle is depicted in Figure 3.01(b), employs simultaneous irradiation of sample and reference standard, either by a true double-beam mode or by alternately deflecting the same beam to either sample or reference surface by means of a rotating mirror. For ideal conditions, sample and standard should be illuminated equally.

In order for the Kubelka-Munk theory to be absolutely valid, a sample should be illuminated by completely diffuse light. For the majority of samples, however, directional light can also be used and changes in the mode or angle of irradiation (different optical geometries) do not result in significant changes of the diffuse reflectance measured[4] (see Table 3.01). However, for samples with a definite structure, a rough texture, or a glossy surface, directional irradiation produces shadow effects and diffuse irradiation becomes essential. This can be achieved by illuminating the sample via the optical sphere in both the substitution and the comparison mode by simply reversing the optical path in either Figure 3.01(a) or 3.01(b). According to the reciprocity relationship $[R_d, (\theta,\phi) = R_{(\theta,\phi)}, d]$[5-7] the results of the measurements should remain unaltered.

Other investigators[8] have recommended a semisphere or circumferential mirror system for the diffuse illumination of roughly textured samples such as textiles, powders with large particle size, etc. Greater efficiency is often claimed when these systems are used.

TABLE 3.01

Diffuse Reflectance of CaF_2 (Measured Relative to an MgO Standard) as a Function of the Optical Geometry[4]

	${}_{45}R'_{\infty_0}$	${}_dR'_{\infty_0}$	${}_0R'_{\infty d}$
CaF_2; R'_∞	0.970	0.978	0.980
	0.941	0.947	0.950
	0.904	0.906	0.912
CaF_2; R'_0	0.885		0.880
	0.856		0.853
	0.827		0.828
CaF_2; R'_0	0.897	0.899	
	0.846	0.845	

(From **Kortüm,** G. and Oelkrug, O., *Z. Naturforsch.*, 19a, 28 (1969). With permission of publisher.)

An actual example of an optical sphere operating with an optical geometry ${}_dR_0o$ is shown in Figure 3.02. The screens S prevent the direct illumination of the sample surface. This design is used in the Zeiss Elrepho Colormeter.

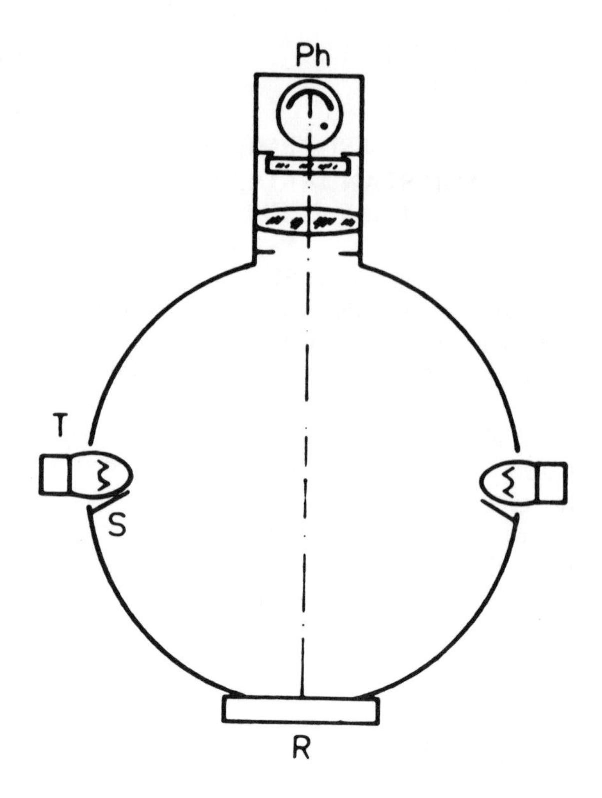

FIGURE 3.02 Optical diagram of the Zeiss Elrepho Reflectance Photometer. Ph, photomultiplier; T, tungsten lamp; S, shield; R, sample.

The principle of radiant flux distribution within an integrating sphere can be illustrated by a summation method for an infinite series of inter-reflecting rays.[9-11] The light intensity reflected from a sample that has been irradiated diffusely via the integrating sphere can be expected to be considerably higher than if directional illumination had been used. The same can be said for the viewing geometry, which employs a sphere for the collection of diffusely reflected light. In both cases a multiple reflection phenomenon, which occurs at the walls of the sphere, is responsible for this amplification effect.

If we assume that a sample is irradiated at a given angle with a radiation flux of intensity S, the reflecting power of the sample would be R_s, and the flux intensity B, which is attributable to specular reflection, would be

$$B_{1s} = R_s S \,\Delta\omega. \qquad (3.01)$$

$\Delta\omega$ represents the area of the exit port through which the measurement has been made with a suitable detector. The same sample is then mounted against the sample port of an integrating sphere, forming part of the inner wall lining of the sphere. The optical sphere possesses an average reflecting power R_{sp} and, as a result of the multiple reflection on the walls of the sphere, the flux intensity B is augmented to

$$B_s = R_s R_w S \frac{1}{1 - R_{sp}} \Delta\omega. \qquad (3.02)$$

The term $1/(1 - R_{sp})$ is a factor often referred to as "sphere efficiency." R_w is the general reflection power of the material used for coating of the sphere wall (generally MgO) and has an order of magnitude of 1. Since R_{sp} is usually about 0.95 it is not uncommon to observe a 20-fold amplification of the radiation flux as a result of this multiple sphere reflection.

The average reflecting power R_{sp} is arrived at by multiplying R_w and R_s with the corresponding surface area a. If one adopts the symbol a_s for the surface area of the sample, a_i and a_e for the areas limiting the emergence and incidence flux (circular areas for detector and entrance ports, respectively), and A for the total inside area of the sphere, then one obtains the following relationship for the average reflection power:

$$R_{sp} = \frac{(A - \Sigma a)\, R_W + a_S R_S}{A}. \tag{3.03}$$

In the substitution mode the sample is then replaced by the reference standard, and Equation 3.02 becomes

$$B_r = R_r R_W S \frac{1}{1 - R'_{sp}} \Delta\omega, \tag{3.04}$$

where R'_{sp} is given by a relationship analogous to Equation 3.03:

$$R'_{sp} = \frac{(A - \Sigma\, a)\, R_W + a_S R_r}{A}. \tag{3.05}$$

The average reflecting power of the sphere depends on the relative difference between R_s and R_r, which may become negligible ($R_s \cong R_r$) when a_s approaches a very small value. From Equations 3.02 and 3.04 the following relationship can then be derived:

$$\frac{B_S}{B_r} = \left(\frac{R_S}{R_r}\right)\left(\frac{1 - R'_{sp}}{1 - R_{sp}}\right). \tag{3.06}$$

It can be seen that the ratio B_s/B_r obtained by the substitution method differs from the ratio of relative reflecting powers by the factor

$$a \equiv \frac{1 - R_{sp}}{1 - R_{sp}}. \tag{3.07}$$

Expression 3.07 is generally known as the "sphere error." Since such errors can become quite significant, an equation for the computation of a corrected relative reflection term R' can be derived from Equations 3.06 and 3.07:

$$R' = \frac{R_S}{R_r} = \left(\frac{B_S}{B_r}\right)\left(\frac{1}{a}\right). \tag{3.08}$$

Table 3.02 gives some sphere errors as determined by Kortüm[11] for a sphere of 12.6 cm diameter, coated with MgO (R_r = 0.98) with an area $a_s = a_i = a_e = 7\ cm^2$ and an R'_{sp} value of 0.9526.

Other methods have also been proposed for a mathematical treatment of the radiant fluxes in optical spheres. Some earlier papers, notably by Taylor[12] and others[13-15] dealt with perfect spheres only, neglecting apertures of any kind. Later, several workers included apertures of various sizes in their discussions.[16,17] Some workers[18,19] have dealt with this problem by using suitable integral equations. Similar results were obtained, however, with any of the different approaches.

In general it can be said that the smaller R_s, the larger the sphere error becomes. Another factor to be considered is the ratio of the surface area of the sample a_s to the total surface area A. In order to keep a small, one should choose relatively small values for a_s and large overall areas A (e.g., a larger sphere can be used). A compromise should be made, however, since an increase of the sphere size will result in a decrease of the reflecting power R.

Excellent reviews and discussions on the theory and sources of error of integrating spheres have been given by Wendlandt and Hecht[20] and by Kortüm.[21]

2. Reference Standards and Sphere-coating Materials

In the past, smoked magnesium oxide was the universally accepted standard for reflectance measurements. This practice dates back to 1931, when the CIE (Commission Internationale de l'Eclairage) recommended that, for colorimetric or photometric reflectance measurements, the brightness or reflectance value of an MgO surface was taken as 100% at 457 nm.[22]

Even though optimum preparation procedures

TABLE 3.02

Sphere Errors Arising from Reflectance Measurements with an Integrating Sphere by the Substitution Method[11]

R_s	0.900	0.800	0.700	0.600	0.500	0.400	0.300	0.200
R_{sp}	0.9515	0.9501	0.9487	0.9473	0.9459	0.9445	0.9431	0.9417
a	1.025	1.055	1.085	1.114	1.144	1.173	1.203	1.233
$1/a$	0.9753	0.9479	0.9220	0.8975	0.8743	0.8522	0.8313	0.8113

(From Kortüm, G., *Kolorimetrie, Photometrie und Spektrometrie,* 4th ed., Springer-Verlag, Berlin, 1962, 345. With permission.)

for the MgO standard have been described, the assumption that MgO has a reflectance of 100% is arbitrary. The actual reflectance is somewhat lower and can vary considerably because of the many parameters that can influence the measurements. The results of several attempts to determine the true value for the reflectance of MgO surfaces have been compiled in Table 3.03.[30] The spread of values in this table (> 2%) may be attributed to differences in the method of measurement,[16] although other parameters, such as layer thickness,[28] particle size,[21,29] method of preparation and aging,[26] or density of packing,[31] may also affect the reflectance values. Obviously it would be difficult to control all these variables sufficiently for the production of an absolutely reliable standard. Data presented by Tellex and Waldron,[28] for example, revealed that even a layer of smoked magnesium about 8 mm thick still possesses some transparency (Figure 3.03). Controversial opinions exist on the best method to coat a sphere or to prepare smoked MgO plate standards. Middleton and Sanders[26] have investigated the different coatings obtained by (a) burning magnesium ribbons and (b) burning magnesium turnings in a silicate dish, as recommended earlier.[32] Although the values did not differ by more than 0.002 reflectance units throughout the visible spectrum, the authors recommended method (a) as the more reliable one. It was shown that up to five successive coatings were needed before no further increase in reflectance values was observed[26] (see Figure 3.04). They also reported[26] a relatively rapid aging process for MgO (see Figure 3.05), which was attributed partly to impurities adsorbed over a

TABLE 3.03

Absolute Reflectance Values for Magnesium Oxide[30]

Authors	R values
Preston[16] (1929)	0.974
Priest and Riley[23] (1930)	0.987 ± 0.001
Taylor[24] (1937)	0.972
Benford, Lloyd and Schwarz[25] (1948)	0.991
Middleton and Sanders[26] (1951)	0.976 ± 0.002
Gordon-Smith[27] (1952)	0.973
Tellex and Waldron[28] (1955)	0.989 ± 0.0014
Budde[29] (1958)	0.980 ± 0.001

(From Budde, W., *J. Opt. Soc. Am.*, 50, 217 (1960). With permission of American Institute of Physics.)

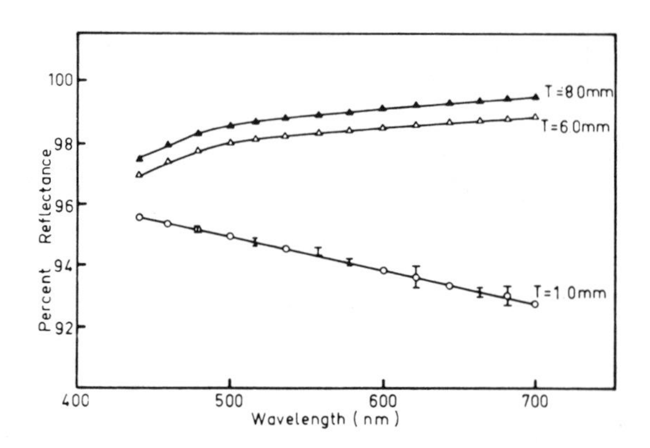

FIGURE 3.03 Coefficient of reflection vs. wavelength for electrostatic-deposited MgO. Thickness as parameter.[28] (From Tellex, P. A. and Waldron, J. R., *J. Opt. Soc. Am.*, 45, 19 (1955). With permission of publisher.)

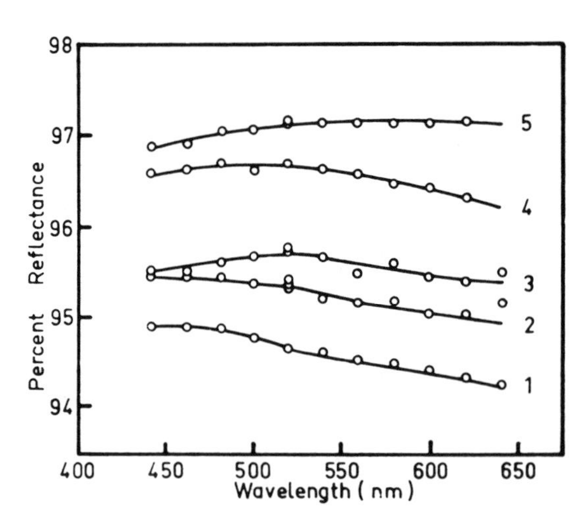

FIGURE 3.04 Improvement of surface by further coating.[26] (From Middleton, W. E. K. and Sanders, C. L., *J. Opt. Soc. Am.*, 41, 419 (1951). With permission of publisher.)

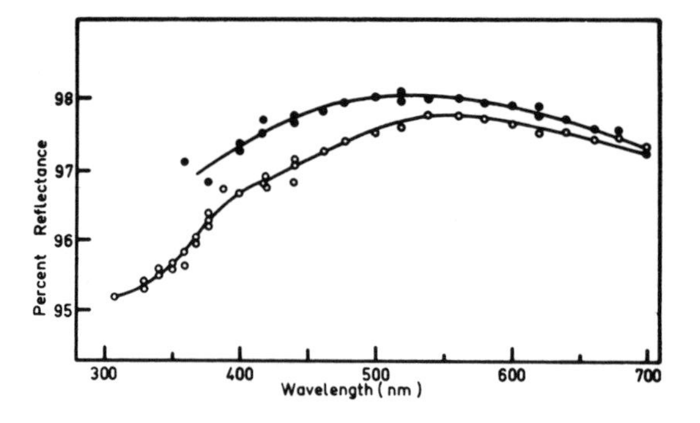

FIGURE 3.05 Effect of aging on freshly prepared MgO. •, December 21, 1950. O, January 9, 1951.[26] (From Middleton, W. E. K. and Sanders, C. L., *J. Opt. Soc. Am.*, 41, 419 (1951). With permission of publisher.)

period of time and partly to decomposition of nitrites (produced during the smoking process) by UV radiation.[23,26] This effect becomes particularly evident in the spectral region below 500 nm (Figure 3.05).

Until recently, the effect of compaction of powders on the reflectance measured has been largely neglected, both in practical and theoretical studies of reflectance phenomena. A detailed investigation of the influence of pressure upon packing of powder samples has been published by Schatz.[31] Studies were carried out on materials such as Al_2O_3, $BaSO_4$, MgO, CrO_3, Fe_2O_3, MnO, Mn_3O_4, NiO, Sm_2O_3, Y_2O_3, ZnO, etc., the first three of which have some importance as reference standards. Some nonoxides were tested also. The general observation was that weak absorbers such as the reference standards show a decrease, and strong absorbers show an increase in reflectance with increased pressure (Figures 3.06 and 3.07). The former effect is more predominant in the visible than in the UV region of the spectrum and increases gradually with an increase in wavelength. Schatz[31] based an attempt to explain this phenomenon on the fact that at high pressures the powder particles move closer to each other. This results in frustrated total reflection[33] and in transmission by the layer of some of the light that otherwise would have been totally internally reflected.

In other words, with increasing pressure the interparticular distance decreases, resulting in an increase in transmitted light and a decrease in reflected light.

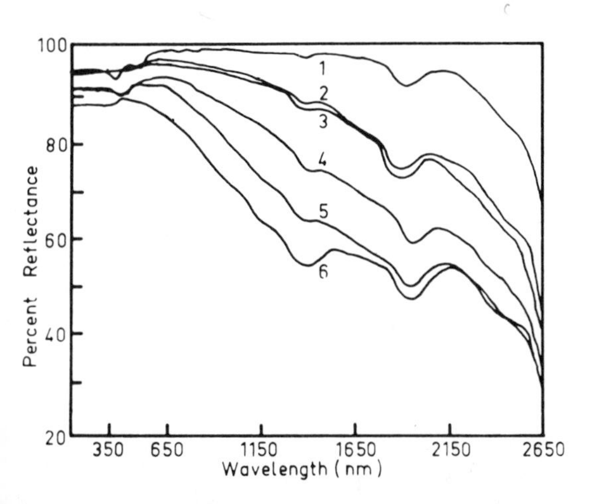

FIGURE 3.06 Spectral total reflectance vs. MgO of compacted MgO powder. Compaction pressure (psi): 1 = 290; 2 = 1,150; 3 = 2,880; 4 = 5,760; 5 = 20,200; 6 = 28,800.[31] (From Schatz, E. A., *J. Opt. Soc. Am.*, 56, 389 (1966). With permission of publisher.)

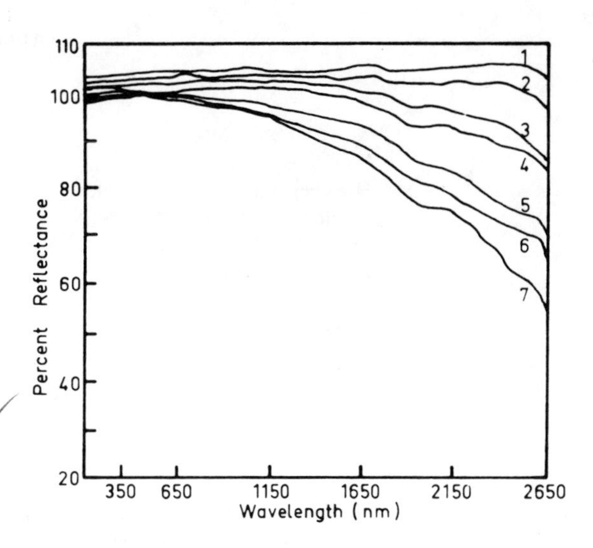

FIGURE 3.07 Spectral total reflectance vs. MgO of compacted Al_2O_3 powder. Compaction pressure (psi): 1. 290; 2. 1,150; 3. 2,880; 4. 5,760; 5. 11,500; 6. 20,200; 7. 28,800.[31] (From Schatz, E. A., *J. Opt. Soc. Am.*, 56, 389 (1966). With permission of publisher.)

In order to overcome some of the previously mentioned disadvantages of a reference standard such as magnesium oxide, the use of barium sulfate as an alternate standard has been suggested repeatedly.[34-36] Extensive studies on $BaSO_4$,[30] however, revealed that despite the good reproducibility of measurement of $BaSO_4$ surfaces for one laboratory or for a single experiment, the results fluctuate considerably from one investigator to another (Table 3.04). Indications are that the fluctuations in reflectance values reported in Table 3.04 are due to factors similar to those discussed for MgO, and it may be reasonable to assume that this holds for practically all powder reference standards. Fluctuations due to different size, shape, and orientation of the powder particles can be considerable (Table 3.05). The degree of impurity is also important in $BaSO_4$, as are the layer thicknesses of the standard surfaces. The hiding power of varying layers of $BaSO_4$ was studied[30] by measuring different thicknesses of one particular grade (0.3 μ particle size) on a dark background (Figure 3.08). At a layer thickness of about 1 mm, no further interference from the background was observed at wavelengths up to 700 nm, and one can reasonably assume that at this particular point an infinite layer thickness has been reached. These results agree fairly well with

TABLE 3.04

Absolute Reflectance Values for Barium Sulfate[30]

Author	R value	Illuminant
Miescher and Rometsch[34] (1950)	0.991 ± 0.002	A
Middleton and Sanders* (1950)	0.975	A
Budde[29] (1958)	0.982 ± 0.001	C
Budde[37] (1958)	0.970 ± 0.001	C
Budde†	0.964 ± 0.001	C
Budde†	0.926 ± 0.001	C
Budde†	0.866 ± 0.001	C
Sanders‡ (1960)	0.957	C
Sanders‡ (1960)	0.964	C
Miller and Sant[38] (1958)	0.945 ± 0.0005	500 W tungsten

*See Reference 34.
†Earlier unpublished measurements listed in Reference 30.
‡Personal communication to Reference 30.
(From Budde, W., *J. Opt. Soc. Am.*, 50, 217 (1960). With permission of American Institute of Physics.)

TABLE 3.05

Reflectance and Chromaticity Coordinates of $BaSO_4$ as a Function of Varying Particle Size and Impurities[30]

$BaSO_4$ samples	Particle size (μ)	Reflectance	Impurities (%)	Chromaticity coordinates	
				x	y
1	1.4	0.970	< 0.055	0.3117	0.3182
2	1.3	0.970	1.26	0.3117	0.3198
3	0.3	0.966	< 0.266	0.3099	0.3164
4	0.3	0.964	0.57	0.3101	0.3166
5	0.2	0.926	0.76	0.3094	0.3167
6	0.2	0.973	< 0.14	0.3097	0.3162

(From Budde, W., *J. Opt. Soc. Am.*, 50, 217 (1960). With permission of American Institute of Physics.)

data reported by Kortüm and Vogel[39] in an investigation of conditions for infinite layer thickness. In general it can be concluded from these studies that $BaSO_4$ has many desirable properties as a reference standard, one of which (in contrast to MgO) is a relatively good stability toward aging effects (< 0.5% after 595 days).[40] Instrument manufacturers also increasingly favor $BaSO_4$ over MgO for sphere coating and standard material.

Magnesium carbonate, which has reflectance properties similar to MgO, is occasionally used as a reference material.[30,41] Its main advantage is the compact block shape in which it is delivered by many instrument companies such as Beckman, Bausch & Lomb, and others. For practical applications other, more durable reference substances (Carrara, Didymium, Vitrolite, Opal Glasses, etc.) are often preferred.[42-45] Vitrolite, which is a white structural glass, is available as standard sample from the National Bureau of Standards[46] and is primarily applied in the "compensated opal" method.[46] Reflectance data for Vitrolite and $MgCO_3$ standards measured vs. MgO are presented in Table 3.06.

Budde[30] has discussed the use of porcelain enamel as a reflectance standard. Such a standard

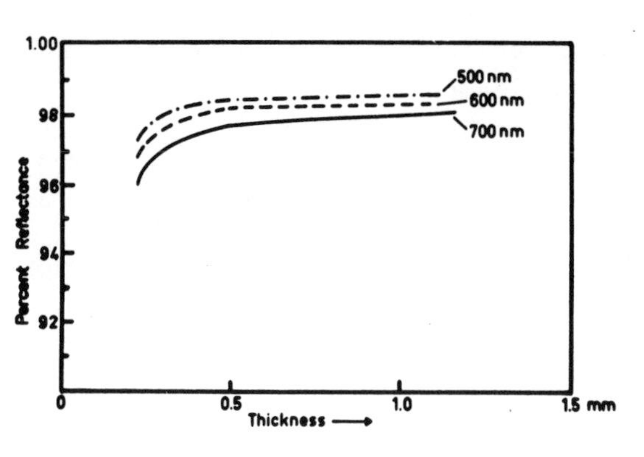

FIGURE 3.08 Hiding power of $BaSO_4$ at three different wavelengths.[30] (From Budde, W., *J. Opt. Soc. Am.*, 50, 217 (1960). With permission of publisher.)

is employed as a built-in swing-in standard in the Zeiss Chromatogram Spectrophotometer. Another white glass standard with a treated dull surface is provided for the Zeiss Elrepho Colormeter. Reflectance values for this standard, measured relative to MgO, are reported in Table 3.07. The absolute values in this table have been measured by Höfert and Loof[47] with the use of a Taylor sphere; they may be the most reliable values available at present. Zeiss is marketing a powder press that enables the production of MgO and $BaSO_4$ standards with a reproducibility of better than ± 0.1%.[48,49]

Very often the standard has to be chosen to fit a particular experimental requirement, e.g., certain adsorbent-adsorbate interactions, such as a Lewis acid-base interaction, ligand exchange functions, complex formations, etc., may have to be considered in the choice of a reference standard, which frequently would consist of the same adsorbent. If desired, the measurements could then be converted to absolute reflectance values by using standard procedures. For quantitative analytical work usually only relative measurements are required for one particular experiment, so a wide range of standards could be chosen. In chromatographic work, for example, the adsorbent used for the separation process is usually recommended as a standard.[50] Materials such as alumina, silica gel, and cellulose have, therefore, widely been used for this purpose. Their spectra, recorded vs. freshly prepared MgO, are presented in Figure 3.09,[51] together with a number of other materials of interest to the analyst. For comparison of data the previously mentioned

TABLE 3.06

Reflectance of $MgCO_3$[41] and Vitrolite[46] Relative to Magnesium Oxide

Wavelength (nm)	$MgCO_3$	Vitrolite
750	–	0.885
700	0.994	0.895
650	0.994	0.901
600	0.995	0.914
550	0.995	0.922
500	0.994	0.919
460	0.989	0.911
400	0.978	0.913
350	0.915	–
300	0.824	–
250	0.781	–

TABLE 3.07

Relative Reflectance of the Zeiss Primary Standard Measured vs. MgO with Absolute Reflectance (Zeiss Information)

Wavelength (nm)	Relative primary standard	Absolute MgO
420	0.992	0.983
460	0.992	0.986
490	0.993	0.986
530	0.994	0.981
570	0.995	0.981
620	0.995	0.988
680	0.996	0.986

disadvantages still exist with the reference materials discussed. The major drawback is the lack of reproducibility of data from one laboratory to another. To overcome these difficulties and uncertainties in the primary standard of reflectance, the CIE recommended in 1959[52] that the "perfect diffuser" be adopted as a reference standard for reflectance measurements of opaque specimens. This recommendation did not bring about the immediate adoption of the perfect diffuser, but at the 16th CIE session in 1967 the committee agreed on the following wording: "The perfect diffuser is recommended as the reference standard. It supersedes magnesium oxide as of January 1, 1969."[53] In practice this would mean that relative reflectances are reported in absolute terms after appropriate recalculation, e.g., with reference to the ideal and hypothetical white

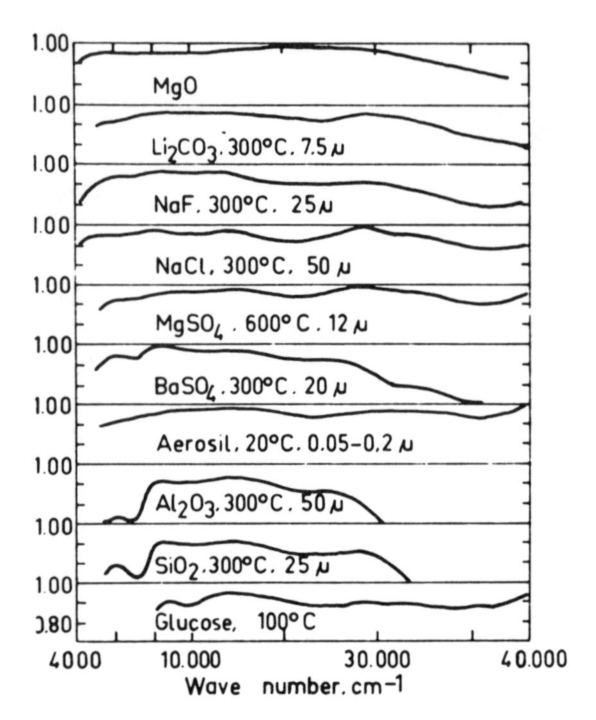

FIGURE 3.09 Dependence on wavelength of the absolute diffuse reflectance of several white standards (maximum grain size in microns) referred to that of freshly prepared MgO.[51] Temperatures given are drying temperatures. (From Kortüm, G., Braun, W., and Herzog, G., *Angew. Chem.*, 2, 333 (1963). With permission of publisher.)

surface that reflects light 100% in a perfectly diffuse manner. Although there are still difficulties to be overcome when this recommendation is put to use, it will become increasingly easier to adopt it as new white standards with better values for their absolute reflectances and improved reflectance spectrophotometers and automatic calculation techniques become available.[54]

Obviously, some materials used for reference standards are also used for the coating of integration spheres. These coating materials should be 100% reflecting at all wavelengths of interest and perform as ideal diffusers. This is as much of an illusion as for the previously discussed standard surfaces, and for practical reasons it has been necessary to resort to the same materials such as MgO and $BaSO_4$ as the most widely used.[28,32,35,55]

There are various methods for the coating of the spheres.[32,55] The substances are applied to the surface either by suitable smoking procedures or in the form of paints. In addition, water glass or certain plastic materials may be used to increase the mechanical stability of the coatings. Detailed coating procedures are usually provided by the instrument manufacturers.

REFERENCES

1. **Goldman, J. and Goodall, R. R.,** *J. Chromatogr.*, 40, 345 (1969).
2. **Sumpner, W. E.,** *Proc. Phys. Soc. (London)*, 12, 10 (1892).
3. **Ulbricht, T.,** *Elektrotech. Z.*, 21, 595 (1900).
4. **Kortüm, G., and Oelkrug, O.,** *Z. Naturforsch.*, 19a, 28 (1969).
5. **Schultz, H.,** *Z. Phys.*, 31, 496 (1925).
6. **Helwig, H. J.,** *Licht*, 7, 99, 119, 140 (1937).
7. **Fragstein, L.,** *Optik*, 12, 60 (1955).
8. **Derksen, W. L., Monahan, T. I., and Lawes, A. J.,** *J. Opt. Soc. Am.*, 47, 995 (1957).
9. **Miller, O. E. and Sant, A. J.,** *J. Opt. Soc. Am.*, 48, 828 (1958).
10. **Longhurst, R. S.,** *Geometrical and Physical Optics*, John Wiley & Sons, New York, 1957, 388.
11. **Kortüm, G.,** *Kolorimetrie, Photometrie und Spektrometrie*, 4th ed., Springer-Verlag, Berlin, 1962, 345.
12. **Taylor, A. H.,** *Scientific Papers Natl. Bur. Std.*, No. 391, 421 (1921).
13. **Karrer, E.,** *Scientific Papers Natl. Bur. Std.*, No. 415, 203 (1921).
14. **Rosa, E. B. and Taylor, A. H.,** *Scientific Papers Natl. Bur. Std.*, No. 447, 281 (1922).
15. **McNicholas, H. J.,** *J. Res. Natl. Bur. Std.*, 1, 29 (1928).
16. **Preston, J. S.,** *Trans. Opt. Soc. (London)*, 31, 15 (1929-30).
17. **Hardy, A. C. and Pineo, O. W.,** *J. Opt. Soc. Am.*, 21, 502 (1931).
18. **Jacquez, J. A. and Kuppenheim, H. T.,** *J. Opt. Soc. Am.*, 45, 460 (1955).
19. **Jacquez, J. A. and Kuppenheim, H. T.,** *J. Opt. Soc. Am.*, 46, 428 (1956).
20. **Wendlandt, W. W. and Hecht, H. G.,** *Reflectance Spectroscopy*, John Wiley & Sons, New York, 1966, 253.
21. **Kortüm, G.,** *Reflexionsspektroskopie*, Springer-Verlag, Berlin, 1969, 142, 225 (English translation, Springer-Verlag, New York, 1969).
22. *Proceedings of the 8th Session of the Commission Internationale de l'Eclairage, Cambridge, 1931*, Cambridge University Press, England, 1932, 23.
23. **Priest, J. G. and Riley, J. O.,** *J. Opt. Soc. Am.*, 20, 156 (1930).

24. **Taylor, A. H.,** *Proc. Phys. Soc. (London),* 49, 105 (1937).
25. **Benford, F., Lloyd, G. P., and Schwarz, S.,** *J. Opt. Soc. Am.,* 38, 445 (1948).
26. **Middleton, W. E. K. and Sanders, C. L.,** *J. Opt. Soc. Am.,* 41, 419 (1951).
27. **Gordon-Smith, G. W.,** *Proc. Phys. Soc. (London),* B65, 275 (1952).
28. **Tellex, P. A. and Waldron, J. R.,** *J. Opt. Soc. Am.,* 45, 19 (1955).
29. **Budde, W.,** *Farbe,* 7, 295 (1958).
30. **Budde, W.,** *J. Opt. Soc. Am.,* 50, 217 (1960).
31. **Schatz, E. A.,** *J. Opt. Soc. Am.,* 56, 389 (1966).
32. Natl. Bur. Std. Letter Circ. LC 547, 1939.
33. **Born, M. and Wolf, E.,** *Principles of Optics,* Pergamon Press, New York, 1959, 46.
34. **Miescher, K. and Rometsch, R.,** *Experientia,* 6, 302 (1950).
35. **Middleton, W. E. K. and Sanders, C. L.,** *Illum. Eng.,* 48, 254 (1953).
36. **Kortüm, G. and Haug, G.,** *Z. Naturforsch.,* 8a, 372 (1953).
37. **Budde, W.,** *Farbe,* 7, 17 (1958).
38. **Miller, O. E. and Sant, A. J.,** *J. Opt. Soc. Am.,* 48, 828 (1958).
39. **Kortüm, G. and Vogel, J.,** *Z. Phys. Chem. (Frankfurt),* 18, 110 (1950).
40. **Laufer, J. S.,** *J. Opt. Soc. Am.,* 49, 1135 (1959).
41. **Jacquez, J. A., McKeehan, W., Huss, J., Dimitroff, J. M., and Kuppenheim, H. F.,** *J. Opt. Soc. Am.,* 45, 971 (1955).
42. **Keegan, H. J. and Gibson, K. S.,** *J. Opt. Soc. Am.,* 34, 77 (1944).
43. **Gabel, J. W. and Stearns, E. J.,** *J. Opt. Soc. Am.,* 39, 481 (1949).
44. **Billmeyer, F. W., Jr.,** *J. Opt. Soc. Am.,* 46, 72 (1956).
45. **Jacquez, J. A., McKeehan, W., Huss, J., Dimitroff, J. M., and Kuppenheim, H. F.,** *J. Opt. Soc. Am.,* 45, 781 (1955).
46. Natl. Bur. Std. Letter Circ. LC 1017, 1948.
47. **Höfert, H. J. and Loof, H.,** *Farbe,* 13, 53 (1964).
48. Zeiss Information No. 50-660/IV-e, Carl Zeiss Inc., Oberkochen, Germany, 1968.
49. **Stenius, A. S.,** *J. Opt. Soc. Am.,* 45, 727 (1955).
50. **Frei, R. W.,** in *Recent Progress in Thin-layer Chromatography and Related Methods,* Vol. II, edited by Niederwieser, A. and Pataki, G., Eds., Ann Arbor Science Publishers, Ann Arbor, Mich., 1970, Chap. I.
51. **Kortüm, G., Braun, W., and Herzog, G.,** *Angew. Chem.,* 2, 333 (1963).
52. *Proceedings of the 14th Session of the Commission Internationale de l'Eclairage, Brussels, 1959,* CIE Publication No. 4, 1960, 36.
53. Proc. 16th Session Commission Internationale de l'Eclairage, Washington, 1967.
54. **Budde, W. and Chapman, S. M.,** *Pulp Paper Mag. Can.,* 69, T206 (1968).
55. **Dimitroff, J. M. and Swanson, D. W.,** *J. Opt. Soc. Am.,* 46, 555 (1956).

IV. INSTRUMENTATION

1. Filter Instruments

The first instrument for the measurement of reflected light used filters in order to obtain light of a defined wavelength range.[1] Filter instruments, both noncommercial and commercial models, have since been designed in large numbers primarily to be used in the paint, ceramics, paper, and textile industries for the defining and matching of colors and for the investigation of specular properties such as whiteness, brightness, etc. Occasionally this technique has been used in the food industry or in the investigation of biological samples, building materials, plastics, etc.

Most of these early instruments operated with an integrating sphere (Ulbrichtkugel) for the collection of diffuse radiation;[1-5] a few others used light pipes, semispheres, mirror systems, etc. to pick up reflected radiation. Specular reflectance was either excluded or included (total reflectance), depending on the location, shape, and distance of the phototube from the sample surface. Several commercial instruments, built specifically for control purposes in the above-mentioned fields, are still available. One of the best-known products, particularly in Europe, is the Elrepho Colormeter (Electric Reflectance Photometer, see Figure 4.01), manufactured by Carl Zeiss Inc., Oberkochen, West Germany. A detailed and critical discussion of this instrument has been given elsewhere.[6-9]

The working principle of the Elrepho Photometer is based on an integrating sphere design and can best be understood on inspection of Figure 4.02. The sample A is mounted at the proper opening of the sphere and is irradiated in a diffuse manner via the optical sphere by two incandescent lamps which cannot be seen in this drawing. Direct radiation is prevented by small screens attached between lamps and sample. A built-in swing-in standard S is irradiated by the same system. The light reflected from the surfaces A and S strikes the two photocells Ph_1 and Ph_2 at a 90° angle to the sample and reference surfaces, respectively. The currents of the two photocells flow in opposite directions. A neutral wedge GK in the measuring beam and a measuring diaphragm MB in the reference beam compensate the two photocurrents. The position of the diaphragm can be read on a scale. The wedge GK can reduce the intensity of the illumination on the photocell by about one tenth. The light filters F are screwed in pairs into the filter changer in front of the photocells. They can be used to give the desired wavelength range. Seven filters are available, covering the effective range 426 to 681 nm with spectral band widths ranging from 15 to 57 nm. The measurements can be carried out by using first a standard of known absolute reflectance ($BaSO_4$) in the sample port A. The measuring diaphragm is then adjusted to this value and with the wedge the two photocurrents are balanced to the null point of the meter N. The standard is then replaced by the sample, the instrument again adjusted to the zero position, and the relative reflectance of the sample read at the measuring drum of the diaphragm. The built-in standard is a piece of

FIGURE 4.01 The Carl Zeiss Elrepho.

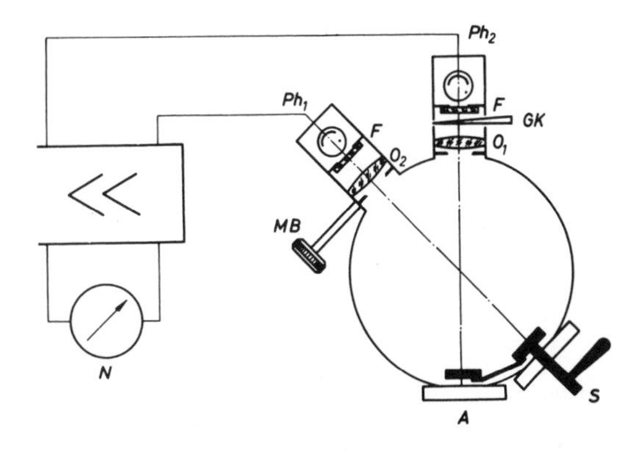

FIGURE 4.02 Optical schematic of the Elrepho.

white glass and has a more stable reflectance value than ordinary reference standards. The advantage of the double-beam arrangement is the same as for regular photometers, viz., to eliminate fluctuations in the energy source. The reproducibility of the measurements is reported as ±0.2%. An accessory permitting scale expansion for the measurement of low-reflectance samples is available. The measuring head is cantilevered over a large stand which also permits the measurement of big samples. The samples are held against the sphere by means of a spring. The smallest sample size that can be accommodated on the standard instrument is a circle with a diameter of 34 mm, or a square of 9.5 x 27.5 mm.

An automatic version of the Elrepho instrument has recently been developed by Zeiss. It is the Colormeter RFC-3 (see Figure 4.03), which provides for digital and printed representation of the data. The color coordinates x, y, and z are computed on the basis of the white reference standard DIN 5033. Two spectral ranges are available: from 400 to 700 nm, with 24 interference filters and a spectral interval of 13 nm, or from 380 to 720 nm, using 18 interference filters and a spectral interval of 20 nm. The instrument features computation and output of the tristimulus values XYZ and the chromaticity coordinates xy for illuminants A, C, and D65 within the 2° or 10° standard colorimetric system according to the weighted ordinate method. Color difference computation may be made by several standard available programs or one formulated to meet the user's needs. The instrument has a wavelength accuracy of ±0.2 nm, and offers reproducibilities of better than ±0.01% for R = 0%, and ±0.04% for R = 100%. Measuring geometries are $_dR_8^\circ$ or $_{45}{}^\circ R_0{}^\circ$. Sample illumination is provided by a 250-W zenon lamp with conversion filter for illuminant D65, and two 40-W filament lamps for spectral irradiation distribution in the sample plane according to illuminant A. The measuring aperture may be switched from 5 to 15 or 30 mm diameter. The instrument also measures transmittance and fluorescence in addition to reflectance. Measurement and output of reflectance and transmittance values take 70 to 90 sec. Computation and output of tristimulus values and chromaticity coordinates for illuminants A, C, and D65 require 20 sec, as do the computation and output of color differences for illuminants A, C, and D65. The instrument has excellent stability (variations of ±0.1% or less at the 0% and 100% readings over an 8-hr period) and, with the variety of output systems offered, is thus ideal for routine analysis with a high degree of automation.

A competitive product, which operates with a similar double-beam arrangement, an optical integration sphere, and a geometry of $_dR_0{}^\circ$ (geometry $_{45}{}^\circ R_0{}^\circ$ is also available) is the Pretema Spectromat FS-38, manufactured by Pretema Ltd., Birmensdorf-Zürich, Switzerland (see Figure 4.04). This instrument likewise uses a small digital computer, programmed to produce an instantaneous readout of CIE color coordinates in printed or digital form. Samples with diameters ranging from 5 to 40 nm can be investigated. The useful spectral range is 390 to 710 nm, and the measurements can be recorded by choice at 33 or

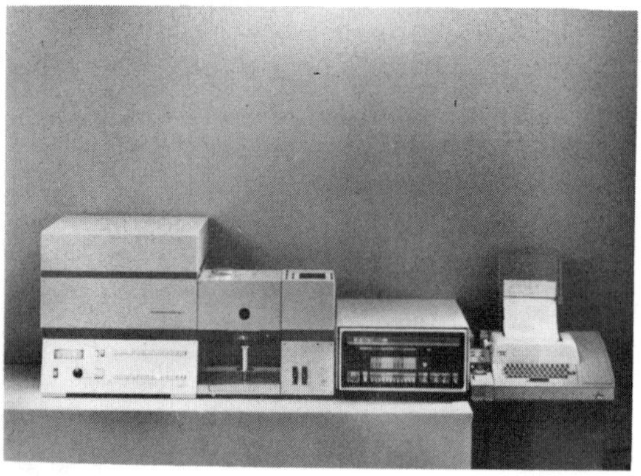

FIGURE 4.03 The Carl Zeiss Automatic Colormeter RFC-3.

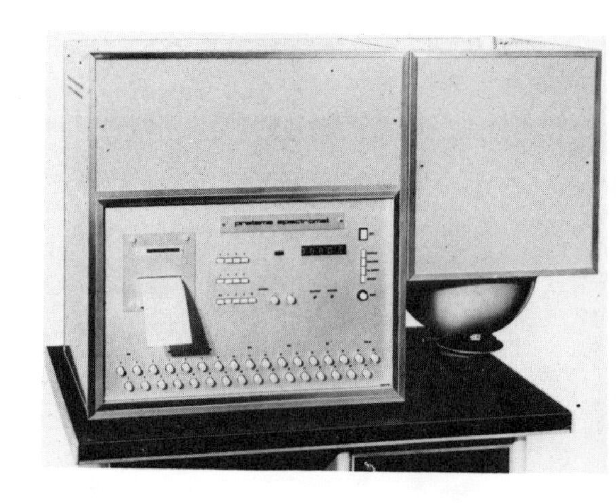

FIGURE 4.04 The Pretema Spectromat FS-38.

16 different wavelengths in 90 or 60 sec, respectively. Additional features of this instrument include modifications for transmission and fluorescence measurements. The latter is possible with a xenon lamp and the sphere is coated with a UV-reflecting aluminum surface. Like the Zeiss RFC-3, this instrument is used widely in industries where large-scale color comparisons in quality and production control are essential.

Van den Akker et al.[10] described and critically discussed an instrument known as the IDL Color-Eye. It is available through Kollmorgen Color Systems, 67 Mechanics Street, Attleboro, Massachusetts 02703. As can be seen from Figure 4.05, the instrument uses an integration sphere, and simultaneous illumination of sample and reference standard is achieved with a rotating mirror, which reflects the polychromatic beam alternately to the reference and standard sample ports. Filters are provided at various intervals ranging from 400 to 700 nm. The instrument permits the measurement of diffuse and/or specular reflectance, absolute or differential. The original Color-Eye of the 1950's was modified in 1963 by the introduction of a larger reflectance sphere. However, both of these designs were manually operated. The present model, the KCS-18 (shown in Figure 4.06), is a fully automated, solid-state instrument with digital readout of wavelength and reflectance or transmittance. Two photometric scales are available, 0 to 199.99% with 0.01% resolution, or 0 to 19.999% with 0.001% resolution. Illumination sources include 2856 K (C.I.E. illuminant A), 3100 K, D6500, and D6500(-UV). Spectrophotometric filters used offer an accuracy within ±0.5 nm of the specified filter wavelength, while the tristimulus filters have been designed to closely match the 1931 C.I.E. observer response. An optical schematic of the instrument is shown in Figure 4.07. The KCS-18 may be used for tristimulus color determination, spectrophotometric analysis, color matching, and colorant formulation with an accurate measurement of small color differences, rapid detection of metamerism, and has full quantitative fluorescence capability. The instrument also is designed for a variety of output systems including teletype, digital printer, spectral plotter, tape punch, and computer interfacing.

Somewhat different from the optical designs for the double-beam filter reflectometers discussed so far is the well-known line of color-difference meters produced by Hunter Associates Laboratory Inc., 9529 Lee Highway, Fairfax, Virginia 22030.

Some of the earliest filter-type reflectometers were designed and described by Hunter in 1934[11] and 1940.[12,13] The modern line of Hunter color and color-difference meters includes Model D25A with standard optical unit, Model D25P equipped with an optical sphere, and Model D25 M/L, which uses a circumferential mirror illumination. All three models operate in the single-beam mode. The electronic unit (solid-state) is the same for all the D25 models. The standard optical attachment D25A (see Figure 4.08) uses an illumination angle of 45°. The diffusely reflected polychromatic light is collected at an average zero-degree viewing angle

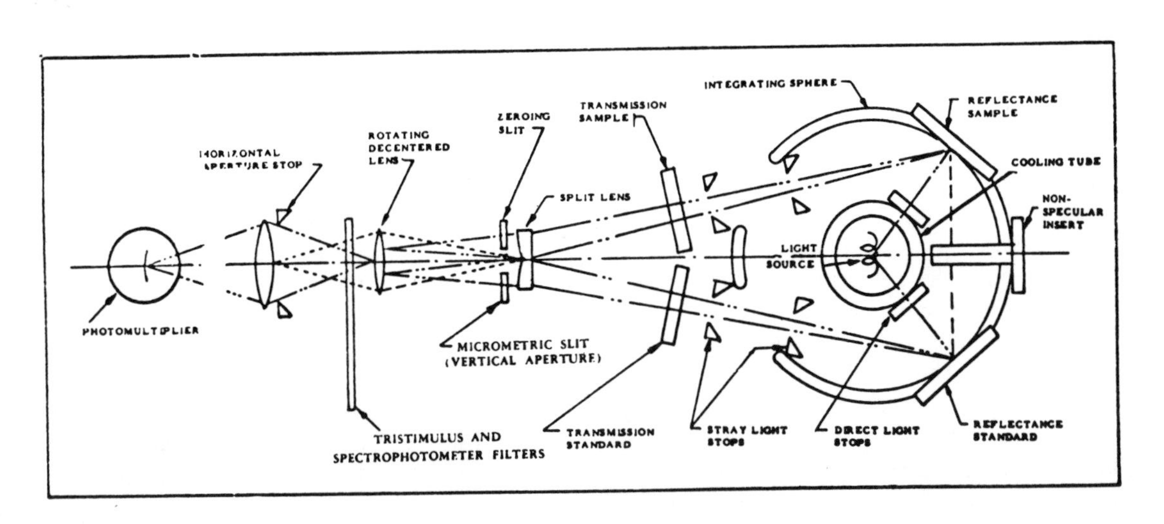

FIGURE 4.05 Optical diagram of the IDL Color-Eye.

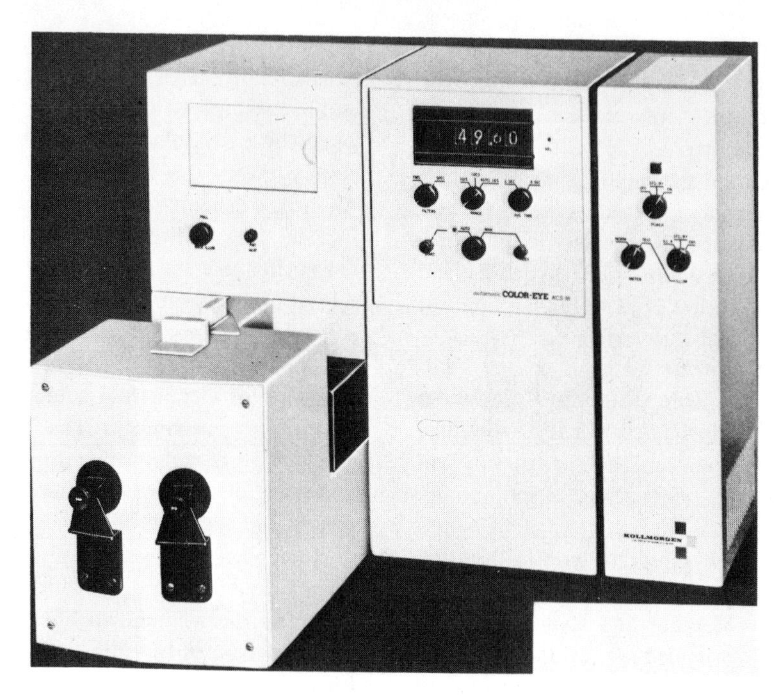

FIGURE 4.06 KCS-18.

by means of a light pipe which is located directly above the specimen. Specular reflectance escapes at a 45° angle of emergence through the viewing port without interfering. The diffuse radiation is split, mixed, and directed by the light pipe to a set of three phototubes. Four broad-band spectral filters (tristimulus X, Y, Z amber, and X blue) can be placed in front of the phototubes.

The optical head D25P is provided as a separate unit to be placed adjacent to the measuring unit D25. Illumination occurs with polychromatic light at a zero-degree angle of incidence. The diffusely reflected radiation is collected in the white-lined sphere (diam. 8 in.) and viewed at the top of the sphere with three phototubes. The set of filters described for the D25A model is used in this attachment also. Specular reflectance can be excluded with this optical arrangement. It is, however, possible to measure total reflectance by swinging either the source or the light sphere about 8° off their common axis.

A rather unusual design is used in the optical head D25 M/L. This attachment is recommended for samples with a relatively irregular and rough texture (e.g., textiles or some powders) where illumination – particularly at a single fixed 45° or zero-degree angle of incidence – would cause large fluctuations of the reflection pattern. The attachment uses a mirror system composed of faceted annular rings. The light from the tungsten lamp is then reflected uniformly from many directions onto the specimen at a 45° angle, causing a circumferential illumination effect, which eliminates fluctuations due to irregularities on the sample surface. To overcome this shadow effect, other instrument designers have illuminated rough-textured samples diffusely through a light sphere. The light reflected from the specimen is collected at a zero-degree angle by a light pipe and directed through four spectral filters to four vacuum phototubes.

A digital version of the D25, the D25D (shown in Figure 4.09) and a companion instrument, the D25D-2, are being marketed also by the company. The D25D offers a variety of optical heads, the DA head for flat opaque surfaces, the DL head for nonuniform surfaces and granular surfaces, the DM head for highly directional surfaces, and the DP head for transparent or translucent liquids and films, metals, and plastics. The D25D offers an instantaneous measurement and display of pushbutton-selected color values in Hunter L, a, b, and C.I.E. X, Y, and Z. Analog-to-digital conversion and automatic computation of tristimulus values are performed by solid-state circuitry and the readout appears within several milliseconds

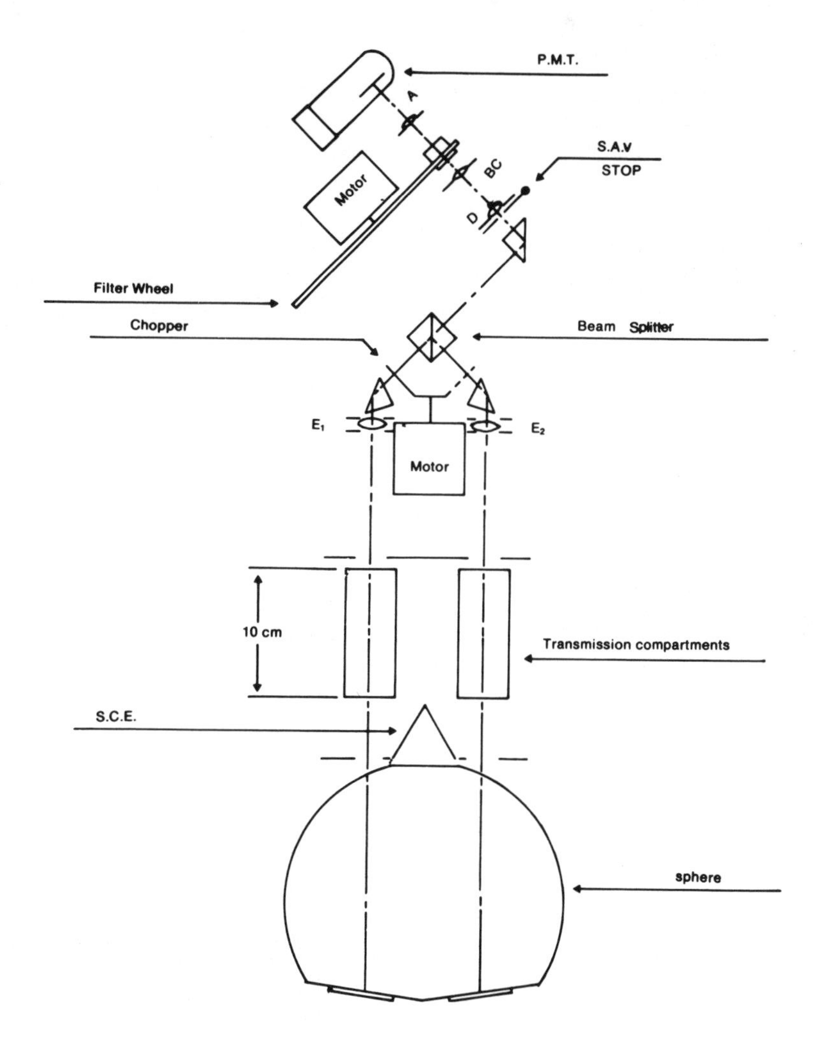

FIGURE 4.07 Optical schematic of the KCS-18.

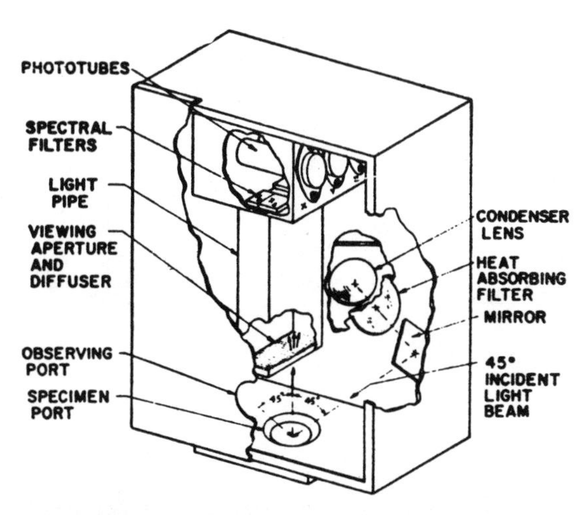

FIGURE 4.08 D25A Optical Unit Arrangement.

FIGURE 4.09 Hunter Color and Color-difference Meter D25D, with standard optical unit D25DA.

after the sample has been positioned. The system is designed to be constantly on full or standby power and requires only 10 min to equilibrate to maximum precision after being switched from standby to full power.

Other Hunter instruments include the D40YZ whiteness reflectometer, which provides measurements of green and blue reflectance and fluorescence of white (and near-white) materials. Another instrument, the D44, is designed for continuous color measurement on production lines and may be used in combination with a digital computer for continuous color control. A special version of this instrument, the D46, measures only a single dimension of the appearance such as blue reflectance. A variety of gloss meters are also available.

Another instrument that falls into the category of filter instruments for reflectance measurements, although it was designed specifically for use in investigations of a microscopic nature, is the Joyce, Loebl Microdensitometer, which is equipped with a reflectance attachment. Various modes of irradiation of the sample surface are possible. Of particular interest in this discussion is the "incident external illumination" mode, which uses a ring condenser that annularly surrounds the objective of a microscope device and is structurally combined with it. The light from the instrumental source is piped to the objective by means of a fiber-optics light guide and is then radiated diffusely onto the sample surface via the ring condenser. The sample is viewed at a zero-degree angle by the objective, and specular reflectance cannot reach the viewing optics. The portion of diffuse radiation that is proportional to the angle of aperture of the objective can reach the viewing optics, no matter whether the light results from diffuse surface reflection or from diffuse depth reflection (in the case of nonmetallic powder substances). We are, therefore, dealing with pure diffuse reflectance. This mode of operation is not suitable for etched and smooth metal surfaces, however, since they would appear as a dark surface due to the gloss portion.

Incident external (diffuse) illumination, on the other hand, is the only satisfactory method for the microscopic reflectance densitometric investigation of structured, nonmetallic and powdered samples. This mode of microdensitometry is hence particularly suitable for investigations and classification of minerals and biological samples. The instrument operates in the double-beam mode, using a beam splitter device, and it permits automatic recording of sample scans.

Considerable developments have been made in this instrument since its introduction, and a complete range of digital systems is now available. The most sophisticated of these, the Computer Controlled Microdensitometer, is shown in Figure 4.10. Widespread application of this instrument is now found where the spectroscopic requirements of the user are best served by programming the computer to move the table according to a scanning pattern suited to his particular needs. The computer also provides data analysis with calculation of results and on-line print-out.

A similar instrument is produced by Photometric Data Systems Corp., 841 Holt Road, Webster, New York 14580. This instrument (Model 1010 Microdensitometer) also has provision for digital data acquisition.

Other filter reflectometers are described in the section for recording instruments used for in situ chromatographic work, e.g., instruments manufactured by Joyce, Loebl & Co. Ltd. and Vitatron Ltd.

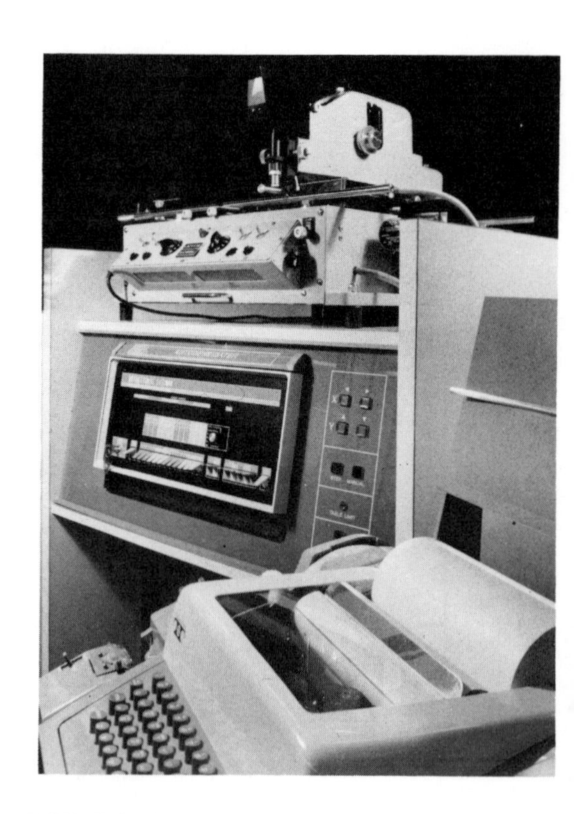

FIGURE 4.10 The Joyce, Loebl Computer Controlled Microdensitometer.

For a more complete survey of instrumentation in this line, a list of commonly used filter reflectometers is given in Table 4.01 along with some optical specifications. Wendlandt[14] has also discussed a variety of filter reflectometers.

2. Monochromator Instruments

A few workers have reported on the design of custom-made reflectance units using a monochromator instead of filters and collecting the diffusely reflected radiation with an integrating sphere or semisphere.[15-18] The spectral ranges explored in this work included UV, visible and near-infrared (220 to 2500 nm). With the advent of commercially available reflectance attachments for practically all standard spectrophotometers on the market, diffuse reflectance spectroscopy has become an increasingly important tool for workers in many fields of analytical chemistry, and the analyst is no longer faced with the problem of having to construct a suitable instrument or attachment for spectral reflectance work. Unlike the filter instruments discussed earlier, which were usually built specifically for diffuse reflectance studies, reflectance attachments were designed at a later date for existing all-purpose spectrophotometers. An exception is the General Electric Hardy Spectrophotometer (General Electric Co., West Lynn, Massachusetts), which was the first commercially available instrument in this field.[19-21]

The devices used for the collection of the diffusely reflected radiation on currently marketed attachments range from full optical sphere to semisphere to circular and ellipsoidal mirror systems and light pipes.

In North America, one of the best-known and most widely distributed spectrophotometers – also in connection with reflectance work – is the Beckman Model DU Spectrophotometer (Beckman Instruments Inc., Fullerton, California). Within minutes, this single-beam instrument can be equipped with a reflectance attachment for diffuse reflectance spectroscopy (see Figure 4.11). A quartz lens is used to make the monochromatic light coming from the slit less divergent. After reflection from the front-surface plane mirror D, the sample E is irradiated at a zero-degree angle of incidence. Some of the diffusely reflected light is collected by the ellipsoidal metal mirror G and transmitted onto the photosensitive surface. The mirror G has a hollow-cone geometry and is centered at an angle of 45° to the sample surface. Light rays emerging from the center of the sample

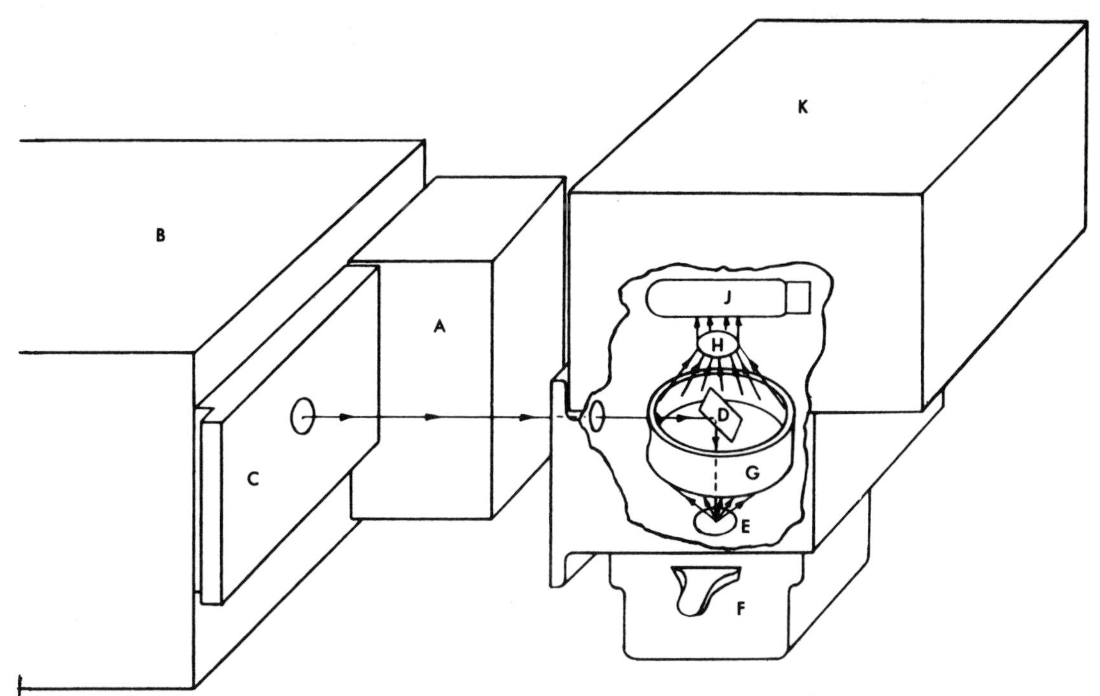

FIGURE 4.11 Optical diagram of the Beckman Model DU Spectrophotometer with diffuse reflectance attachment. A, source; B, monochromator; C, mounting block; D, deflection mirror; E, sample surface; F, sample drawer; G, circular ellipsoidal mirror; H, frosted glass diffusing screen; J, phototube; K, phototube housing.

TABLE 4.01

Filter Instruments for Reflectance Measurements

Manufacturer	Model	Mode of operation	Optics	Optical geometries	Range of operation (nm)	Miscellaneous
Gardner Lab. Inc.	Model 100M-1	double-beam	lens system	45/0	visible	One beam goes to reference detector
Hunter Assoc. Lab. Inc.	D25A	single-beam	light pipe	45/0	visible	All three available with automatic readout of color coordinates
	D25P	single-beam	sphere	0/d; 8/t	visible	
	D25 M/L	single-beam	circumferential mirror	d/0	visible	
Industrial Development Lab. Inc.	IDL Color-Eye	double-beam	sphere	–	400-700	
	IDL Colorede	double-beam	sphere	–	400-700	Semiautomatic
Joyce, Loebl & Co. Ltd.	Microdensitometer attachment	double-beam (chopper mirror)	annular ring condenser	d/0	visible, UV	Small sample area
Magnuson Engineering Inc.	Agtron M400A	single-beam	collimating lens	45/0	visible	(Food industry)
Manufacturing, Engineering and Equipment Corp.	Colormaster Model V	double-beam	lens system	45/0	visible	Digital readout
National Instrument Lab. Inc.	Colorcord II A	double-beam	sphere		visible	
Photometric Data Systems Corp.	Model 1010 microdensitometer	double-beam	objective	d/0	visible, UV	
Pretema Ltd.	Spectromat FS-38	double-beam	sphere	d/0; 45/0	390-710	Automatic computation of x, y, z coordinates
Carl Zeiss, Inc.	Elrepho	double-beam	sphere	0/d; d/0	400-700	
	RFC-3	double-beam	sphere	d/8; 45/0	400-700	Automatic computation of x, y, z coordinates

or from the reference surface at angles of 35 to 55° to the perpendicular are measured. The sample and reference material are accommodated in a drawer F located at the bottom of the attachment. The various reflectance cells described later in this chapter were originally designed to fit into this sliding drawer of the Beckman DU accessory but were later used with other attachments without any alterations.

The reflectance accessory for the Models DK-1A or DK-2A Spectrophotometers (spectroreflectometers) is shown in Figure 4.12. As can be seen from Figure 4.12, this accessory uses an integrating sphere in which both sample and reference surfaces are irradiated alternately with monochromatic light. The light passes from the prism monochromator to an oscillating mirror, which produces the double-beam effect. In the normal sample position only diffuse radiation is measured, and specular reflectance is attenuated. Total reflectance (specular plus diffuse radiation) can be determined by mounting the sample and reference materials at an angle of 5° to the incident beam. The range of the reflectance accessory is from 210 to 2700 nm. A photomultiplier tube views the UV and visible range, and for the near-infrared region a lead sulphide cell functions as a detector.

Still another diffuse reflectance attachment was recently introduced for use in conjunction with the Beckman Models DB and DG Recording Spectrophotometers (Figure 4.13). Figure 4.14 shows the light path used in this attachment. Diffuse reflectance attachments for the Bausch & Lomb instruments Spectronic 20 and Spectronic 505 (Bausch & Lomb Co., Rochester, New York) have also gained wide acceptance. The Spectronic 20 accessory is illustrated in Figures 4.15 and 4.16. The monochromatic light-beam is reflected by a mirror and passes through a lens and filter system onto the sample or reference standard placed at the top of the integrating sphere. The diffusely reflected light is viewed by a photomultiplier tube placed at a 90° angle to the sample surface. A 2 x 8 mm spot of light strikes the sample surface, hence permitting the investigation of relatively small samples. A modified attachment with a light-beam area of 12 x 12 mm can be purchased separately and is recommended for rough textured materials such as textiles. A band width of 20 nm is employed at any point in the spectrum 400 to 700 nm. The monochromator consists of a replica grating.

The reflectance attachment for the Bausch & Lomb Model Spectronic 505 double-beam Spectrophotometer is depicted in Figure 4.17. The accessory shows clearly the shape of the integrating sphere equipped with an end-on photo-

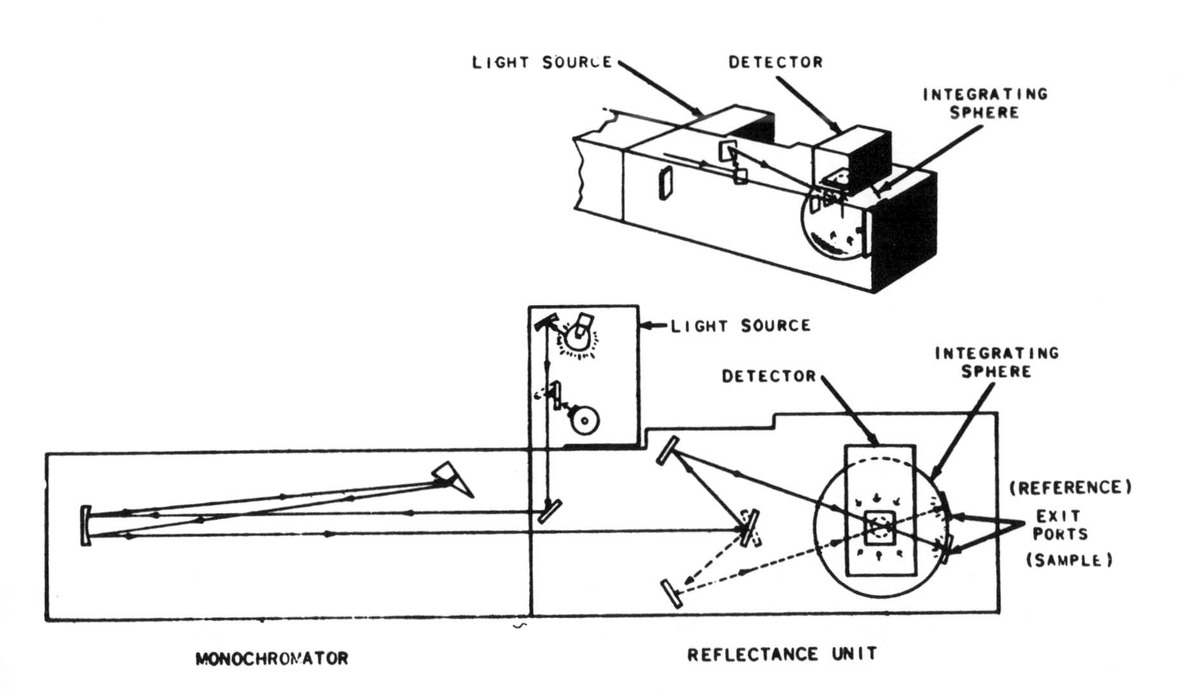

FIGURE 4.12 Schematic diagram of the Beckman Model DK-2A Spectroreflectometer.

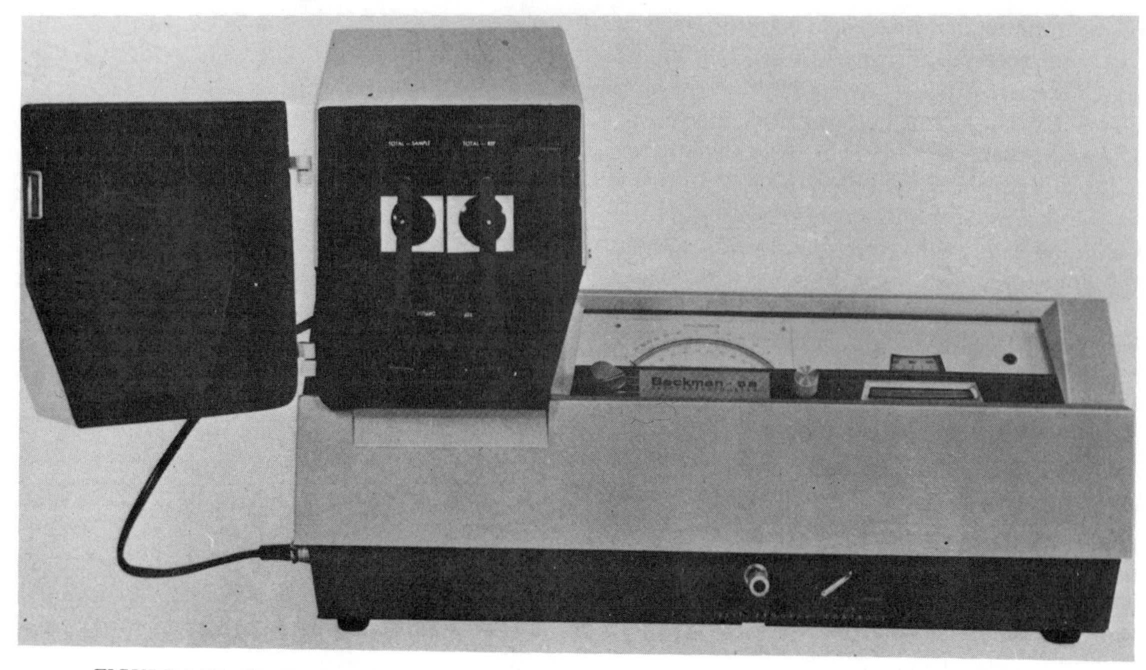

FIGURE 4.13 The Beckman Model DB Spectrophotometer equipped with a diffuse reflectance attachment.

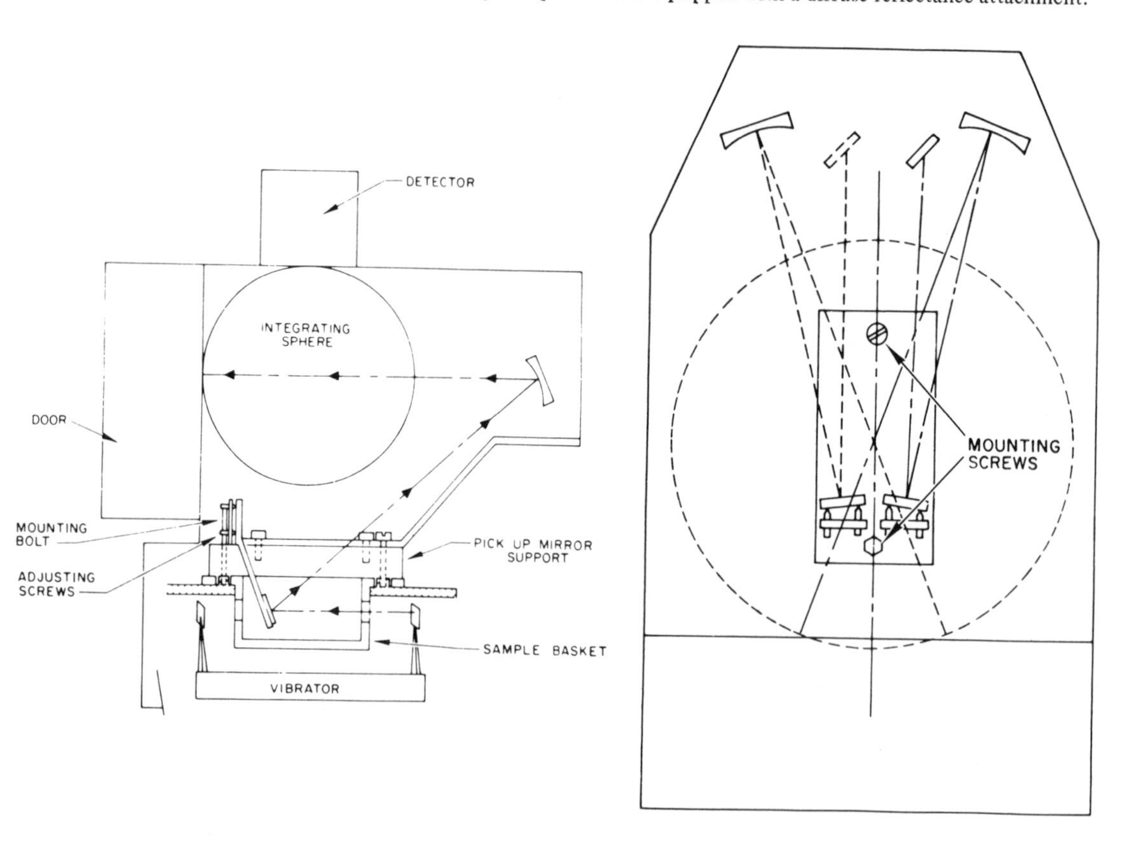

FIGURE 4.14 Top view and side view of light path used in Beckman DB diffuse reflectance attachment.

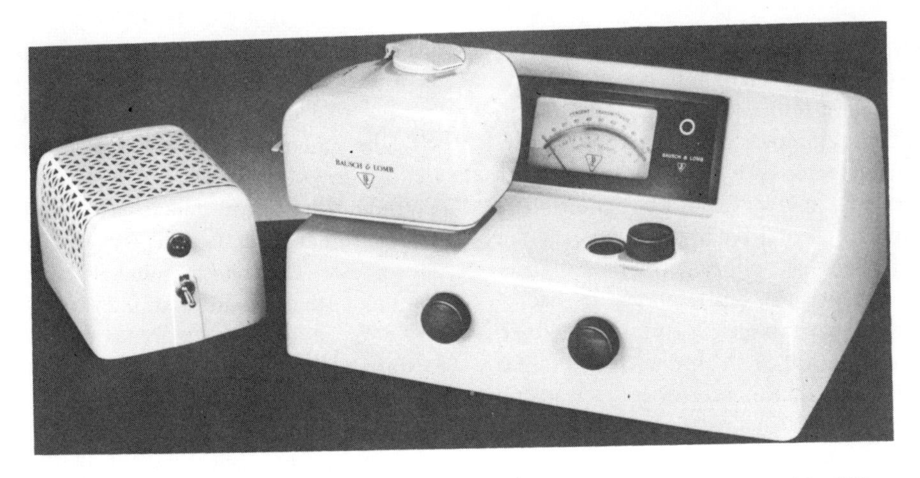

FIGURE 4.15 The Bausch & Lomb Spectronic 20 Spectrophotometer with diffuse reflectance attachment.

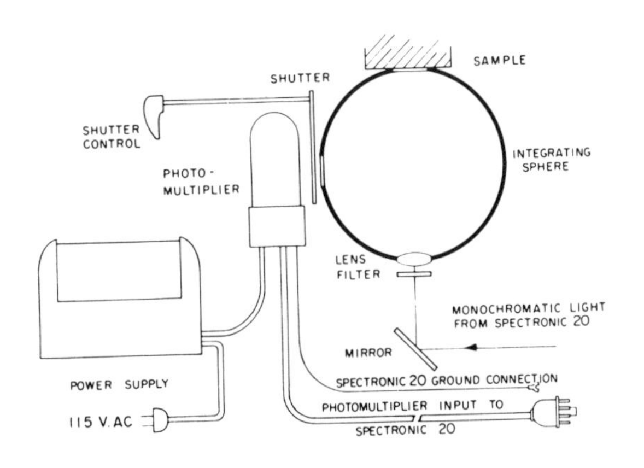

FIGURE 4.16 Diagram of the reflectance attachment to the Spectronic 20.

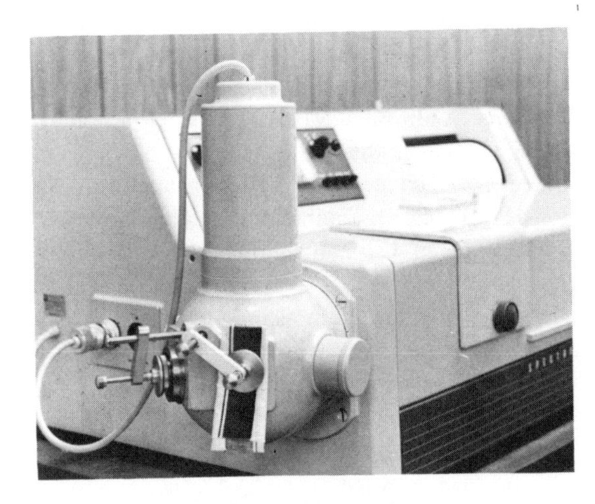

FIGURE 4.17 The Bausch & Lomb Spectronic 505 Spectrophotometer equipped with diffuse reflectance attachment.

multiplier tube. The spring-loaded sample and reference holders can be seen outside the sphere, loaded with the standard powder sample holders manufactured by Bausch & Lomb. Light from the monochromator is focused by a lens system (built inside the attachment) onto the sample and reference material, respectively. The circular incident light spot is about 16 mm in diameter. No oscillating mirror is used, as in the Beckman DK model, but before amplification the signals are chopped with an electronic chopper. The diffuse radiation is detected by the photomultiplier located at the top of the optical sphere. Light traps located at an angle of 90° to the reference and sample surfaces can be used to eliminate the specular reflectance component. The traps can be replaced by a reflecting surface for the measurement of total reflectance. The band-width selector has to be set at 50 Å to compensate for energy losses due to scattering. The standard attachment can be used in the spectral region 400 to 700 nm. Another sphere with different UV-reflecting coating is available; this permits an extension of the useful spectral range down to about 220 nm.

The Spectronic 600 is an instrument designed mainly for use in routine measurements at a few selected wavelengths where stability and reliability requirements are high. Two spectral ranges are available, 350 to 650 nm (tungsten lamp) and 200 to 650 nm (combination light source, tungsten and deuterium). The instrument is a double grating spectrophotometer (1200 grooves/mm gratings, blazed at 300 nm) and the wavelength range may be extended from 650 to 800 nm by a suitable choice of photomultipliers. Bandpass is constant at 5 or 50 nm, with a resolution of 0.5 and 5.0 nm, respectively, constant over the wavelength range. The standard detector is a 1P28 photomultiplier with readout on a meter. Other photomultipliers and a recorder or digital readout attachment are available as options. The photometric presentation is linear in transmittance, but nonlinear in absorbance, with photometric scale expansion of 0 to 20, 0 to 100 and 0 to 200%T. Absorbance is 0 to 2.0 full-scale. Two reflectance accessories are available for the Spectronic 600 – one for operation from 400 to 700 nm, the other for measurements from 220 to 650 nm.

A set of three different reflectance attachments is available for adaptation to the Cary Models 14 and 15 Spectrophotometers (Cary Instruments, Applied Physics Corp., 2724 South Peck Road, Monrovia, California 91016). One attachment, which works on the principle of an integration sphere, is depicted in the actual installation position (Figure 4.18). Sample and reference surface are illuminated by a directional chopped monochromatic beam at a zero-degree angle of irradiation. The radiation from sample and reference is diffusely reflected within the sphere and viewed by a phototube located at the top of the sphere (see Figure 4.19). In the second mode of operation with the sphere, sample and reference are illuminated by an external source of radiation (see Figure 4.20), and the diffuse radiation is fed through the monochromator and detector of the spectrophotometer. The working range is given as 250 to 750 nm, with a resolution of 0.25 to 0.7 nm.

Another type of accessory uses a circular mirror mounted above the sample for the collection of the diffuse radiation within cones of angles of 45° ±7° (see Figure 4.21). This arrangement is similar to the collection devices used in the Beckman DU and Unicam Spectrophotometers. The standardization of the reference beam with respect to the reference sample is achieved with the screen attenuator. The horizontal port location makes the measurement of powdered samples convenient. A working range of 220 to 700 nm with a resolution of 0.13 to 0.7 nm is reported for this attachment.

The third type of accessory is the cell-space diffuse reflectance attachment. It uses a sample sphere and light-integrating reference box. Chopped monochromatic radiation is reflected by mirrors onto the sample and reference material.

FIGURE 4.18 The reflectance attachment to the Cary Model 14 Spectrophotometer.

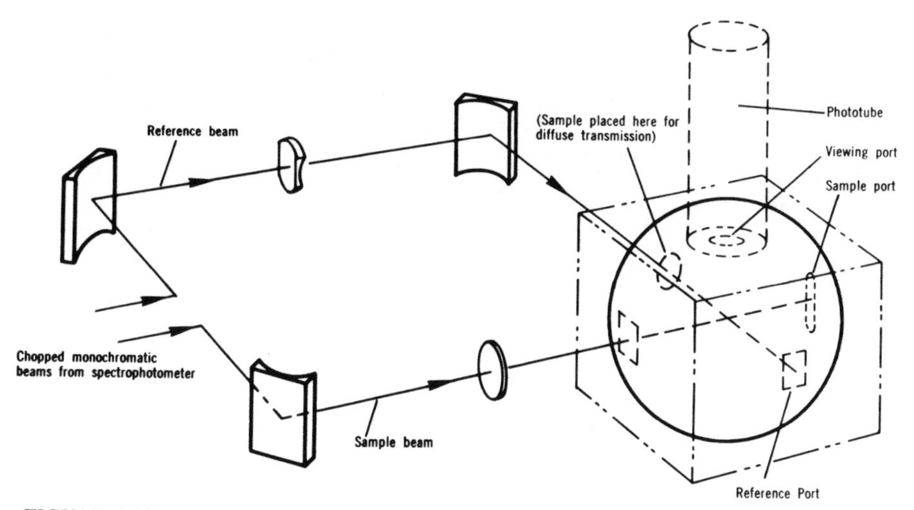

FIGURE 4.19 Optical diagram (dispersed illumination) of reflectance accessory of Cary 14 Spectrophotometer, using an integrating sphere for measurement of diffuse reflectance or diffuse transmittance.

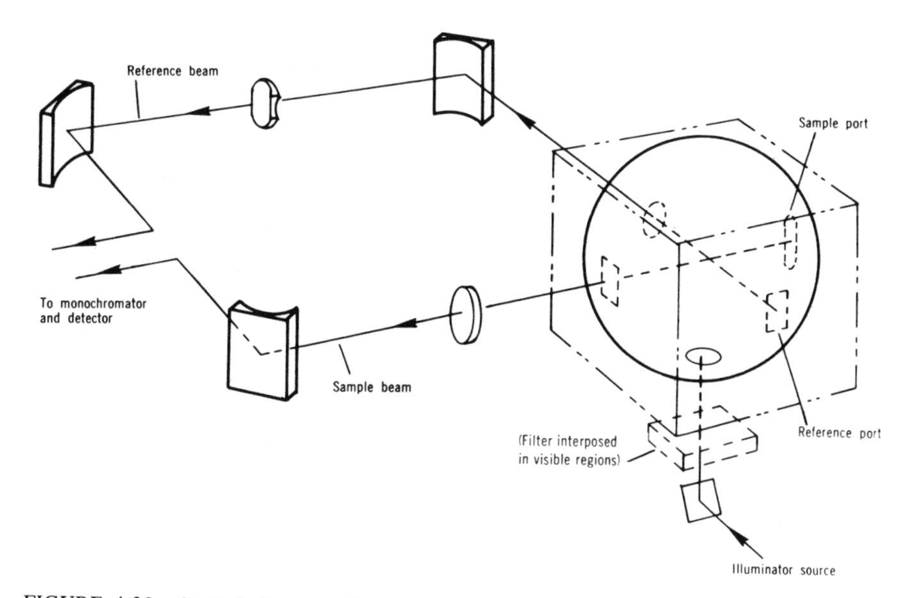

FIGURE 4.20 Optical diagram of reflectance accessory to the Cary 14 Spectrophotometer, using an integrating sphere and an external source of radiation.

The diffusely reflected light passes through the exit ports to the detectors.

A new reflectance accessory for adaptation to the Cary Models 14, 14R and 14RI is now available. With an altered integrating sphere design it can cover the spectral range 220 nm to 2.5 μ, depending on the detectors used. The working principle of this attachment was discussed in detail elsewhere.[21] It is mounted in the same position as the previously discussed attachment (see Figure 4.18). The monochromatic chopped reference and sample beams enter at the top of the sphere and simultaneously illuminate the sample and reference material. The diffuse radiation is collected via the integrating sphere at the detector located at the bottom port between the sample and reference position. The double beam compensates for source fluctuations, and hence errors of <1% over the entire spectral range are reported. The samples are placed in a horizontal position favorable for the

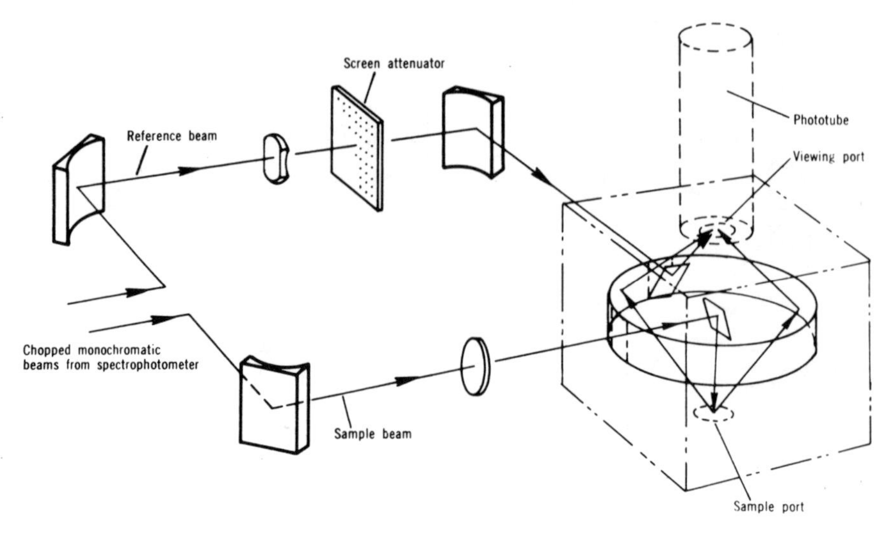

FIGURE 4.21 Optical diagram of reflectance accessory to the Cary 14 Spectrophotometer, using a circular mirror system (ring collector).

measurement of liquids and powder samples. The accessory can easily be converted to the specular reflectance mode. The use of this diffuse reflectance attachment is recommended also in conjunction with a computer system for the computation of tristimulus color measurements.

An attachment for the measurement of diffuse reflectance is available for the Perkin-Elmer Spectrophotometers Models 350 and 450 (The Perkin-Elmer Corp., Norwalk, Connecticut). Rather than using one integrating sphere with two sample ports for double-beam operation, as is the case with most other attachments, the Perkin-Elmer accessory uses two spheres, both coated with magnesium oxide. Each sphere has an individual end-window photomultiplier detector, and hence interaction between sample and reference reflectance is eliminated. The optical diagram for one of the reflectance spheres is shown in Figure 4.22. Incident monochromatic radiation at an angle of incidence of 16°15′ is reflected by either sample or reference, further reflected by the magnesium oxide surface of the sphere, and finally collected at the photomultiplier detector. A choice of either black or white port covers enables the operator to reject or accept the specular component. With the standard instrument, a wavelength range from 220 to 750 nm can be covered. Extensions to 2.5 μ are possible with the use of a lead sulfide cell. The precision for reflectance measurements carried out with this attachment is reported to be equal to one obtained by the transmission mode for the same instrument (±0.25% in the 1 to 100% reflectance range). The same type of attachment has previously been discussed in great detail.[23] These instruments are now regarded as obsolete by the manufacturer.

The Hitachi-Perkin-Elmer Model 139 single-beam UV-visible Spectrophotometer (Perkin-Elmer Corp., Coleman Instruments Division, 42 Madyson Street, Maywood, Illinois 60153) can also be equipped with a diffuse reflectance accessory (Figure 4.23). The optical schematic (Figure 4.24) and the photograph (Figure 4.23) clearly show the existence of an integrating sphere in this attachment. Attached to the sphere is a revolving mount

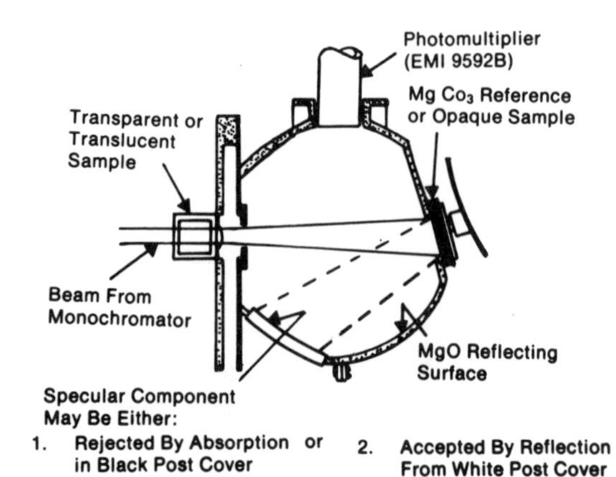

FIGURE 4.22 Optical diagram for a diffuse reflectance sphere of the Perkin-Elmer reflectance accessory to the Models 350 and 450 Spectrophotometers.

FIGURE 4.23 The Hitachi Perkin-Elmer Spectrophotometer with diffuse reflectance attachment installed and diffuse reflectance attachment separate.

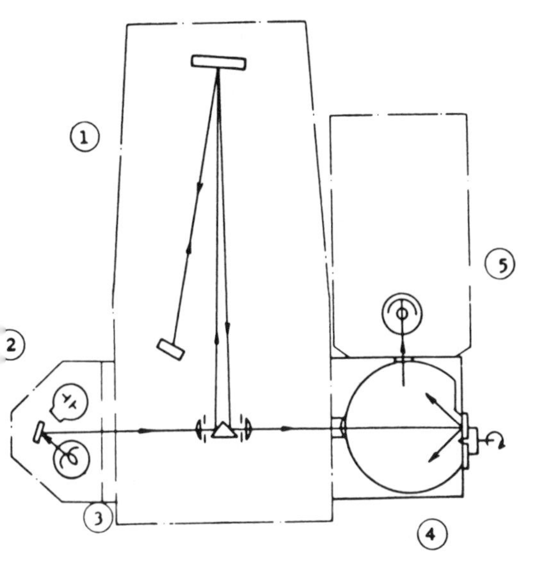

IGURE 4.24 Optical diagram of the reflectance attachment for the Hitachi Perkin-Elmer Model 139 Spectrophotometer.

ith spring-clamp holders for sample and reference aterial, respectively. Monochromatic light is adiated perpendicularly to the sample surface and he diffusely reflected energy is collected in the phere and viewed at 90° to the light path and ample with a suitable photodetector. The optical eometry could hence be described as $_0{}^\circ R_d$. The pectral range of this attachment is 400 to 760 m. It can be extended down to 250 nm with the use of a quartz plate. Powder cells, which are delivered with the attachment (Figure 4.23), are 22 mm in diameter and 10 mm in depth.

A new spectrophotometer which features high performance in the UV, visible and near-infrared regions is the Hitachi-Perkin-Elmer Model 323. The instrument operates from 185 to 2500 nm using a double-beam, electrical, direct ratio system. A deuterium lamp is the light source for the UV-range, while a tungsten lamp is used at longer wavelengths. Resolution is 0.2 nm in the UV. The lamps and detectors are switched over automatically as the spectrum is scanned from the UV to the visible.

The photometric system consists of a double beam with two optical choppers and electronic direct ratio recording using a fast response X-Y recorder. Continuously variable, bilateral slits (0.005 to 2 mm) are automatically controlled for constant energy. The prism monochromator is fused quartz with an apex angle of 30° and is optically transparent from 170 to 2600 nm. A range of scale expansions (X2, 4, 10 and 20) and scan speeds is available. An integrating sphere attachment is available as an option.

Another well-known instrument of Japanese origin, the Shimadzu Model QV-50 single-beam Spectrophotometer (obtainable from American Instrument Co., Inc., 8030 Georgia Avenue, Silver Spring, Maryland 20910), is now available with a diffuse reflectance attachment. An integrating sphere is used and three samples can be placed in a horizontal sample drawer similar to the sliding sample drawer in the Beckman DU. Samples up to 25 mm in diameter can be accommodated. The optical performance is similar to that discussed previously for other single-beam instruments covering the UV and visible portion of the spectrum.

Pye Unicam Ltd., York Street, Cambridge, England, have developed diffuse reflectance attachments for the Unicam Models SP500, SP700 and SP800 Spectrophotometers. All three models have an ellipsoidal mirror system for the collection of the diffuse radiation. The SP500 is a single-beam instrument and can be operated with the reflectance accessory in the spectral range 350 to 1000 nm. The attachment and an optical diagram are shown in Figures 4.25 and 4.26. The optical arrangement is similar to the one used in the Beckman Model DU accessory. A monochromatic light-beam is deflected by the mirror onto the sample at a zero-degree angle of incidence and a portion of the diffusely reflected light is collected in the concave mirror and fed into the detector system.

For reflectance measurements by the double-beam mode, the Unicam Model SP700 with attachment SP735 can be used. The attachment and an optical diagram are illustrated in Figures 4.27 and 4.28. The two monochromatic light-beams emerging from the SP700 beam splitter unit are reflected by the mirror M_2 into the apertures located in the ellipsoidal mirror M_4 and by means of an optical system (lenses L_2, reflecting mirrors M_3) are focused on the sample and reference surfaces, respectively, at a zero-degree angle. The diffuse radiation with reflectance angles between 35 and 55° is collected by the ellipsoidal mirror and focused onto the detector. The useful spectral range is 200 to 2500 nm with a lead sulfide cell.

The diffuse reflectance accessory for the Model SP800 (Figures 4.29 and 4.30) can be used for measurements in the UV and visible region of the spectrum. As can be seen in Figure 4.30, it operates essentially as a double-beam instrument but the reference beam strikes only the detector. The monochromatic sample beam is focused onto the sample with the illumination and observation geometry being ${}_{0}{}^{\circ}R_{d}$. Light reflected again between 35 and 55° angles is collected by the ellipsoidal mirror and viewed in the detector assembly.

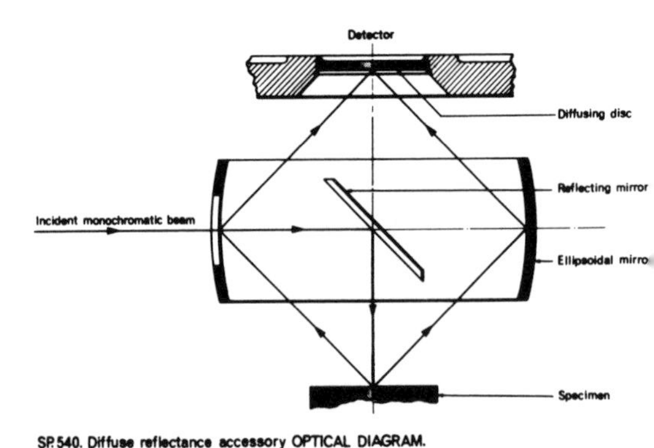

FIGURE 4.26 Optical diagram for the SP540 diffuse reflectance accessory.

FIGURE 4.25 The Unicam SP540 diffuse reflectance accessory for use with the SP500 series 1 and 2 Spectrophotometers.

FIGURE 4.27 The Unicam SP735 (series 2) diffuse reflectance accessory for use with the SP700 Spectrophotometer.

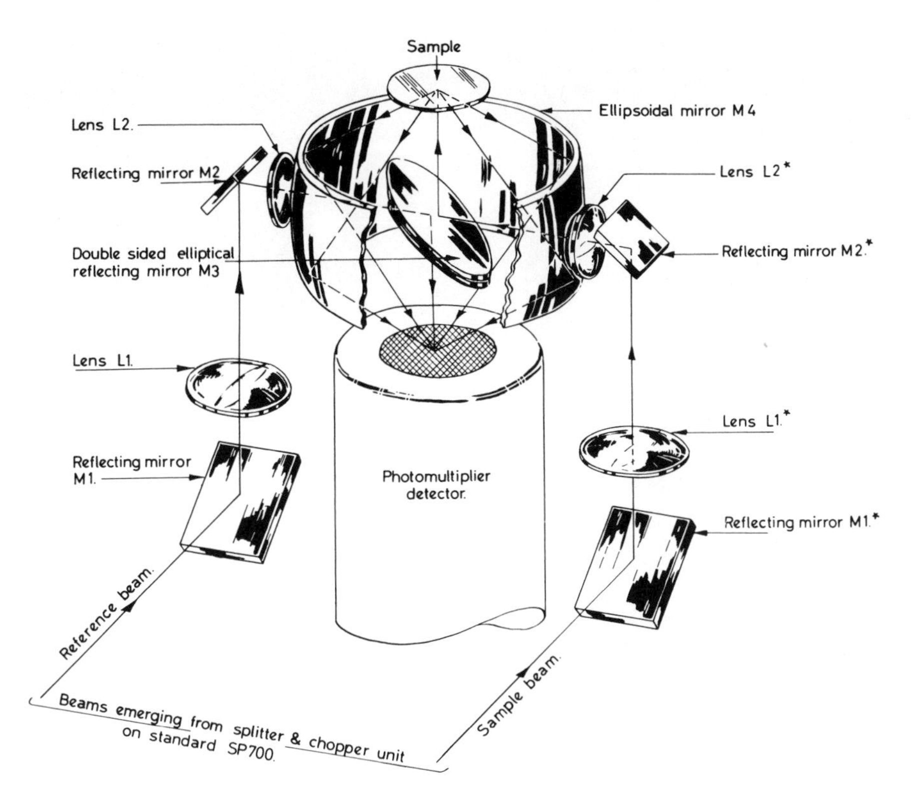

FIGURE 4.28 Optical system of the SP735 Accessory.

A sophisticated line of reflectance spectrophotometers is available from Zeiss (Carl Zeiss, Inc., Oberkochen, West Germany). Three types of reflectance accessories can be chosen for work with the Zeiss Model PMQ single-beam Spectrophotometer. The attachment RA-2 (see Figures 4.31 and 4.32) uses monochromatic light which emerges from the exit slit and is irradiated on the sample at a 45° angle with the use of a lens L and mirror system M. The field lens L at the exit slit E of the monochromator forms an image of the monochromator prism on the sample, which means that changes in slit width will only influence the intensity and spectral purity of radiation on the sample. The diffusely reflected light D is then viewed at right angles to the sample with a lens system L built in front of the cathode C of the photomultiplier P. The specular component F is eliminated in this process. The amount of light entering the photomultiplier is controlled by the shutter LS. The spectral range of this apparatus includes the UV and visible and near-infrared portion of the spectrum. The near-infrared portion requires a change of detectors. Samples and reference standard are pressed in horizontal position against the sample port by means of a spring-loaded cup.

The geometry used for the illumination and observation with the attachment RA-2 ($_{45°}R_{0°}$) will lead to systematic errors with samples of rough texture, such as textiles or certain large-grain powder samples. This is due to irregularities in the surface.[9,24–26] When dealing with this type of surface, it is recommended to use diffuse radiation and to measure at a zero-degree angle ($_{d}R_{0°}$). This can be done with the attachment RA-3, which operates on the basis of an integrating sphere (Ulbrichtkugel).

As can be seen from Figures 4.33 and 4.34, two modifications are possible with the RA-3 attachment. Modification (1) (Figure 4.33) uses diffuse sample illumination by polychromatic light from an incandescent lamp. The light is diffusely scattered by the integrating sphere and radiated on the sample. Diaphragms keep the direct reflectance component away from the sample and the mirror system. The reflected light is observed at a zero-degree angle and deflected by the swiveling mirror through an intermediate optical system and into the monochromator. The spectrally dispersed

FIGURE 4.29 The Unicam SP890 diffuse reflectance accessory for use with the SP800 Spectrophotometer.

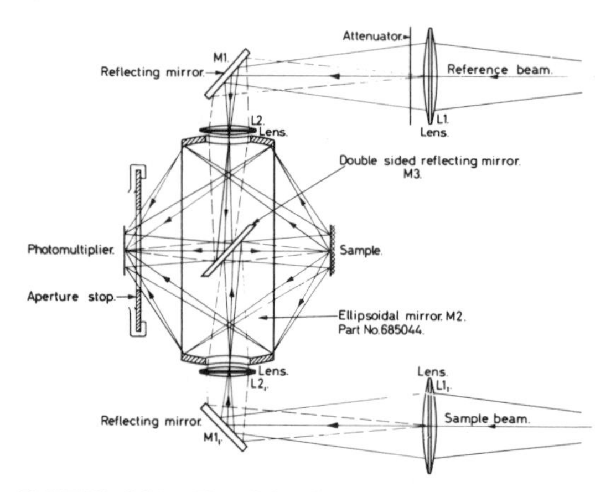

FIGURE 4.31 The Zeiss Spectrophotometer PMQ II with the Model RA-2 reflectance attachment installed (attachment to the left of the monochromator).

radiation is then directed toward the detector. The measuring set up (1), which uses the geometry $_dR_{0°}$, corresponds to that of the Elrepho filter instrument discussed in the previous section and can be used for the calibration of color standards also. The spectral range is fixed by the incandescent lamp at 380 to 2500 nm. In modification (2) (Figure 4.34), the illumination and observation

FIGURE 4.30 Optical diagram of the SP890 Accessory.

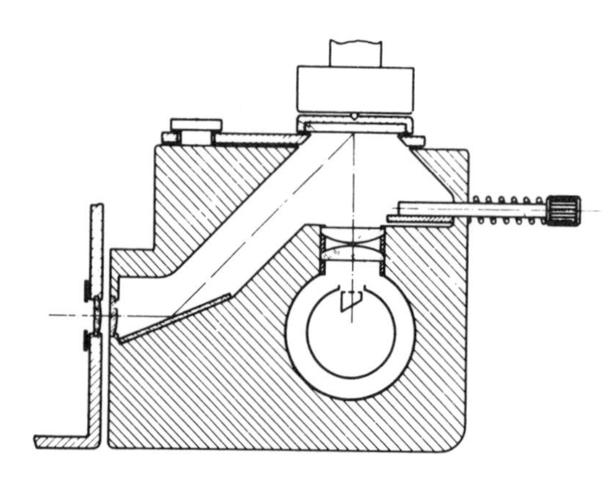

FIGURE 4.32 Optical diagram of the Zeiss Model RA-2 reflectance attachment.

geometry have been switched to $_{0°}R_d$. This version is suggested for investigations of light-sensitive samples because of the narrow band of monochromatic light that impinges on the sample. The change from one modification to another can be made within minutes by replacing the lamp with a photomultiplier and the detector with a lamp unit that contains a hydrogen and tungsten lamp. Modification (2) can also be used for work in the UV region of the spectrum if used in conjunction with the hydrogen lamp. Quartz optics and a suitable photometer sphere coating are provided. (Recently built accessories are coated with $BaSO_4$.) In modification (2) the sample is illuminated by monochromatic light at a zero-degree angle of incidence and the resulting diffuse radiation is collected in the photomultiplier via the integrating sphere. The Model RA-3 Reflectance Spectrophotometer is shown in Figure 4.35.

A third reflectance attachment is available for very small samples. It operates essentially on the

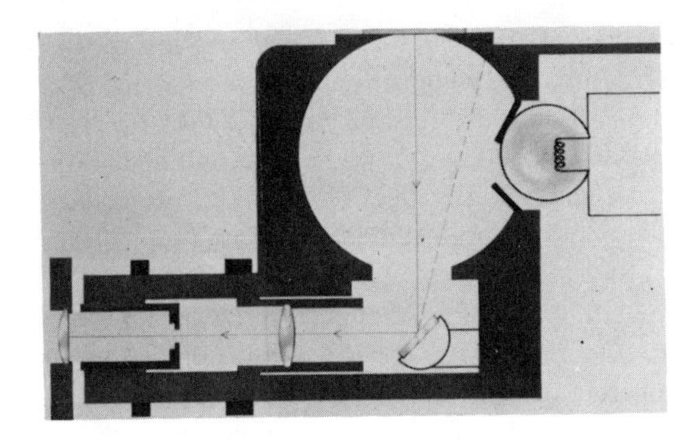

FIGURE 4.33 Optical diagram of the Zeiss Model RA-3 reflectance attachment.

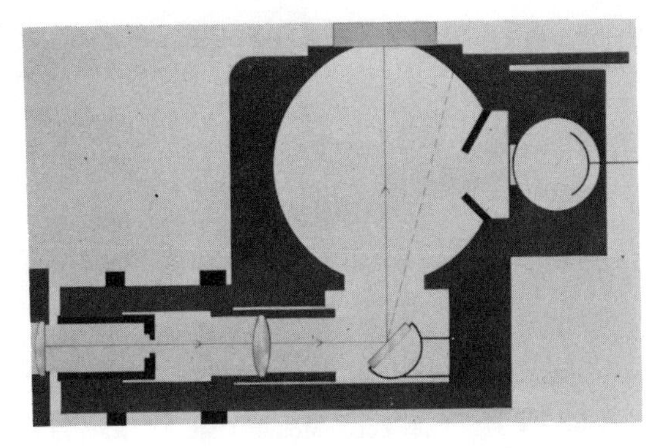

FIGURE 4.34 Optical diagram of the Zeiss Model RA-3 reflectance attachment. Geometry $_{0°}R_d$.

same basis as the RA-3 attachment. The sample is illuminated with monochromatic light at a zero-degree angle of incidence. The reflected light is directed via the integrating sphere to the photomultiplier. Quartz optics and a bonded MgO (currently $BaSO_4$) sphere coating suitable for UV reflectance work are used. The total spectral range of the attachment is 250 to 2500 nm, depending on the detector. A swivel standard is provided for setting the reference value, and the sample need not be removed during the entire measuring operation. Samples with a minimum surface area of 0.1 mm^2 can be accommodated. The same attachment can easily be converted to a $_dR_{0°}$ illumination, observation geometry similar to the RA-3 accessory. The detector is simply replaced by an incandescent lamp that illuminates the sample diffusely. The reflected radiation is picked up at a zero-degree viewing angle and fed into the monochromator compartment. (The accessory can also be used in the transmittance mode.) The application possibilities of this microreflectance attachment are numerous. It is well suited for microanalytical work, e.g., the examination of a single crystal surface, spots from TLC plates, etc.

An attachment for measuring diffuse reflectance with the new double-beam Recording Spectrophotometer Zeiss Model DMR-21 is shown in Figure 4.36. Like the attachment for the earlier Model RPQ 20-A, it operates with an integrating sphere and the spectral range can be chosen from 200 to 620 nm with the photomultiplier tube and up to 2500 nm with the use of a lead sulfide detector. The optical geometry is $_8R_d$; the illuminated sample area is 20 x 20 mm. Either monochromatic or polychromatic sample illumination may be used.

A real breakthrough in the field of diffuse reflectance spectroscopy and color measurements in general has been achieved with the Zeiss Model DMC-25 Recording Spectrophotometer (Figure 4.37). Although specifically designed to satisfy the needs for complex color analysis, it is remarkable for its versatility. Optical geometry and chromaticity of the illumination can be changed rapidly from one mode to another. Spectral reflectance can be recorded simultaneously with the logarithm of the Kubelka-Munk function, or, if desired, a simultaneous record of transmittance and absorbance can be obtained. The recording is

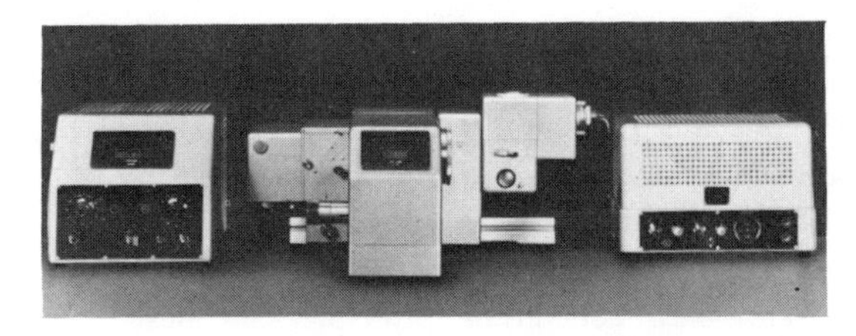

FIGURE 4.35 The Zeiss Model RA-3 reflectance accessory to the Model PMQ II Spectrophotometer.

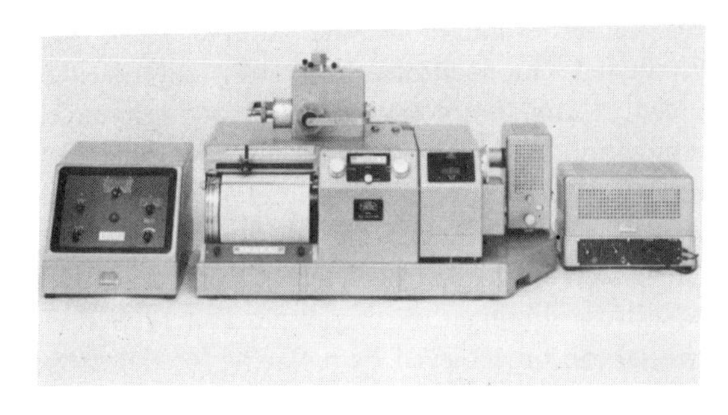

FIGURE 4.36 The Zeiss Model DMR-21 double-beam Spectrophotometer equipped with the reflectance attachment (monochromatic sample illumination).

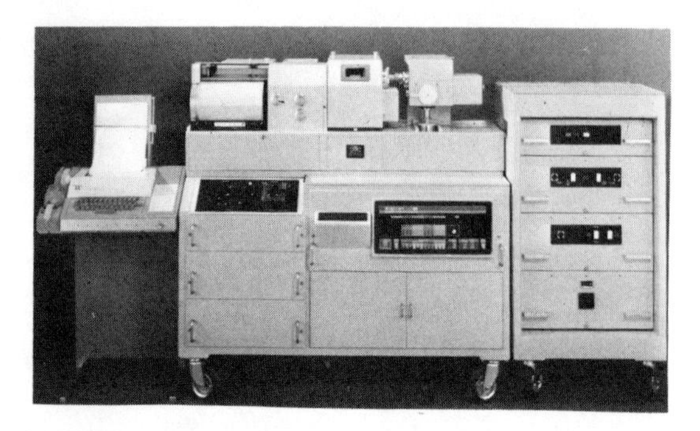

FIGURE 4.37 The Zeiss Model DMC-25 Recording Spectrophotometer.

done with an analog twin recorder. The measurements are printed in digital form at preselected wavelength intervals. A built-in digital computer (Davidson & Hemendinger, Tatamy, Pennsylvania) also determines the X, Y and Z values for any of the illuminants A, C, D or E. The results are printed at the completion of the recording.

The basic operation principle can best be understood from Figure 4.38. Light from the light source L passes through the monochromator M at preselected wavelength intervals and is alternately deflected and transmitted by a chopper mirror DS. This causes standard S and sample P to be illuminated monochromatically in alternating sequence similar to the sample and reference illumination in the Beckman Model DK-2 double-beam Recording Spectrophotometer attachment for diffuse reflectance. The diffusely reflected light is received by the detector E. Polychromatic and diffuse illumination of sample and reference material can be achieved by replacing the detector with a tungsten or xenon lamp. The detector would then occupy the position L and the optical path is reversed. Some of the unique technical

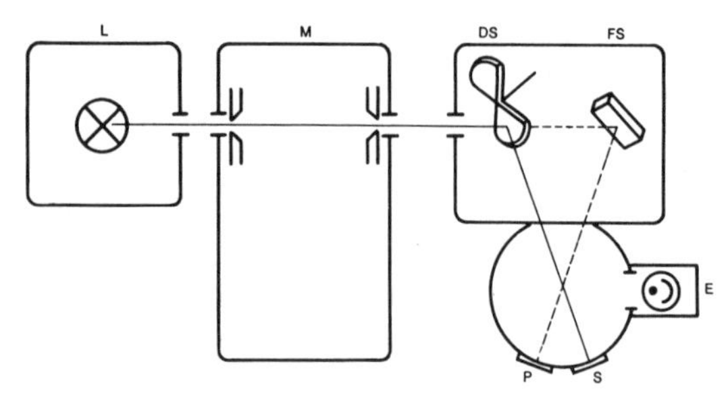

FIGURE 4.38 Optical diagram of the Zeiss Model DMC-25 Spectrophotometer. L, light source; M, monochromator; DS, chopper mirror; FS, fixed mirror; P, sample; S, standard; E, detector.

features of the DMC-25 are a xenon arc for polychromatic illumination and interchangeable prism sets for the spectral ranges 200 to 650 nm and 380 to 2500 nm. A wavelength reproducibility and accuracy better than 0.1 nm is reported. For diffuse and directional mode measuring geometries of ${}_{d}R_{8°}$ and ${}_{8°}R_{d}$ are available. Gloss traps for the elimination of specular reflectance can be inserted into the integrating spheres. Photospheres with diameters of 100 mm and 130 mm are available. For the directional oblique illumination mode, a measuring geometry ${}_{45°}R_{0°}$ is used. Sample surfaces can vary between 15 x 20 mm and 50 x 38.5 mm, but a further reduction of the sample area can be effected if needed. Cell holders for liquid or solid samples of various properties are provided.

The interchangeable detector assemblies include a photomultiplier with S-10 response and one with S-5 response for monochromatic illumination of 380 to 720 nm and 200 to 650 nm, respectively. For polychromatic illumination a photomultiplier with S-20 response is used for the spectral range 380 to 720 nm and a lead sulfide cell for the range 700 to 2500 nm. Since seven measuring ranges are available for diffuse reflectance measurements, the instrument is useful for differential reflectance spectroscopy.

A diffuse reflectance attachment is also available for the Hilger & Watts Uvispek Spectrophotometer (Rank Precision Industries Ltd., Camden Road, London, N.W.1, England). It operates on the single-beam principle and is applicable to the UV and visible regions of the spectrum.

Although in the infrared region above 2.5 μ diffuse reflectance spectroscopy has rarely been applied, Dunn et al.[27,28] have worked on an ellipsoidal mirror reflectometer for diffuse reflectance measurements in the 4 to 40 μ region. A similar instrument was later manufactured by Block Engineering Inc., Silver Spring, Maryland, and discussed by Kneissl[29] at the 1970 Pittsburgh Analytical Conference in Cleveland.

The reflectometer system consists principally of a radiation source, the transfer optics, a Fourier Transform Spectrometer, an ellipsoidal mirror and a large area detector. As seen in Figure 4.39 the flux from a blackbody source is focused through the Fourier Transform Spectrometer onto a stop located at the exit port of the spectrometer. The emerging beam is folded twice and finally focused by a spherical mirror. The spherical mirror produces an image of the exit port aperture on the

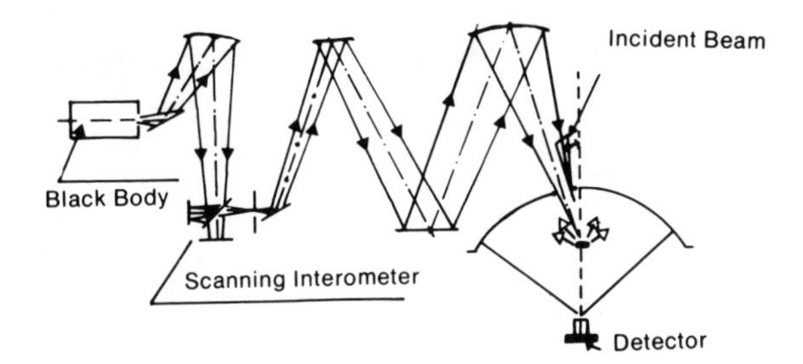

FIGURE 4.39 Optical schematic of the infrared reflectance spectrophotometer (Block Engineering Inc.).[29]

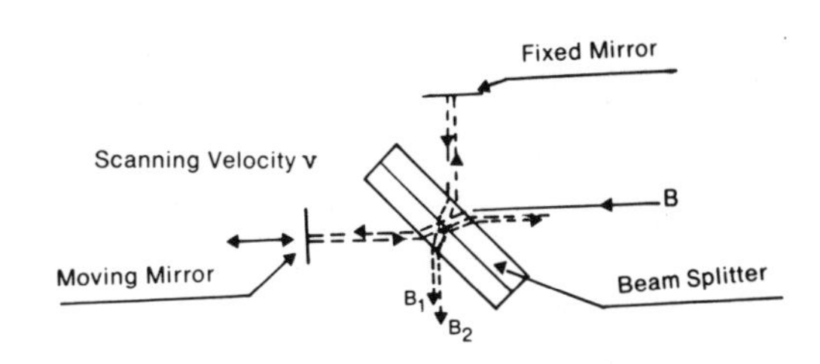

FIGURE 4.40 Optical schematic of the basic interferometer.[29]

sample. The sample is located at the first focal point of an ellipsoid (diam. 12 in.). The flux reflected by the sample is collected by the ellipsoid and focused onto the detector, which is mounted at the second focal point of the ellipsoid.

A blackbody radiator operated at 1000°C was chosen as the source instead of a Globar since it has a much better thermal stability due to its high thermal inertia and its feedback-controlled power supply, which uses a platinum resistance thermometer as the sensing element. The blackbody also provides about 25% more radiant power than a Globar at the same temperature.

The spectrometer employed in this system is a fast-scanning Michelson Interferometer built by Block Engineering Inc. Since the operation of the instrument is radically different from that of dispersive spectrometers, a short description seems to be in order.

A schematic of the basic interferometer is shown in Figure 4.40. The beam splitter divides the incoming beam into two halves and recombines them after one half has been reflected by a fixed mirror and the other half by a scanning mirror. The scanning mirror moves with a constant velocity V. The portion of the beam which is reflected off the scanning mirror experiences a phase-shift. Since this phase-shift is different for each wavelength, the modulation frequency caused by successive constructive and destructive interference between the recombined beams is a unique function of wavelength. Since each wavelength is modulated with its own characteristic frequency, the interferometer eliminates the need for a mechanical chopper. Using the interferometer as an optical chopper allows differentiation between the flux emitted and the flux reflected by the sample. This becomes of increasing importance the further one advances into the infrared region of the spectrum.

The interferometer spectrometer has two basic advantages over a dispersive instrument. First, since it does not employ any slits, the through-put (a measure of the flux accepted by an optical instrument) for a given resolution is higher than for a grating or prism spectrometer. Second, because the detector "views" all spectral elements during the entire scan, an increase in the signal-to-noise ratio is achieved.

The ellipsoidal mirror reflectometer, which has been described in detail,[28] employs an aluminum coated metal ellipsoid with focal lengths of 4.25 inches and 10.25 inches, respectively. A schematic of the ellipsoidal mirror reflectometer is shown in Figure 4.39. A converging beam enters the ellipsoidal mirror through an entrance hole and is incident on the sample under a polar angle θ. The reflected flux is collected by the ellipsoidal mirror and focused onto a detector placed at its second focal point. The mirror forms a magnified image of the irradiated sample area on the detector. The size of the irradiated sample area is therefore limited by the size of the sensitive element of the detector. Usually total reflectance is measured with this instrument, but the specular component can be eliminated by inserting a radiation trap at the position in the mirror where the directly reflected portion impinges.

The solid angle under which the irradiated sample area "sees" the entrance port determines the solid angle of the incident beam. Of course, one would like to use as large a beam as possible to obtain a high signal-to-noise ratio. In this instrument the size of the entrance port was chosen so that the flux escaping was less than 3% under the assumption of a perfectly diffuse sample.

The detector for the reflectometer had to fulfill several partially conflicting requirements. First, the responsiveness of the detector had to be uniform over its sensitive area. This is necessary because the flux distribution will vary greatly with the directional distribution of the reflected flux from the sample. Averaging devices such as integrating spheres can be used to eliminate this problem. However, they are highly inefficient in the use of the radiant power and are only practical in special regions where the measurement is not energy-limited. Second, a large sensitive area was required because the ellipsoidal mirror produces a magnified image (linear magnification of about 2.5) of the sample's irradiated area. Third, the time constant of the detector had to be short enough so that it could respond to frequencies from 250 to 25 Hz, the modulation frequencies corresponding to the wavelengths from 4 to 40 μ, respectively. Finally, the specific detectivity had to be high enough to allow measurements at the longest wavelengths where the least energy was available. A recently developed pyroelectric detector came closest to fulfilling all these requirements. A special large area pyroelectric detector with a sensitive element of 12 x 12 mm was built by Barnes Engineering. The heart of the detector is a piezoelectric crystal that also exhibits spontaneous polarization. These crystals are called pyroelectric, since the value of spontaneous polarization is temperature-dependent, and a change in charge will appear whenever the temperature is changed. The detector has a built-in FET preamplifier which lowers the output impedance to about 10 kΩ.

The entire reflectometer is mounted in a sealed housing to allow effective purging of the optical path. Purging becomes important in the far-infrared because absorption by water vapor and carbon dioxide can become appreciable and lower substantially the signal-to-noise ratio.

The sample holder can accept three samples, any of which can be rotated into the measurement position remotely without breaking the seal.

The system was supplied with three ellipsoidal mirrors having entrance ports at 15, 30, and 45°, respectively. Each mirror was kinematically mounted to allow for easy interchange without the necessity of realignment. These interchangeable mirrors, together with a multiple-tilt mirror and three focusing mirrors, mounted permanently in the cabinet, permit the investigator to perform reflectance measurements at angles of incidence of 15, 30, and 45° with a minimum of wasted time.

Digilab Inc., a subsidiary of Block Engineering, Inc., 237 Putnam Avenue, Cambridge, Massachusetts 02139, has developed a line of infrared spectroreflectometers based on the principles described. The most versatile of these is the FTS-14, shown in Figure 4.41. The spectrophotometer unit contains the Digilab Model 296 high-resolution Fourier transform optical system with an infrared source and accommodations for double-beam sampling. A small, general purpose computer is built into the system and has been configured to work with analytical instrumentation. The FTS-14 may be operated from 1 to 10,000 μm by exchanging beam splitters and detectors. Conversion to operate in the far-infrared takes about 15 minutes and, after purging the instrument for an hour, high quality far IR data may be measured rapidly and accurately. In addition to analysis of data, the computer operates the spectrophotometer and constantly monitors its operation and performance. Considerable versatility and the handling and output of data have also been built into the system.

Other Digilab instruments include the FTS-12, a single-beam, medium resolution spectrophotometer which, as it is small and portable, is useful for remote sensing and rapid scan spectroscopy. The FTS-16 is a ratio-recording, far-infrared spectrophotometer, which uses sequential sampling between sample and reference cells. The instrument incorporates a rapid-scanning interferometer for the far-infrared and an on-line data system. The instrument delivers high resolution spectra at all frequencies due to its rapid scanning and precision motor drive. The FTS-20 is an infrared spectrophotometer designed to measure spectra over a wide frequency range at high resolution (10^{-1} cm^{-1}).

Another infrared instrument designed for operation in the region from 500 to 5000 cm^{-1} is the Willey 318 total reflectance infrared spectrophotometer. An integrating sphere is used to collect diffusely reflected light as well as transmitted light from the sample, which may be as small as 25 mm or as large as a wall or a tree. The 318 also may be used to measure the diffuse

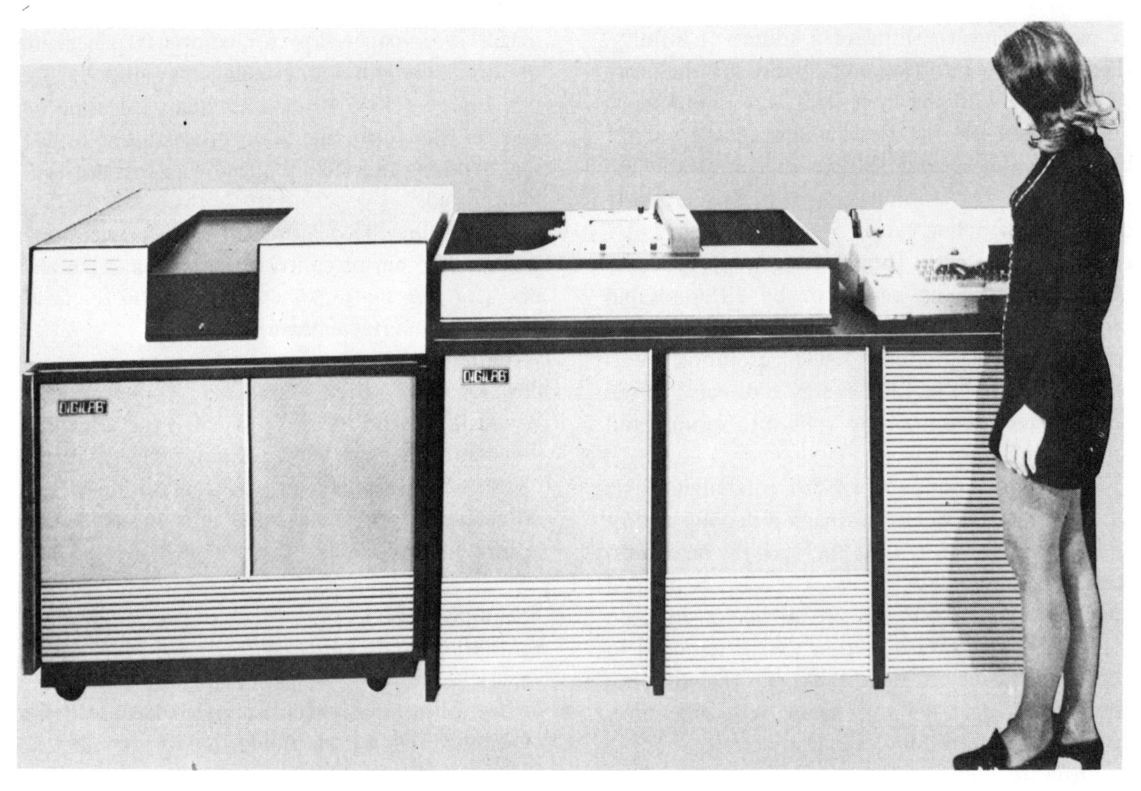

FIGURE 4.41 Digilab FTS-14 Fourier Transform Infrared Spectrophotometer.

reflectance of very rough surfaces that formerly were measured by their specular reflectance. According to the manufacturer, the combination of an integrating sphere with double-beam operation guarantees photometric accuracy.

The 318 has the following major optical components: laser and laser detector, infrared source, collimator mirror, Michelson interferometer, rotating mirror chopper, integrating sphere, and infrared detector. Three distinct beams are provided for in the optical paths of the instrument: a laser measuring beam, a reference beam and a sample beam. The laser beam and source beam are present simultaneously and share the same path.

The Michelson interferometer is used to give an improved signal-to-noise ratio and greater measurement speed. Sensitivity is increased by orders of magnitude as compared to conventional prism or grating instruments. Optimum resolution of 10 cm^{-1} in the dual beam mode of operation for a full range scan takes about seven minutes, but rapid scans of lower resolution require proportionally less time.

The laser measuring beam, which incorporates a helium-neon laser, is used to determine accurately the position of the scanning mirror. The laser beam is focused on a small mirror, then through a hole in a larger mirror (part of the source beam path) to the beam splitter where it is split into two components. One component is reflected off the scanning mirror while the other is reflected off a fixed mirror. The components are then recombined by the beam splitter and focused onto the laser detector through a hole in the source collimator mirror. Each time the scanning mirror moves one-half wavelength of laser light, a sinusoidal signal is sensed by the detector. When the wavelength of the laser light is known precisely, the scanning mirror travel may be accurately determined. The accurate frequency measurement capability of the instrument is achieved by having the laser and source beams share the same optical path through the interferometer.

The infrared source is a NERNST glower, which produces a bright beam of infrared light. The light from the source is reflected off a collimator mirror in parallel rays to simulate a source at infinity, then separated into two components by the beam splitter. As with the laser light, the components are reflected off the fixed and scanning mirrors back to the beam splitter where they are recombined. The scanning mirror is moved slowly during the sampling cycle to produce a constantly changing pathlength, thus causing the amplitudes of the recombining beams to be different and producing an interferogram. Decreasing the mirror scanning travel produces lower resolution interferograms. About half the energy is directed to the chopper mechanism to be split into sample and reference beams.

The chopper consists of a disc rotated at 13 Hz which passes between the main reflecting mirror and the three mirrors used to focus the beam into sample and reference paths. The disc is divided into three equal sections – a mirror, an infrared blackbody, and a third section that is removed to allow unobstructed light passage. The chopper alternately separates the beam into reference, sample and null (room temperature reference).

From the chopper, the sample beam is focused by a series of mirrors into the transmittance sample chamber and then into the integrating sphere where it encounters the sample being measured. Similarly, the reference beam is directed by means of a series of mirrors through the sample compartment and into the integrating sphere where it strikes the reference material. The integrating sphere is designed to collect both diffuse and specular reflectance so that total reflectance may be measured. Reflected beams are bounced and integrated within the sphere before being directed to the detector, and any scattered rays are detected as the inside of the sphere appears as a diffuse white surface to the infrared. When the specular port covers are removed, a mirror-like sample may be measured for diffuse reflectance, as the specularly reflected rays are passed through the specular ports out of the sphere leaving only the diffuse rays to be detected. The various optical paths for the laser, sample, and reference beams are depicted in Figure 4.42.

The analog signals from the infrared detector are filtered and amplified, then routed to an analog-to-digital converter, while the laser detector is simultaneously amplified and passed to the data preprocessor where the signals are combined. The laser signal then becomes the time base for the measurements. The output is recorded on magnetic or paper tape for computer processing, or direct computer interfacing may also be used (see Figure 4.43). All data are analyzed using fast Fourier transforms and other processing to provide fast handling and yield reduced data that are easily interpreted.

The Willey 318 features all solid-state circuitry and a minimum of controls to speed and simplify operator use. Options include provision for liquid and powder surface measurement.

Other less common reflectance measuring devices have been discussed by Judd and Wyszecki[30] and by Wendlandt and Hecht.[21] Wendlandt[14] has also discussed the major commercially available monochromator devices for reflectance work; a summary of such instruments is given in Table 4.02.

3. Sample Holders

a. Commercial and Microcells

The variety of samples being investigated by diffuse reflectance spectroscopy makes the design of a universal sample holder an impossible task. Clamping the sample to the sample port either in horizontal or vertical position with sample face up or down is usually accomplished with spring-loaded cups in one fashion or another. Solid and compact samples do not usually pose a problem and modifications to suit a particular application are easily designed. The problems become more difficult if one wants to measure liquids or loose powders. For liquids, often the same cells used for transmission measurements can be used, but using the same cells for powders or slurries creates problems with the removal of samples and cleaning of the cell. If samples are positioned horizontally with face up, liquids can be examined simply in small sample cups, beakers, or planchets. With powders the loose packing in cups or planchets creates the problem of maintaining a uniform surface, which is easily disturbed when samples and standards are moved back and forth. In many attachments samples have to be placed vertically or horizontally with face down and sample covers become necessary.

Few commercial sample holders are available for powder materials. Bausch & Lomb manufacture windowless cells, which can be packed with some powders by means of a sample press. Similar equipment is provided by Zeiss, who now also manufactures a wider line of sample holders for a variety of samples to be used with the new

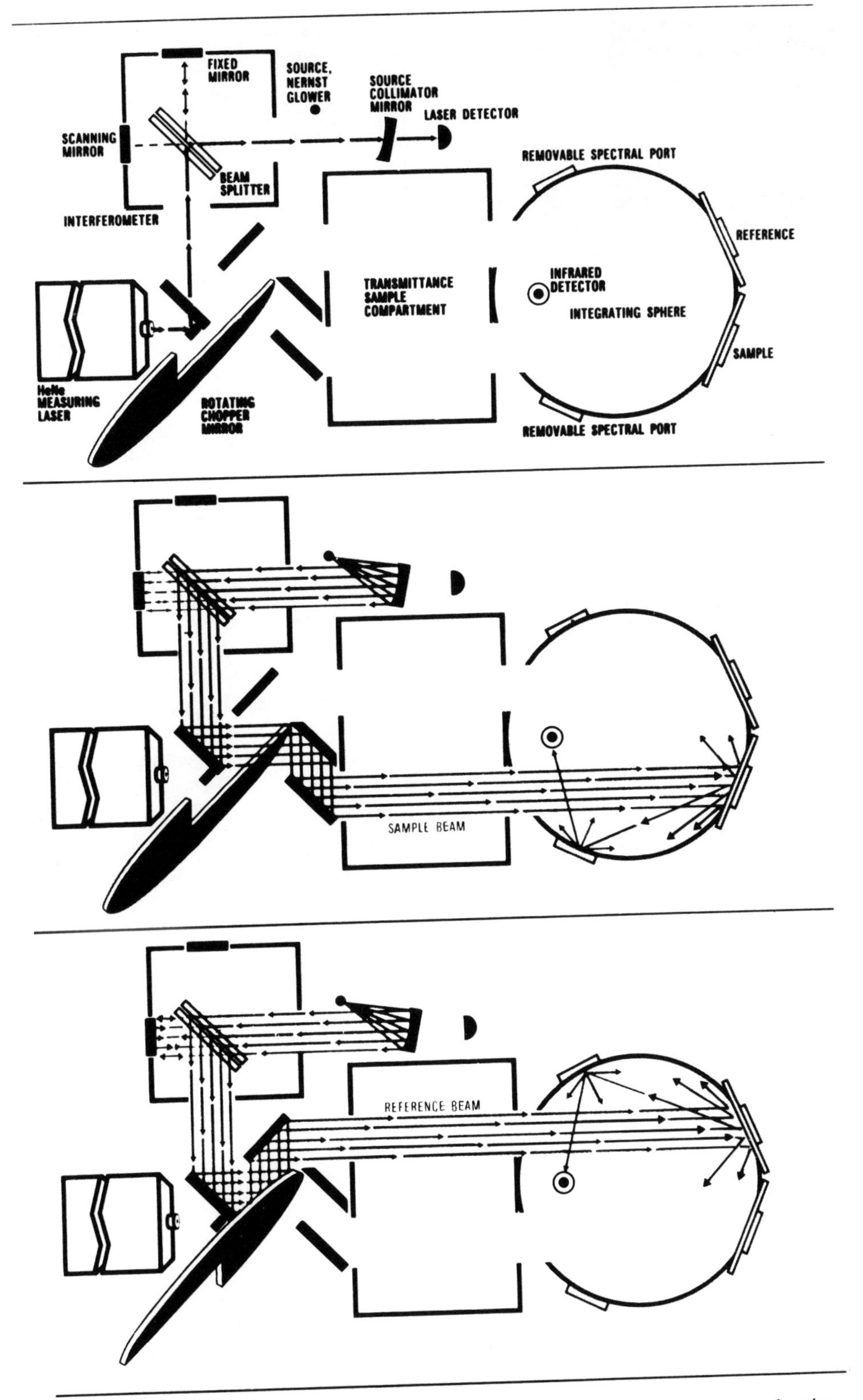

FIGURE 4.42 Optical diagram of Willey 318 Total Reflectance Spectrophotometer showing laser beam, sample beam and reference beam.

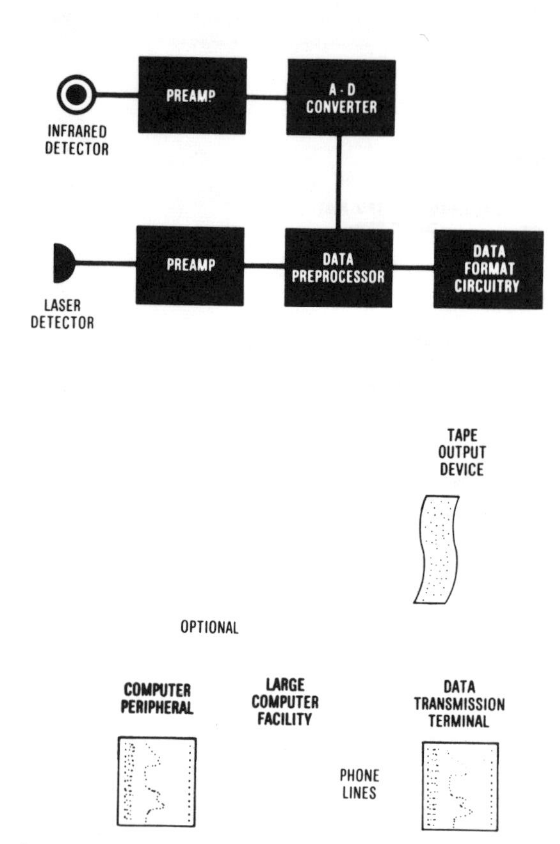

FIGURE 4.43 Schematic diagram of the data-handling facilities available with the Willey 318.

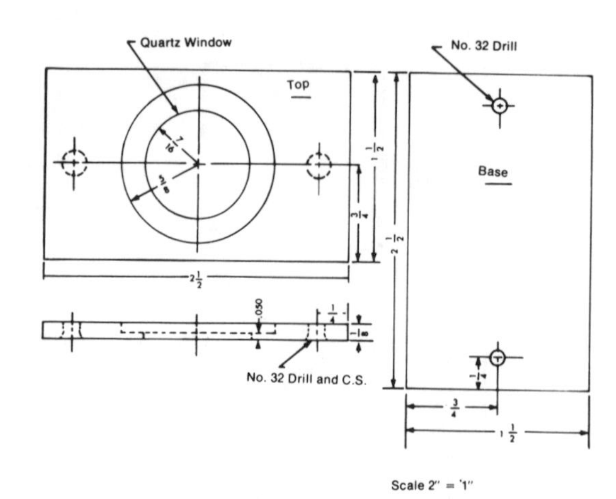

FIGURE 4.44 Sample holder for spectral reflectance measurements (dimensions in cm).[32] (From Barnes, L., Goya, H., and Zeitlin, H., *Rev. Sci. Instrum.*, 34, 292 (1963). With permission of American Institute of Physics.)

Model DMC-25 Color Spectrophotometer. The Bausch & Lomb cells can only be used with relatively adhesive powders such as $MgCO_3$ or $BaSO_4$ and are primarily recommended for the preparation of reflectance standards. Cells of 22 mm diameter and 10 mm depth are provided for the Hitachi Perkin-Elmer instrument Model 139. A number of noncommercial holders were discussed by Tonnquist.[31]

A typical example for a low-cost reflectance cell designed for the measurement of powders and slurries in the UV, visible and near-infrared region of the spectrum was designed by Barnes et al.[32] (see Figure 4.44). The cell consists of a Bakelite® or Lucite® cover plate and base. The quartz window (1.1 x 0.13 cm) is cemented into the hole (diam. 1.1 cm) with epoxy resin. A standard aluminum planchet (1¼ x 1/16 in., or ca. 3.2 x 0.16 cm) serves as the sample holder and is held tightly by means of two screws in the circular indentation of the cover plate between the base and the top. The cell was originally designed for use with the Beckman Model DU attachment but can be used in practically every other commercial instrument with little or no modification. Dimensions can easily be modified to suit special needs; instead of screws, masking tape can be used to hold the cell together. The only relatively costly part is the quartz disk. Larger series of similar cells can therefore be produced with a minimum of time and money involved. The danger of breaking the cell is small. This cell has been used extensively for the investigation of adsorption phenomena of many systems. Since it can be sealed easily with silicone grease on the edge of the planchet, it will remain leakproof for a prolonged period of time, e.g., humidity from air can be kept away from the sample during the measuring process.

b. Semi-microcells

A universal drawback of the cells discussed so far is the large amount of sample needed for packing the holders. The amounts range from a few grams, e.g., for the Bausch & Lomb sample holders, down to 0.5 and 1 g for the holder described by Barnes et al.[32] When the problem arose of accommodating spots scraped off thin layer chromatograms for qualitative and particularly quantitative measurements, a new line of reflectance cells had to be designed to enable the measurement of a few milligrams of powder sample. In this section three types of reflectance cells are described; each is capable of accommodating less than 100 mg to as little as 10 mg of sample.

TABLE 4.02

Monochromator Instruments for Reflectance Measurements

Manufacturer	Model	Mode of operation	Optics	Optical geometries	Range of operation (nm)	Miscellaneous
American Instr. Co. Inc.	Shimadzu Model QV-50	single-beam	sphere	0/d	UV-visible	
Bausch & Lomb Co.	Spectronic 20	single-beam	sphere		400–700	
	Spectronic 505	double-beam	sphere		220–700	Light traps for specular reflectance
Beckman Instruments Inc.	Model DU	single-beam	ellipsoidal mirror	0/d	210–700	
	DK-2	double-beam	sphere	0/d, 5/t or d/0	210–2,700	
	DB, DG	double- or single-beam	sphere	0/d	210–700	
Cary Instruments, Applied Physics Corp.	Models, 14, 15	(1) double-beam (chopper mirror)	sphere	0/d, d/0	250–750	
	(3 attachments)	(2) double-beam (chopper mirror)	circular mirror system	0/d	220–700	
		(3) double-beam (chopper mirror)	one sphere one ref. box	0/d	220–700	
	(new attachment) Models 14, 14R, 14RI	double-beam	sphere	0/d	220–2,500	
General Electric Co.	Hardy Spectro-photometer	double-beam	sphere		380–700 (or 380–1,000)	First model built in 1931
Perkin-Elmer Corp.	Model 350	double-beam	two spheres	16°15′/d	220–750	Traps for specular reflectance
	Model 450	double-beam	two spheres	16°15′/d	(220–2,500 on request)	
Perkin-Elmer Corp. Coleman Instruments Div.	Hitachi Perkin-Elmer Model 139	single-beam	sphere	0/d	400–760 (or 250–760)	
Pye Unicam Ltd.	SP500	single-beam	ellipsoidal mirror	0/d	350–1,000	
	SP700	double-beam (beam splitter)	ellipsoidal mirror	0/d	200–2,500	

TABLE 4.02 (Continued)

Manufacturer	Model	Mode of operation	Optics	Optical geometries	Range of operation (nm)	Miscellaneous
	SP800	double-beam	ellipsoidal mirror	0/d	200–750	Reference beam strikes detector only
Rank Precision Industries Ltd. (Hilger & Watts Ltd.)	Uvispek	single-beam	sphere	0/d	UV-visible	
Carl Zeiss, Inc.	PMQ II RA-2	single-beam	lens system	45/0	220–2,500	
	PMQ RA-3	single-beam	sphere	0/d; d/0	380–2,500 (or 220–2,500)	(d/0 for rough textures)
	Micro attachment	single-beam	sphere	0/d; d/0	250–2,500	Minimum sample area 0.1 mm^2
	DMR-21	double-beam	sphere	8/d	220–2,500	Gloss traps
	DMC-25	double-beam	sphere	d/8; 8/d; 45/0	200–2,500	Digital readout of KM function and x, y, z coordinates

1. Glass Window Cell.[33] This cell consists of white paperboard to which a 3.7 x 2.5 x 0.1 cm microscope cover glass is affixed with two pieces of masking tape. The white backing paper is cut to 4.0 x 3.0 x 0.1 cm, which permits its introduction into the sample sliding drawer of the reflectance accessory built for the Beckman DU Spectrophotometer. A sketch of this cell is presented in Figure 4.45. The analytical sample S consists of the adsorbent and the resolved compound and is carefully compressed between the cover glass C and the paperboard P, until a uniformly thin layer of densely packed powder is obtained. If properly packed, the layer varies in thickness between 0.3 and 0.5 mm depending on the material. The main advantage of this type of cell is its low cost and simplicity.

2. Quartz Window Cell.[34] This cell consists of a circular quartz plate, (diam. 22 mm) superimposed on a 40 x 40 x 3-mm plastic plate which is affixed to the backing paper with two pieces of masking tape. A circular window (diam. 19 mm) in the upper surface of the plate opens into a concentric circular well (diam. 24 mm) that is deep enough to accommodate the quartz disk. A sketch of the cell is given in Figure 4.46. The analytical sample, after being introduced into the cell, is carefully compressed between the quartz disc and the paperboard by rotating the disc until a thin layer is obtained. Between 60 and 90 mg of sample is required.

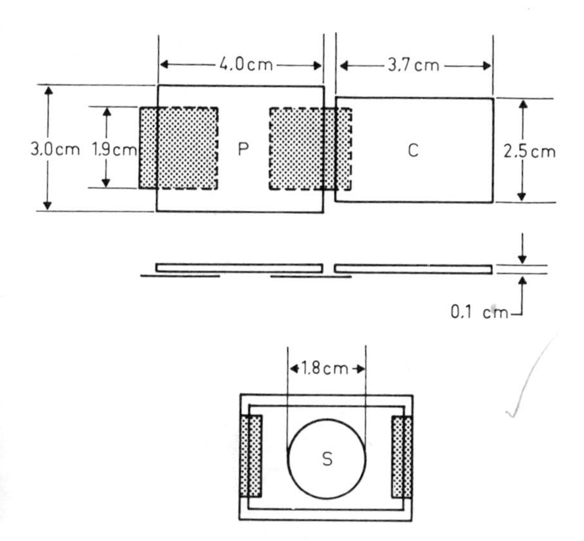

FIGURE 4.45 Glass window cell. P, paperboard; C, glass cover; S, sample.[33] (From Frei, R. W. and Frodyma, M. M., *Anal. Chim. Acta*, 32, 501 (1965). With permission of Am. Elsevier.)

3. Windowless Cell.[35] This cell consists of a 35 x 40-mm plastic plate affixed to white paperboard of the same dimensions with two pieces of masking tape. As shown in Figure 4.47, the plastic plate has a circular opening (diam. 21 mm) in its center. The cell is packed by introducing the sample into the opening and then compressing it with a fitted tamp made of an aluminum planchet affixed to a cork stopper.

Obviously the glass window cell can be employed only with samples that absorb in the visible region, whereas the other two are suitable for use in both the visible and UV regions of the spectrum. Of these two, the windowless cell, although it is simpler in design and costs almost nothing to make, provides less protection to the analytical sample.

c. Variable Temperature Cells

Valuable additional information can often be obtained if reflectance spectra are recorded at *elevated temperatures.* For work in high-temperature reflectance spectroscopy, a special set of sample holders is needed either for maintaining the sample at one particular temperature (usually between 100 and 300°C) and recording the

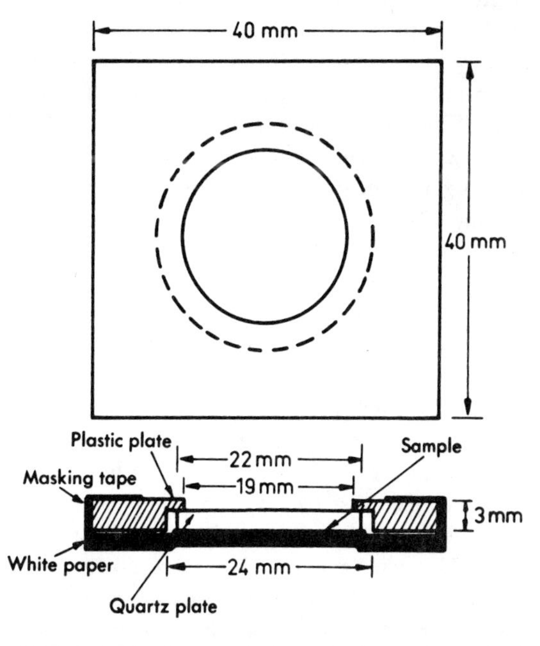

FIGURE 4.46 Quartz window cell. Dimensions of cell elements and sketch of assembled cell.[34] (From Frodyma, M. M., Lieu, V. T., and Frei, R. W., *J. Chromatogr.*, 18, 520 (1965). With permission of Am. Elsevier.)

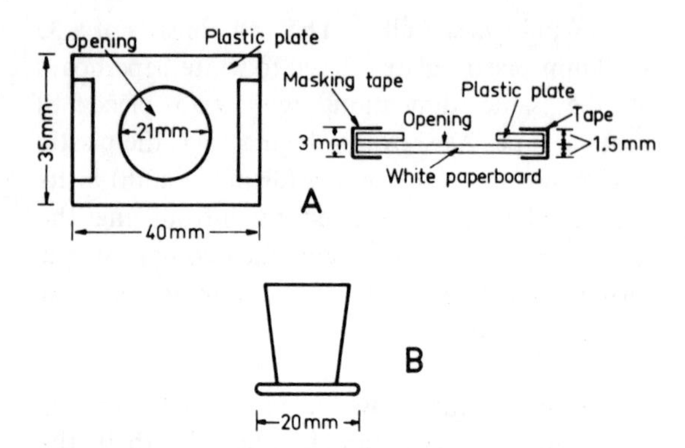

FIGURE 4.47 Windowless cell. (A) Dimensions of cell elements and sketch of assembled windowless cell. (B) Tamp used to pack cell.[35] (From Frodyma, M. M. and Lieu, V. T., *Anal. Chem.*, 39, 814 (1967). With permission of American Chemical Society.)

spectrum or for gradually increasing the temperature at a uniform rate and measuring the sample at one preset wavelength. The latter method is often referred to as dynamic reflectance spectroscopy. A number of reflectance studies have been carried out at elevated temperatures (see Chapter 6, Section 2a) and some sample blocks suitable for controlled heating of the samples with an embedded filament were described.[36-41] Asmussen and Andersen[37] reported on a sample holder with a cylindrical nickel-plated brass heating block (diam. 60 mm, height 85mm). The temperature was regulated through the filament of a small bulb mounted in a chamber at the bottom of the cylinder. A sample cell 35 mm in diameter and 0.5 mm deep was contained on top of the heating block. The temperature of the powder sample was measured with an embedded thermocouple.

Wendlandt et al.[38-41] described a number of cells used for their numerous studies of high-temperature and dynamic reflectance spectroscopy. The first of their heated sample holders consisted of an aluminum block 60 mm in diameter and 11 mm thick. A circular indentation (25 x 1 mm) with two circular ridges (to prevent the compressed powder sample from falling out of the holder when mounted in vertical position) was cut into the aluminum block to serve as a sample holder. In some cases the sample was covered with a glass plate. The heating was done with a Nichrome® wire, wound in a spiral on an asbestos board and covered with a layer of asbestos paper. This heating element was fixed close to the sample cavity. The sample temperature was monitored by a Chromel-Alumel thermocouple contained in a ceramic insulating tube and fixed directly behind the sample-holder indentation. A thermal spacer made of insulating material prevented heat transfer to the optical sphere. Modifications of this sample cell were described later.[21,41] One of the changes involved the heater element, which was inserted in cartridge form directly behind the sample well. One thermocouple was then placed adjacent to the heater and served to control the temperature programmer, and the other was positioned in the bottom of the sample cavity for detection of the sample temperature.

Another sample holder described by Wendlandt and Hecht[21] consisted of an aluminum block 50 mm in diameter and 25 mm in height. A circular sample indentation (25 x 1 mm) was machined into the block and heated with a 35-W stainless steel sheathed heater assembly embedded in the main block. The thermocouple arrangement was as described previously. Thermally unstable samples were covered with pyrex or quartz glass.

In 1968 a cell that can be used for high-temperature reflectance studies of very small samples was described.[41] The holder is based on a design reported previously[33] (Figure 4.45). A drawing of this sample holder is shown in Figure 4.48. A small aluminum block (40 x 50 mm) is heated internally by a circular heater element. Electrical connections to the heater and the

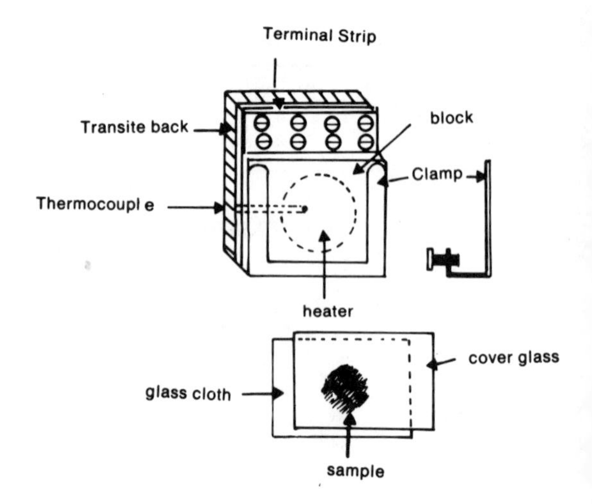

FIGURE 4.48 Heated sample holder.[41] (From Wendlandt, W. W., in *Modern Aspects of Reflectance Spectroscopy,* Wendlandt, W. W., Ed., Plenum Press, New York, 1968, 58. With permission.)

thermocouple are made with the terminal strip on top of the assembly. Both the heating block and the terminal strip are mounted on a 5 x 5-cm transite block. Similar to Figure 4.45, the sample is compressed between a glass-fiber cloth mounted on the heating block and a cover glass. The cover glass is tightened to the assembly by means of a metal clamp. For dynamic reflectance spectroscopic (DRS) work, any of these cells can be used by hooking them up with a heater controller,[42] a strip-chart recorder to record the sample temperature, and an X-Y recorder for the DRS curve, recording the change of sample reflection at one particular wavelength as a function of temperature.

Wendlandt and Dorsch[43] described a controlled atmosphere sample holder in 1970 for temperature-controlled reflectance measurements suitable for use in high-temperature reflectance spectroscopy (HTRS) and in dynamic reflectance spectroscopy (DRS). The cell has an operating range from room temperature to approximately 500°C. The cell consists of a silver heater block (25 mm diameter) heated by two 2.6 ohm Nichrome wire heaters. The surface of the block has a 1 x 10 mm indentation into which the sample is placed. The silver block is enclosed in the main body of the sample holder and insulated from it by a layer of ceramic fiber insulation. The sample side of the cell is covered with a quartz plate and an 'O' ring between this cover plate and the sample holder provides a gas tight seal. Inlet and outlet lines are provided so that gas may be flushed through the cell to provide a controlled atmosphere. The sample holder has been attached to a Deltatherm III differential thermal analysis instrument in place of the DTA furnace and sample holder so that the temperature programmer and readout of the instrument may be used with the DRS sample holder. The temperature may thus be held constant or varied at heating rates of from 0.625 to 10°C/min.

To measure a DRS curve, the sample is placed in the cell and tamped lightly into position with a smooth metal rod. The quartz cover plate is clamped in place and the sample holder is then attached to the sample port of the integrating sphere. The gas inlet and outlet hoses are attached and a stream of gas is passed through the chamber. During the heating cycle, the gas flow may be continued or interrupted as desired. As the temperature is changed, the DRS curve of the sample is recorded. Using a two channel recorder, the temperature may be recorded simultaneously on the second channel.

As the temperature of the sample may also be held at any desired temperature within the operating range, the cell is also suitable for use in high-temperature reflectance spectroscopy (HTRS). In this case, the spectrum of the sample would be recorded at a given temperature, then the temperature would be changed and the spectrum recorded again. The procedure is repeated until the desired temperature range has been covered. For operation at temperatures above 300°C, Wendlandt and Dorsch[43a] recommend that an external cooling fan be used to keep the main body of the sample holder at a reasonable temperature and prevent damage to the rubber 'O' rings, which provide the seal.

Reflectance measurements at *low temperature* can also be useful at times, e.g., in cases where stability is a problem or for kinetic studies of systems in the adsorbed state, etc. A typical example for a low-temperature diffuse reflectance cell is the combined reference and sample-material holder designed by Symons and Travalion,[44] specifically for use with the Unicam SP 540 reflectance accessory. Similar sample holders could be used with the Beckman DU or Shimadzu single-beam Spectrophotometers. The holders are recessed to contain powder samples. They are constructed of copper rods (diam. 1.25 in.) connected to each other with copper tubing and with connections from the base of the rods to a Dewar flask by means of flexible tubing. Liquid nitrogen is pushed through the cooling system by low-pressure nitrogen gas. With this technique a temperature of 77°K can be achieved.

d. Tablet Holder

A special cell was designed by Urbanyi et al.[45] for the reproducible measurement of tablets in the Beckman DU Spectrophotometer with standard reflectance attachment. The holder (Figure 4.49) permits the accurate and reproducible centering and leveling of individual tablets in the reflectance attachment for the measurement of diffuse reflectance. The holder is made of oxidized steel, with dimension "A" being the diameter of the tablet. The tablet to be investigated is placed in the holder and the leveling screw is adjusted so that the top edge of the tablet is level with the holder, which is mounted in the front compartment of the

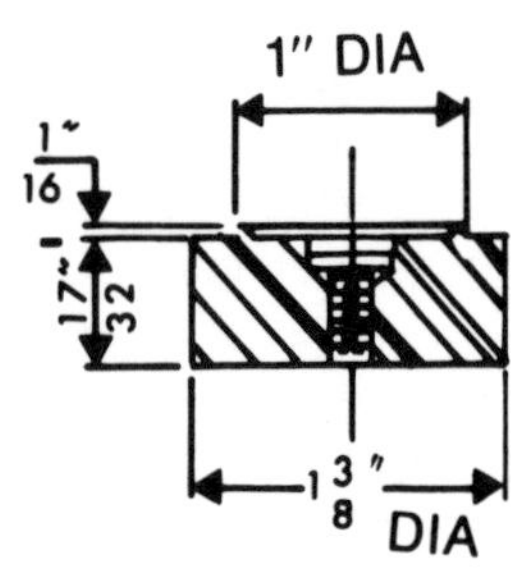

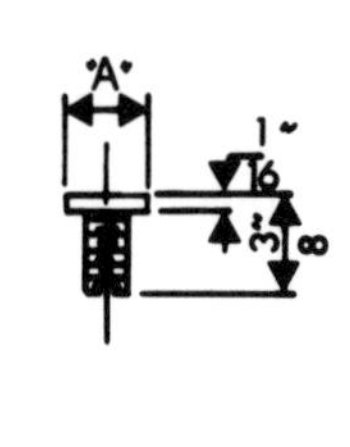

FIGURE 4.49 Tablet holder for Beckman DU reflectance attachment.[44]

reflectance attachment. The same holder can be adapted to other spectrophotometers with minor modifications.

4. Instruments for *in situ* Measurement of Chromatograms

The technique of direct scanning of paper or thin-layer chromatograms by transmitting light through the chromatogram and measuring the absorbance has been commonly used for more than a decade. This technique, generally known as densitometry, has been discussed in a number of books and reviews.[46-49] Analogous to scanning spots by densitometry, reflectance spectroscopy can be used for direct scanning of separated materials on chromatoplates and sheets. The reason in situ reflectance spectroscopy has lagged behind in its development compared to transmission densitometry can be attributed to a lack of knowledge and familiarity with the general technique and instrumentation in this field.

a. Noncommercial Designs

Klaus[50,51] has described possibilities for the use of the Zeiss Model PMQ II Spectrophotometer equipped with the standard optical sphere attachment for a direct evaluation of thin-layer plates by spectral reflectance. De Gallan et al.[52] have discussed the construction and use of an accessory suitable for the scanning of chromatoplates over the sample aperture of the Zeiss diffuse reflectance attachment and have demonstrated its use in the visible and UV region of the spectrum. Other noncommercial instrument modifications suitable for in situ reflectance measurements were described by Gordon[53] and Hamman and Martin.[54] Beroza et al.[55] devised a simple and low-cost instrument for the automatic recording of spectral reflectance from thin-layer chromatograms. Fiber optics are used in this instrument, and both single- and double-beam operation can be carried out. The apparatus is shown in Figure 4.50. In the double-beam mode, the instrument uses two fiber optics assemblies (Dolan-Jenner Industries, Inc., Melrose, Massachusetts). The spots are scanned with the scanning head and a reference head scans the adjacent blank area. Each assembly operates with a Y-shaped bundle of randomly arranged glass fibers of 24 in. length. The two arms of the Y have complementary functions in that one conducts the light from an incandescent bulb to the sample surface (diffuse polychromatic irradiation) and the other picks up the diffuse reflectance and conducts it to a cadmium sulfide cell. Each CdS cell forms one leg of a Wheatstone bridge (Figure 4.51) with a ten-turn 500-ohm potentiometer and two 100-ohm resistors serving as the other two legs. The unbalanced output is attenuated by a 10-kΩ linear taper potentiometer and fed to a strip-chart

FIGURE 4.50 Apparatus for determining diffuse reflectance of TLC spots with fiber optics.[55] (From Beroza, M., Hill, K. R., and Norris, K. H., *Anal. Chem.*, 40, 1611 (1968). With permission of American Chemical Society.)

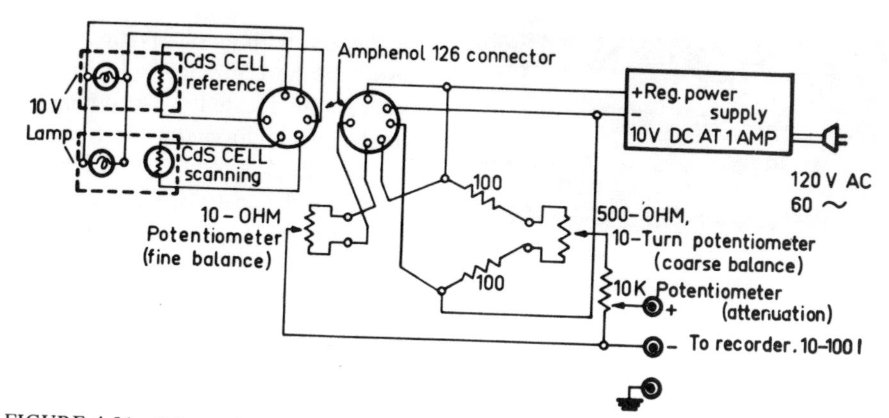

FIGURE 4.51 Schematic diagram of scanner bridge circuit.[55] (From Beroza, M., Hill, K. R., and Norris, K. H., *Anal. Chem.*, 40, 1611 (1968). With permission of American Chemical Society.)

recorder. A 10-v regulated solid-state power supply (Lectroteck Model 1015) provides the power for the bridge and light bulb.

The thin-layer plate scanning assembly has been adapted from the one used for the Aminco Bowman Spectrofluorometer (American Instrument Co. Inc., 8030 Georgia Avenue, Silver Spring, Maryland 20910) by removing the optics and the lid. The fiber optics scanning and reference heads are fastened side by side at the end of a freely pivoting aluminum rod supported by laboratory clamps and rods. During the scanning operation the heads remain stationary while the plate moves under them at a speed of 3 in./min. The distance between the head and the TLC plate is maintained with the use of a cover plate (1.5 mm thick) on which the scanning head slides. Scanning friction is reduced by means of a Teflon® strip attached to the head with masking tape. The reference head is clamped slightly higher and does not touch the plate assembly.

For scanning the spots with monochromatic light, the original tungsten light source is replaced by a beam of monochromatic light from a Beckman Model DU Spectrophotometer. Single-beam operation is used in this case and the reference head is removed. The cell holder and photocell compartment of the spectrophotometer are removed to expose a circular port, and the arm of the scanning assembly normally leading to the light source is fitted into the port with a one hole rubber stopper and is carefully aligned to receive the light output of the Beckman instrument. With this arrangement and a maximum slit width setting on the spectrophotometer, the bridge circuit can be balanced from 480 to 700 nm when the scanning head is placed over a white reflecting surface.

The advantage of the fiber optics is the possibility of bringing the light to a small defined area on the plate. By keeping the light-collecting fiber optics close to the plate, the loss through scattering of diffusely reflected energy can be minimized. Specular reflectance poses no problem with a fiber optics assembly. The validity of the Kubelka-Munk function and hence the proof that the collected radiation consists only of diffusely reflected light was determined on the basis of reflectance data gathered from some pesticides spots. The instrument described by Beroza et al.[55] is being developed commercially by Kontes Glass Company (Vineland, New Jersey). The authors have been informed that the commercial instrument contains some improvements over the original instrument and that a prototype is now ready for display.

A TLC scanning device operating in the "flying spot" mode has been described by Goldman and Goodall.[56] The scanner was adapted to the Joyce, Loebl Chromoscan.

b. Commercial Instruments

One of the first commercially available attachments for the direct mechanical scanning of thin-layer plates by reflectance spectroscopy was the Zeiss Chromatogram-Spectrophotometer PMQ II. It has been described in detail[47,57-63] and is shown in Figure 4.52.

For the quantitative determination of chromatographically separated substances by means of

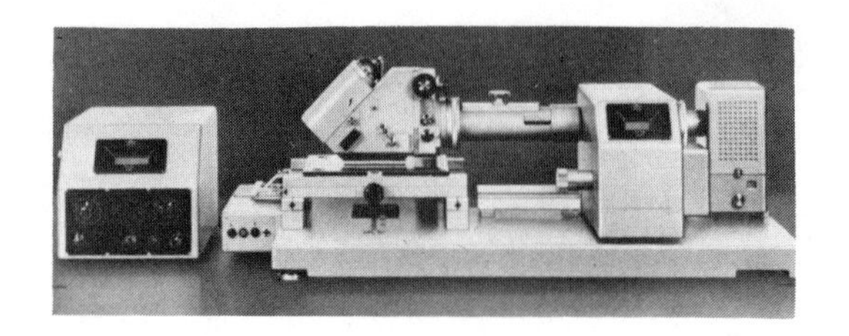

FIGURE 4.52 The Zeiss Spectrophotometer PMQ II equipped with TLC attachment for diffuse reflectance spectroscopy.

their light absorption, the chromatogram is illuminated with monochromatic light at a zero-degree angle with respect to the chromatogram surface. (The radiation from the light source is spectrally dispersed in the monochromator and then strikes the sample.) The diffusely reflected nonabsorbed light is measured at an angle of 45° (Figure 4.53). In this attachment the use of an integrating sphere (Ulbrichtkugel) has been abandoned; instead, a light pipe is used for the collection of the diffuse radiation. The geometries of illumination and observation are $_{0^\circ}R_{45^\circ d}$. Direct reflectance escapes through the same opening in the light pipe, which is used as entrance port for the incident monochromatic light-beam. For measurements the thin-layer or paper chromatogram is placed horizontally on the mechanical stage (Figure 4.54), which can be moved in two directions at right angles to each other. The stage is moved by a synchronous motor at one of ten preselected speeds in the direction of the y-axis and adjusted manually in the direction of the x-axis. Provided the zone is visible, the position of the spot in relation to the measuring aperture can be checked with an observation device. Zones that absorb in the UV range and are not visible to the eye can be located by scanning (at a suitable wavelength) a chromatogram trace lying above the starting point.

The mechanical stage and the lower surface of the measuring head are parallel in any position, and, with the use of a scale, the distance between the two can be adjusted to fit a particular chromatoplate. The local resolution in the direction of the motorized feed (y-axis) depends on the dimensions of the scanning spot in this direction. This dimension can be varied between about 0.02 and 2 mm. The maximum length of the scanning spot at right angles to the direction of travel is 14 mm, but the length of this edge can be reduced to match variations in the diameter of the spot. Change of focusing gives any defined scanning spot in the different spectral ranges. With the most recent models, focusing of the rectangular scanning spot (slit image) is performed automatically by means of a quartz achromatic optical system. It is also possible to use a circular scanning spot to cover the entire chromatogram spot. The maximum diameter of the scanning spot is 30 mm; this can be varied for optimum conditions.

A built-in swing-in standard which can be moved into the path of the beam allows test values to be calibrated without affecting the position of the object under investigation. For the measurement of transmitted light, the detector is placed underneath the sample. Since the monochromator gives any desired radiation between 200 and 2500 nm for the test, it is possible to determine spectra (light absorption of reflectance as a function of the wavelength) for qualitative identification or

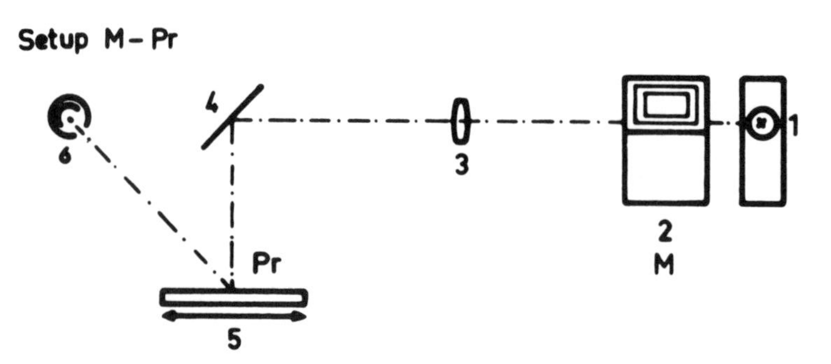

FIGURE 4.53 Simplified optical diagram of the Zeiss Chromatogram Spectrophotometer. Setup M-Pr (M, monochromator; Pr, probe-sample). 1, Light source: hydrogen lamp, incandescent lamp or high-pressure xenon lamp; 2, monochromator; 3, intermediate optical system (simplified); 4, deviating mirror; 5, chromatogram; 6, detector.

FIGURE 4.54 Optical head and X-Y stage of the Zeiss Chromatogram Spectrophotometer.

FIGURE 4.55 The Camag-Z-Scanner, mounted to the Zeiss PMQ II Spectrophotometer.

characterization of separated substances. With small alterations in the optical system, the same instrument can be used for fluorescence measurements.

Camag Ltd., Muttenz, Switzerland, has started production of the Camag-Z-Scanner (Figure 4.55) that can be used in conjunction with the Zeiss Spectrophotometer PMQ II for the direct measurement of spectral reflectance and fluorescence from thin-layer and paper chromatograms. Figure 4.56 shows a close-up of the assembly set up for reflectance work at an optical geometry of ${}_{0°}R_{45°}$. The Z-Scanner is placed directly adjacent to the monochromator unit of the Zeiss instrument and aligned on the same triangular rod used on the optical bench system. As can be seen from the optical diagram (Figure 4.57) the thin-layer or paper chromatogram is placed in a vertical position to the optical light-path system. In Figure 4.57a the monochromatic light-beam impinges on the chromatogram at a zero-degree angle. The diffuse radiation is picked up by a suitable lens and photodetector assembly, which has to be close to the sample surface to minimize radiation loss.

The system can be reversed if measurements at a geometry ${}_{0°}R_{45°}$ are undesirable. In Figure 4.57b the sample surface is irradiated polychromatically at a 45° angle and the chromatogram is viewed at a zero-degree angle. The diffusely reflected radiation is dispersed in the monochromator and measured with the photodetector. In both modes (${}_{0°}R_{45°}$ and ${}_{45°}R_{0°}$) specular reflectance can be excluded. In the visible region a tungsten lamp is used as an energy

FIGURE 4.56 Rear view of the Camag-Z-Scanner set up for reflectance work at an optical geometry ${}_{0°}R_{45°}$.

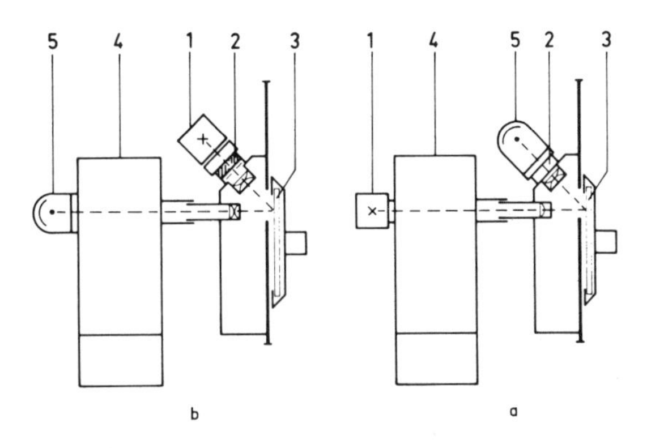

FIGURE 4.57 Optical diagram of the Camag-Z-Scanner. a. Geometry ${}_{0°}R_{45°}$. b. Geometry ${}_{45°}R_{0°}$. 1, deuterium or tungsten lamp; 2, lens system; 3, chromatoplate; 4, monochromator; 5, photomultiplier.

source. The use of quartz optics and a deuterium lamp enable the extension of the spectral range of this attachment down to 220 nm.

For in situ fluorescence measurements the assembly shown in Figure 4.57b is used in conjunction with a blocking filter and a mercury lamp. For quantitative scanning the chromatogram plate holder moves horizontally at one of two preselected speeds. The vertical position of the scanning track is set by a precision adjustment. Since the plate holder has a removable bottom plate, the chromatogram can be observed from the back thus facilitating correct and precise positioning. Width and length of the scanning slit can be selected from a wide variety of dimensions; the maximum available slit length is 16 mm.

The Camag-Z-Scanner can also be used for the manual recording of spectra in the region 220 to 750 nm. In order to compensate for changes in photomultiplier response at different wavelengths, background corrections are made at certain intervals. This can be done with a built-in blocking device, which permits a reproducible shifting of the plate holder from the spot area to an empty region on the chromatogram and vice versa. With the previously discussed Zeiss attachment, this function is performed by the swing-in reflectance standard. Using the same chromatographic adsorbent as a reference standard, however, is in certain cases more advantageous.

In principle it is believed that the Camag-TLC-Scanner, designed for direct fluorometric measurements with the Turner Model 111 instrument can also be modified for reflectance work in the UV and visible region of the spectrum.[64]

The Vitatron Densitometer TLD 100 (Vitatron Ltd., Dieren, Holland) is another single-beam instrument designed for the in situ measurements of reflectance, transmittance and fluorescence on paper and thin-layer chromatograms (Figures 4.58, 4.59). An interesting feature is the "flying spot" scanner, which is quite different from previously discussed designs that scan the spots with an optical slit. The principle of "flying spot" scanning was first discussed by Goldman and Goodall.[56] To measure the total concentration of the chromatogram spot an oscillating movement is given to the plate with a stroke length slightly larger than the spot diameter. At the same time the plate is moved at a uniform speed in the Y-direction. The integrated values for the optical density, measured over the stroke length, are

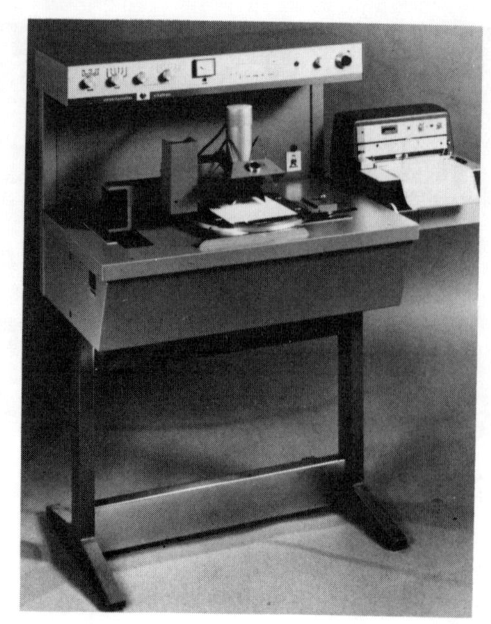

FIGURE 4.58 The Vitatron Densitometer TLD 100.

FIGURE 4.59 Close-up of the vibrating sample stage and optical head of the Vitatron Model TLD 100 Densitometer.

registered on a chart recorder and the peak areas are computed with an integrator. With this scanning technique instantaneous spot areas under observation are very small and appear homogeneous. This approach is therefore expected to be particularly suitable for irregularly shaped spots (tailing spots).

An optical and mechanical diagram is shown in Figure 4.60. Two light sources are available, one

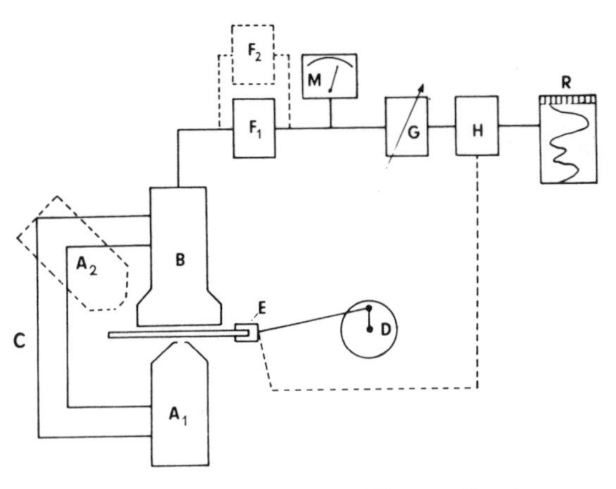

FIGURE 4.60 Schematic of the Vitatron Densitometer TLD 100. A_1, light source for transmittance; A_2, light source for reflectance or fluorescence; B, photomultiplier unit; C, optical base; D, drive units; E, sample table and holder; F, measuring amplifier; M, maximum indicator; G, variable amplifier; H, electronic speed corrector; R, recorder.

for transmittance and one for diffuse reflectance measurements with an irradiation angle of 45° to the sample plane. The radiation is rendered monochromatic with suitable filters. A quartz-iodine lamp of 12 v and 55 w is used for reflectance and transmittance work and an Hg lamp can be mounted instead for direct fluorometry. The diffusely reflected radiation is collected by a photomultiplier assembly positioned at a zero-degree angle to the chromatogram surface. The manufacturers claim a better performance for the instrument if it is operated in the transmission mode. This may be true for the "flying spot" method, since light scattering is kept at a minimum under these conditions. On the other hand, it should be kept in mind that the reported useful optical range (320 to 700 nm) could be extended down to 220 nm if the reflectance mode is used, but not in the transmission mode in connection with thin-layer chromatograms. This can be done simply by replacing the iodine lamp with a deuterium energy source. Quartz optics are already built into the instrument. The Vitatron TLD 100 is not suitable for the measurement of spectra.

Several instruments, which operate on the double-beam principle, are now available for in situ quantitative evaluation of chromatograms by diffuse reflectance spectroscopy. One of them is the Chromoscan with chromatogram accessory (manufactured by Joyce, Loebl & Co. Ltd., Princesway, Team Valley, Gateshead 11, England). The Chromoscan was one of the first commercial instruments used in conjunction with a scanning unit (Figure 4.61) for the quantitative evaluation of chromatograms by transmission densitometry and in situ fluorometry. Recently the unit has been modified to enable the direct measurement of spectral reflectance also. The actual operation principle of the Chromoscan used without the TLC unit is demonstrated in the optical scheme (Figure 4.62). The Chromoscan is a filter instrument. The light passes from the external lamp housing L through a filter-holder assembly f_1, a system of lenses and aperture holders, to a rotating chopper mirror which alternately passes or deflects the light to produce a double-beam system. In the vertical deflection the beam passes through a base line and control wedge and a secondary filter mount f_2 to the photomultiplier R. The light that passes the chopper undeflected (if used in the reflectance mode) impinges on the sample surface at a zero-degree angle. The diffuse radiation is viewed at a 45° angle by the photomultiplier R, which is preceded by a third filter mount f_3. No secondary filters f_2 and f_3 are used for reflectance work.

The thin-layer attachment works on essentially the same basis as the main instrument, but it

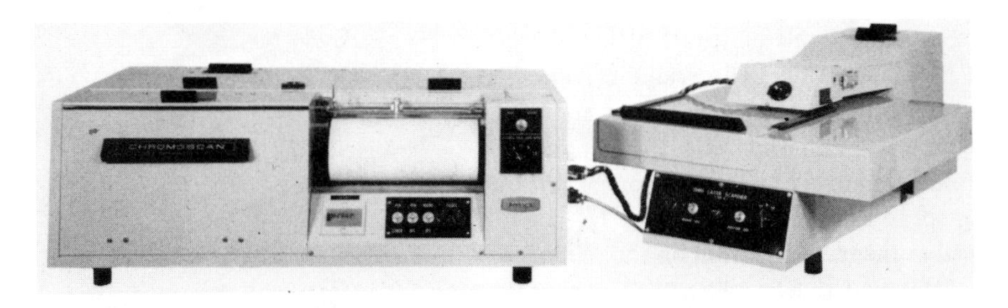

FIGURE 4.61 The Joyce, Loebl Chromoscan with chromatogram scanning unit.

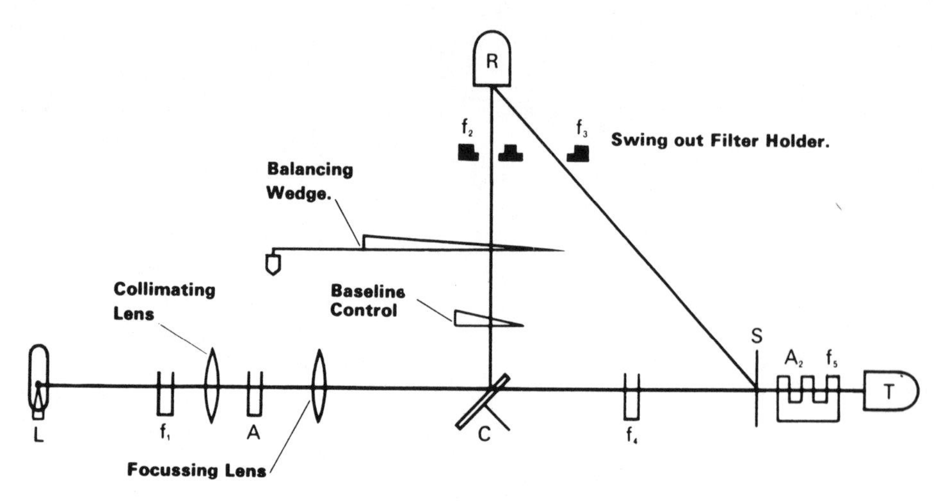

FIGURE 4.62 Optical diagram of the Chromoscan operated in the reflectance mode. L, lamp housing; F, filters; A, aperture holder; C, rotating chopper mirror; S, solid sample; R, photomultiplier in reflectance; T, photomultiplier in transmittance.

provides its own optical system, e.g., main apertures, filters, etc., are mounted in the TLC unit. The chopper phase is advanced by 180° to that of the parent instrument, so that its photomultiplier receives light signals alternately with the photomultiplier in the attachment. The thin-layer accessory also provides its own scanning beam and can therefore be changed instantaneously from reflectance to transmission operation with a switch controlling a mirror position in front of the accessory filter mounts.

A series of filters, which enables the choice of a proper spectral region between 400 and 750 nm, is available for reflectance work. Although no special reference is made to in situ UV reflectance spectroscopy, the authors think that the instrument could be adapted to the UV region of the spectrum with minor changes only. All optics are made from quartz, and the lamp house is so designed that alternative sources such as a deuterium lamp could be fitted for operation at lower wavelength. The instrument is not suitable for the recording of spectra. Similar to the design used by Zeiss, it is the sample table rather than the optical head that is movable. The scanning table can be adjusted manually in one direction, and a mechanical device is used in the other direction to move the chromatogram spot past the optical slit. The scanning speed can be chosen at 1/9, 1/3, or the same speed as the recorder drum on the parent instrument. Plates and sheets 200 x 200 mm or smaller can be placed on the sample table.

Modifications of the Chromoscan densitometer have been built and used by Goldman and Goodall[56,65] for a more theoretical treatment and comparison of reflectance and transmission phenomena on chromatograms. A detailed discussion of instrumental parameters is also included in this study.

A versatile double-beam (or single-beam) spectrophotometer is manufactured by Farrand Optical Co. Inc., 117 Wall Street, Valhalla, New York 10595. The Farrand VIS-UV Chromatogram Analyzer is shown in Figure 4.63; it has been discussed in detail by Cravitt.[66] A schematic block

FIGURE 4.63 The Farrand VIS-UV Chromatogram Analyzer.

diagram of the instrument is presented in Figure 4.64. As the name says, the instrument can be operated in the spectral region 200 to 800 nm, but experience has shown that for reflectance work below 220 nm no useful data usually can be obtained due to scattering and loss of radiation energy. Both double- and single-beam mode are possible and the light can be rendered monochromatic in the exciter leg either with a 14,400 lines/inch grating with a dispersion power of 11 nm/mm or, if desired, with a set of filters. As can be seen in the optical diagram (Figure 4.65) the excitation beam is split into a reference and analyzer beam. The double-beam operation in conjunction with a motorized monochromator drive enables automatic recording of UV and visible reflectance spectra. A sliver of light, whose size is adjustable by proper selection of slits, illuminates the sample at a zero-degree angle to the chromatogram surface. The central portion of the light-beam on the sample is viewed only by the analyzer optics; both sides of the beam are viewed by the reference channel and the two reference signals are averaged.

The diffuse radiation, which is observed at a 45° angle for the reference beam and at a 25° angle for the sample beam, is detected by two similar photomultipliers; the ratios of the analyzer and the reference branch signals are recorded on a conventional strip-chart recorder. This operation of the reference beam is rather unique. By referencing both sides of the spot the manufacturer claims to achieve a better compensation for background variations in the chromatographic substrate. For quantitative work the instrument can be changed to single-beam operation. Light sources as well as optics can be chosen to fit the spectral region of interest. The instrument can be modified in a matter of minutes for in situ

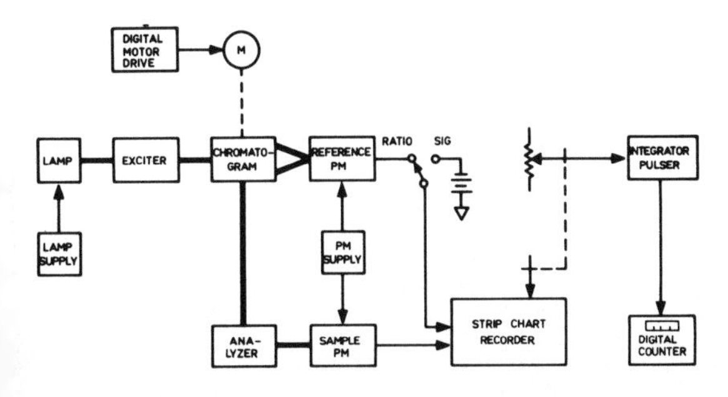

FIGURE 4.64 Schematic block diagram of the Farrand Chromatogram Analyzer.

fluorescence measurements. The scanning table is in a light-proof position inside the instrument housing. Automatic scanning in the Y-direction is possible at nine scan rates, varying from 10 to 300 mm/min over a distance of 170 mm. Manual adjustment is made along the x-axis. Chromatograms (TLC or PC, maximum size 200 x 200 mm) can be placed on the sample table.

The TLC attachment for the Farrand Mark I Spectrofluorometer permits work in the reflectance mode also. The attachment is shown in Figure 4.66. A single-beam operation is used for the measurement. The beam from the exciter monochromator exit slit is reflected by a first mirror through a first quartz lens and the scanning slit to the chromatogram surface at a zero-degree angle. The reflected light is viewed at a 45° angle by a suitable mirror and lens system and focused through a final slit to the photodetector. The quartz optics permits measurements in the UV region. The scanner device has specifications similar to the one discussed above.

An interesting new instrument produced by the Farrand Optical Company is the MSA Microscope Spectrum Analyzer (shown in Figure 4.67), which was described by Gurkin and Kallet in 1971.[67] The instrument is compatible with most microscopes and uses two monochromators – one for the illumination of the sample and the other attached to the eyepiece. The MSA thus provides a significant advance over previous instruments that depended on filters and thus limited frequency selection. The acquisition of narrow band filters for the UV-range is in itself a problem. The MSA may thus be used to obtain quantitation and flexibility in analysis not possible with previous techniques.

The MSA system may be used for fluorescence or absorbance measurement and consists of analyzing and illuminating sections. The analyzing section consists of an eyepiece-monochromator assembly and a photomultiplier-photometer. A beam splitter in the eyepiece reflects energy to the monochromator and transmits a portion of the beam to the eye. This assembly may be clamped to the standard barrel of a monocular, binocular, or trinocular microscope, usually without any modification of the microscope being required. Seven different bandwidths varying from 1.0 to 20 nm may be chosen by changing the exit slits. A Farrand F/8.0 Czerny-Turner (205 to 780 nm) grating monochromator is used, but instead of the

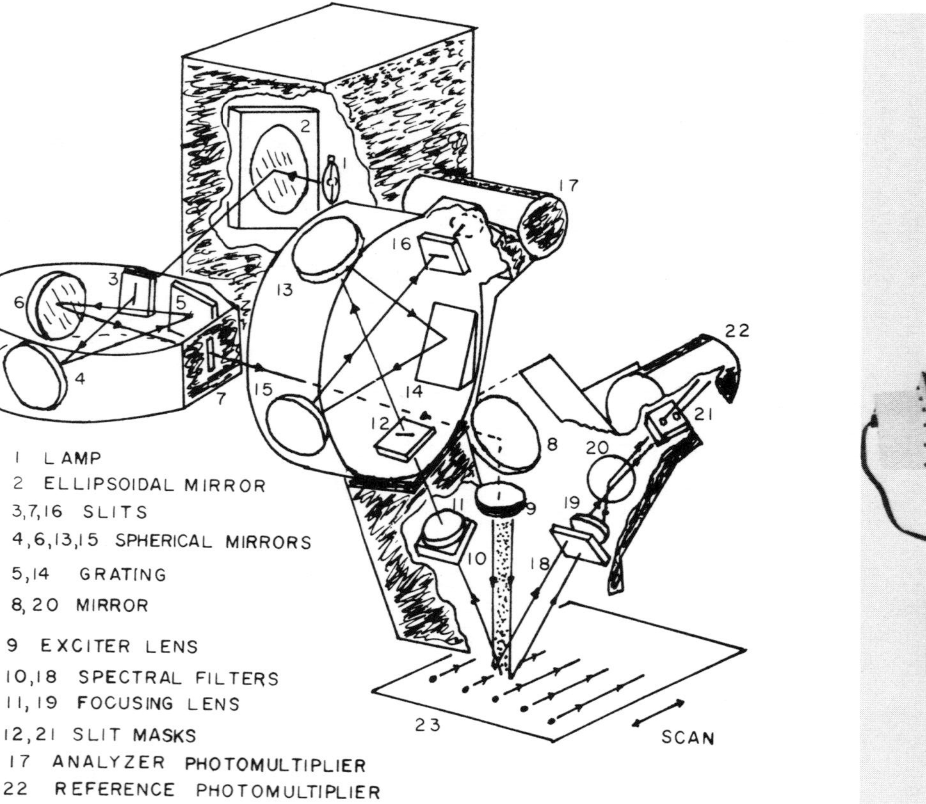

FIGURE 4.65 Optical diagram of the Farrand Chromatogram Analyzer.

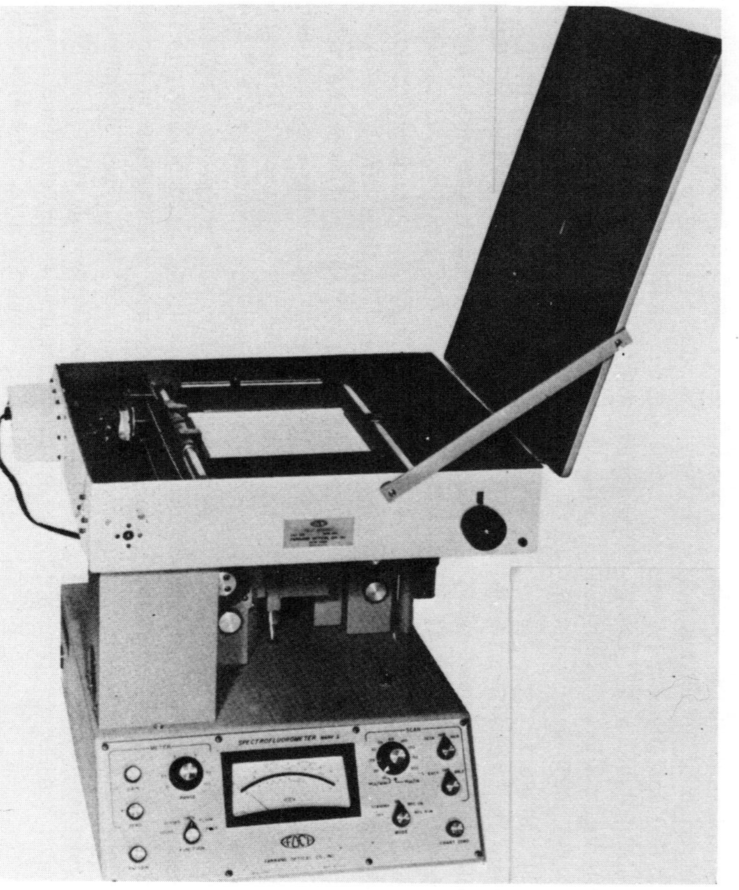

FIGURE 4.66 The Farrand Mark I Spectrofluorometer equipped with the chromatogram scanner attachment.

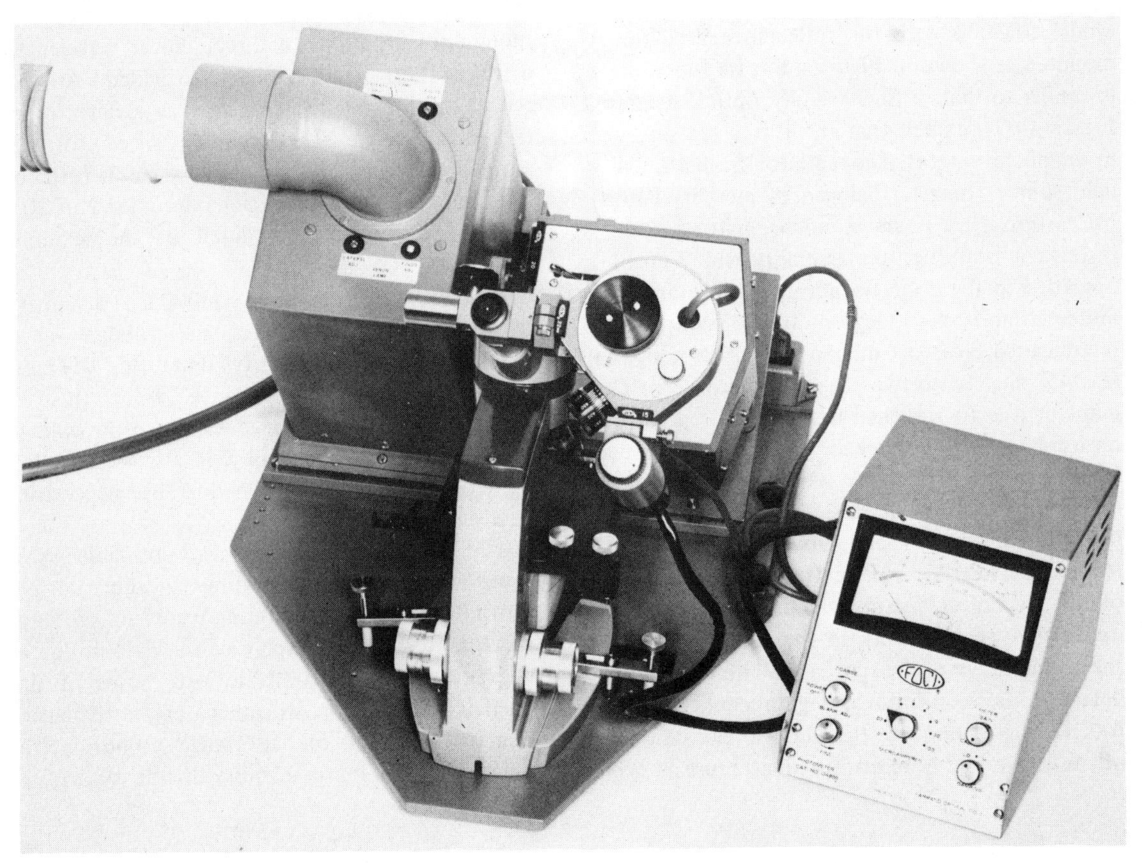

FIGURE 4.67 The Farrand MSA Microscope Spectrum Analyzer.

usual entrance slit, one of eight different size pinholes ranging from one to hundreds of microns diameter is used. The solid-state photometer package includes the meter, detector signal amplifier and photomultiplier power supply and provides an outlet for a recorder.

A 150-W xenon high pressure lamp is the light source provided with the illuminating section. An off-axis ellipsoidal mirror is used to focus the arc at the entrance slit of a Farrand F/3.5 grating monochromator (range 200 to 700 nm). Bandwidths from 1.0 to 20 nm are again provided by interchangeable slits. The microscope condenser is used to focus the light from the monochromator onto the specimen. The lamp housing and monochromator are mounted on an adjustable base.

As the analyzing and illuminating sections are independent, either one or both may be used as required. The MSA has the unique feature that the user may visually locate the specimen of interest exactly in the field of view of the analyzing monochromator, even though the selected pinhole represents a field diameter as small as 1 μ.

The MSA should find particular use in both biochemical research and also in clinical use, or in any other field of research where the measurement of spectra and quantitation of specimens of microscopic size is of interest. Options for the instrument include a motorized wavelength drive and recorder for the measurement of spectra.

Two densitometers marketed by Schoeffel Instrument Corp., 24 Booker Street, Westwood, New Jersey 07675, have recently been modified to permit in situ scanning of chromatograms by the reflectance mode. These are the Models SD2000 (the Chrometrician) and SD3000. The basic principle of operation is similar for both models, but the SD2000 is a lower-cost filter instrument recommended for routine use in medical laboratories. The Model SD3000 uses a prism monochromator to render the radiation monochromatic, and can, with the many possible modes of operation, serve as a research instrument. The

Model SD3000 with the reflectance attachment mounted is shown in Figure 4.68. Its functioning is similar to that of SD2000 (see optical diagram, Figure 4.69), except that the filters are replaced by monochromators. The radiation from a suitable light source (tungsten-halogen or mercury lamps) passes through a prism monochromator and then through a beam splitter assembly which enables operation in the double-beam mode. The angle of illumination is zero degrees to the sample and reference surface; the diffuse reflection is viewed at a 45° angle with two separate phototubes. The detector legs are designed to include filters for the measurement of fluorescence quenching. No filters are needed for reflectance work. The instrument can also be used in the single-beam mode and for in situ transmission and fluorescence measurements on TLC, PC, and electropherograms. Some of the technical features include an all-quartz optics that permits work in the UV as well as in the visible range of the spectrum. With a change in detector, the near-infrared region could be used too. The light source in the SD3000 can be a Xe- or Xe-Hg lamp. A miniature quartz prism type monochromator with a direct, linear wavelength drive of 200 to 700 nm is provided. An S-5 ten-stage photomultiplier serves as a detector in the same range. The scanning speed for the chromatoplate can be selected at 8, 4, 2, 1, 0.5 or 0.25 in./min and plates of a maximum size of 20 x 20 cm can be accommodated on the scanning table.

The basic readout unit SD3000 has a density computer which computes the voltage signal analogs of % T to density units (log 1/Transmission). For the reflectance work this is, however, not too advantageous, since the voltage signal would be an analog of % R, and the function log 1/% R vs. concentration does not necessarily result in linear relationships.

The Model SD3000 is also available as a computing/readout/recording system, or a computing/readout/integrating-recording system (see Figure 4.70). A display of the data in digital or printed form is available also. Some of the specifications of this instrument are: attenuation accuracy ±0.25% on all ranges; photometric accuracy ±0.5%, reproducibility of scan ±0.2%.

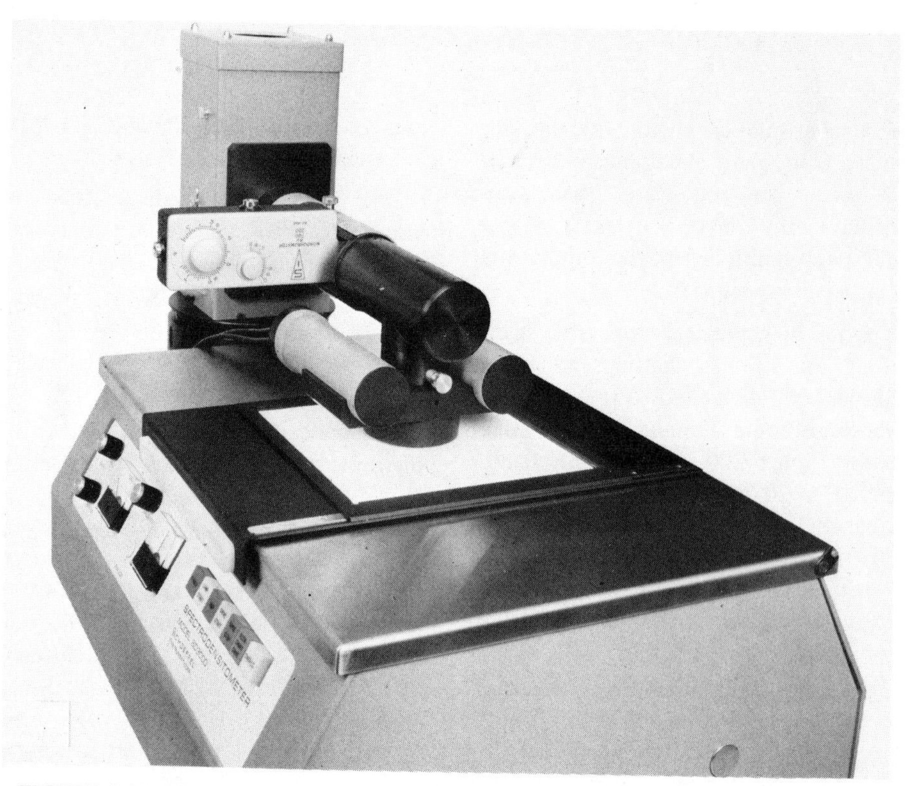

FIGURE 4.68 The Schoeffel Model SD3000 Spectrodensitometer with reflectance measuring head.

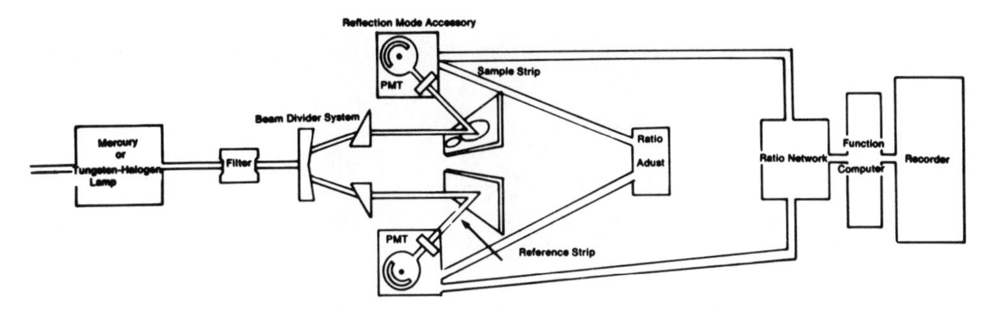

FIGURE 4.69 Optical diagram of the Schoeffel SD2000 Spectrodensitometer.

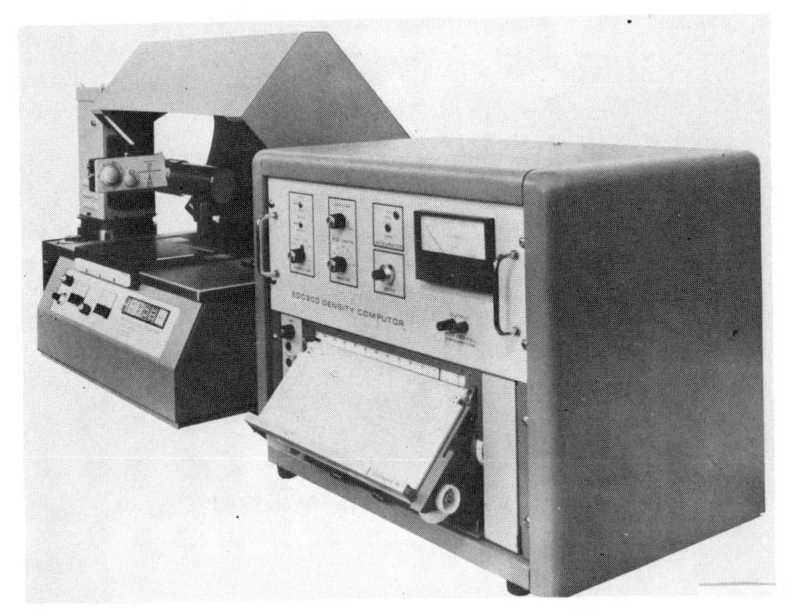

FIGURE 4.70 The Schoeffel Spectrodensitometer SD3000, equipped with a density computer.

Nester/Faust have recently developed a chromatogram scanner (Uniscan 900) which permits recording of chromatograms in the visible reflectance mode (Figure 4.71). Reflectance work in the UV region is not possible with this instrument and the manufacturers recommend the use of fluorescence quenching for UV-active non-fluorescent compounds. For scanning of narrow specimens such as paper strips, films, or small plates, where simultaneous scanning of the matrix with a reference beam is not possible, single-beam operation can be chosen. For normal operation, a double-beam mode is preferred to compensate for layer fluctuations. Rather than using a beam splitter, two separate energy sources (incandescent lamps) are used for the sample and reference beams. The geometry of illumination and observation is $_{45^\circ}R_{0^\circ}$ and a solid-state matched pair detector with ±1% linearity is provided.

For rendering the illuminated radiation monochromatic, a set of five standard color-developing filters is supplied. Collimating slits are provided for enhancement of resolution. The instrument can also perform in situ fluorescence measurements. Upon request, an electronic digital integrator is supplied with this instrument.

A rather unusual feature is the scanning head, which is moved at speeds of 3.5, 1.25, or 0.42 in./min in either direction. For previously discussed instruments; the scanning table rather than the optical head is moved. The maximum

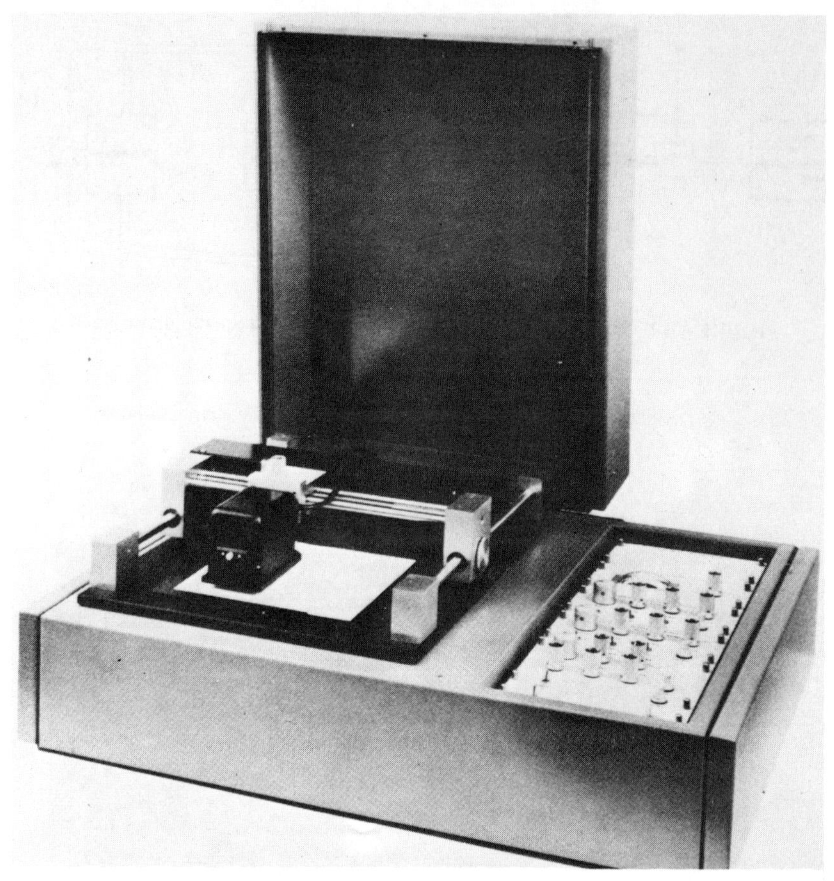

FIGURE 4.71 The Nester/Faust Uniscan 900.

plate size that can be accommodated and scanned by the measuring head is 10 x 20 in.

A TLC attachment is also available for the Aminco-Bowman Spectrofluorometer, which in the near future will permit reflectance measurements to be carried out. Its function is similar to the Farrand Mark I accessory described earlier. Fiber optics may be used to pick up diffuse reflectance.

The American Instrument Co. Inc. introduced an instrument several years ago (Aminco Dual-wavelength Scanner) specifically designed for in situ reflectance, transmittance, fluorescence and fluorescence quenching measurements on chromatograms. The basic principles of the measurement are depicted in Figure 4.72. In the design, both single- and double-beam operation were provided, and measurements could be made in the UV and visible regions of the spectrum. Time-shared reference and sample light beams struck the surface of the chromatogram at a zero-degree angle via a mirror with the reflected light being received by the photomultiplier tube at a 45° angle from the slit image center. Due to the size and proximity of the cathode, diffusely reflected radiation between the angles of 30 to 60° was measured.

The light was rendered monochromatic with an Aminco 0.25 meter Czerny-Turner type duochrometer. A 200-W tungsten-iodide lamp provided the energy source in the region 325 to 825 nm, while mercury and xenon lamps were used in the UV-region. Scanning speeds available were 2, 4, and 8 in./min on the y-axis with manual adjustment of the x-axis. The lightproof measuring compartment accommodated plates up to a maximum size of 8 x 8 in. As the instrument was undergoing redesign at the time of preparation of this book, no pictures or description of the new model were available.

A TLC attachment has been built by Perkin-Elmer (Norwalk, Connecticut) to fit the Model

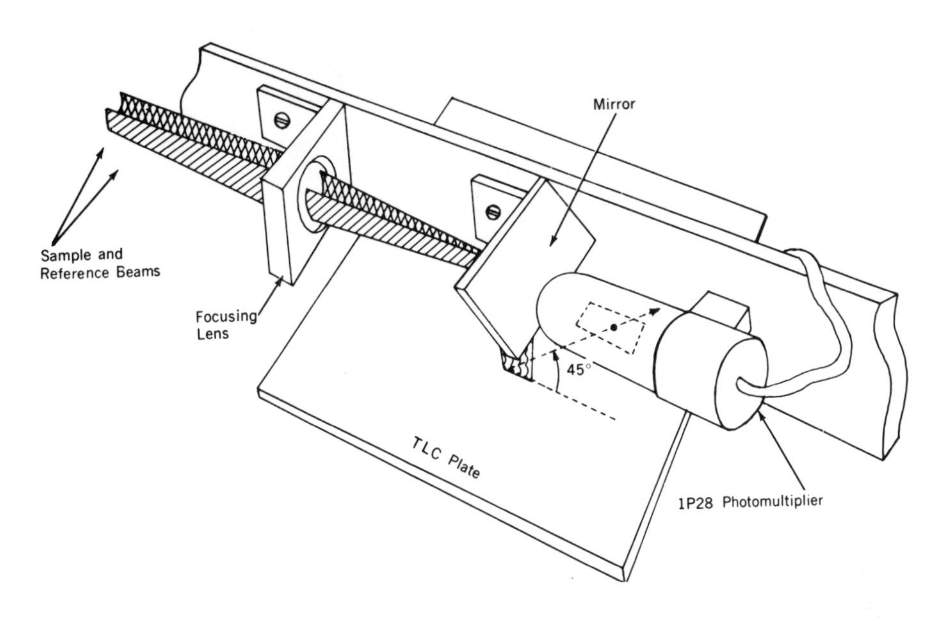

FIGURE 4.72 Optical principle of the Aminco Dual-wavelength Chromatogram Scanner.

MPF Spectrofluorometer. Some specifications are given in Table 4.03, along with a list of the previously discussed chromatogram scanners. Densitometer attachments for scanning thin-layer plates are also available for the Pye Unicam SP500 and a scanning densitometer attachment is also being manufactured for the Pye Unicam SP1800 series double-beam recording UV instrument.

Lefar and Lewis[68] discussed in detail some commercially available TLC scanners in a publication on the current state of thin-layer densitometry (including reflectance mode), and instrumentation for thin-layer chromatography has also been reviewed by Lott and Hurtubise.[69]

c. Automated TLC

An automated system for thin-layer chromatography (see Figure 4.73) using a reflectance spectroscopic detector has recently been introduced by the J. T. Baker Chemical Company. A flow diagram of the instrument is shown in Figure 4.74. The chromatograms are developed on adsorbent strips which have been precoated on the Chromotape® film. Samples are applied by pneumatic pressure from open-ended capillary tubes contained in a rotating sample tray. Each five-in. section of tape contains two parallel bands of adsorbent coating (Duogram® strips). Thus, two chromatograms may be developed on each five-in. section of tape. Prior to the spotting operation, the tape passes through an oven that may be operated at temperatures ranging from 50 to 120°C to activate the adsorbent.

After sample application, which is done with reproducibility of ±1% of the nominal one microliter volume of the applicators and activated by photoelectric detectors that sense when the Duogram strips are in position, the chromatogram is developed. Solvents are delivered from a reservoir to a syringe that applies them to a sponge at the lead end of each adsorbent strip. The syringe may be adjusted or changed to deliver different amounts of solvent and is automatically emptied and filled once during each cycle. Delivery rates are also adjustable. Chromatograms are developed in a temperature-controlled sandwich-type chamber 64 in. in length which may contain up to 16 pairs of Duogram strips at any one time. As the chromatogram leaves the developing chamber, the sponges are removed automatically before the strips pass into a drying oven.

Chromogenic developing agents may either be included in the developing solvent or may be applied at a spraying station after the chromatogram leaves the drying oven. Another oven following the spraying station dries the chromatograms prior to measurement and may be used to develop colors if necessary. After color

TABLE 4.03

Chromatogram Scanners for Reflectance Measurements

Manufacturer	Model	Mode of operation	Optics	Optical geometries	Range of operation (nm)	Monochromatic source
American Instr. Co. Inc.	Chromatogram attachment to Aminco-Bowman spectrofluorometer	single-beam	lens system or fiber optics	0/>0	UV-visible	Monochromators
	Dual-wavelength chromatogram scanner	double- and single-beam	lens system	0/45(d)		Monochromators
Camag Ltd.	Camag-Z-scanner (Attachment to Zeiss PMQ II)	single-beam	(light pipe) lens system	0/45; 45/0	220–2,500	Monochromator
Farrand Optical Co. Inc.	VIS-UV Chromatogram Analyzer	double- and single-beam	lens system	0/45(d)	200–800	Monochromator and filters
Joyce, Loebl & Co. Ltd.	Chromoscan (attachment)	double-beam (chopper mirror)	lens system	0/45(d)	400–750	Filters
E. Leitz, Inc.		single-beam	lens system	45/0	UV-visible	
Nester/Faust MFG Corp.	Uniscan 900	double- and single-beam	lens system	45/0	visible	Filters
Perkin-Elmer Corp.	Attachment to MPF-2A	single-beam	lens system	0/>0	UV-visible	Monochromators
Schoeffel Instrument Corp.	SD2000	double- and single-beam	lens system	0/45(d)	visible	Filters
	SD3000		lens system	0/45(d)	UV-visible	Prism monochromator
Vitatron Ltd.	Model TLD 100	single-beam	(light pipe) lens system	45/0	320–700	"Flying spot" scanning filter instrument
Carl Zeiss, Inc.	Chromatogram Spectrophotometer (attachment to PMQ II)	single-beam	light pipe	0/45(d)	220–2,500	Monochromator
Zeiss Jena DDR	ERI-10	single-beam	lens system	0/45(d)	visible	Filters

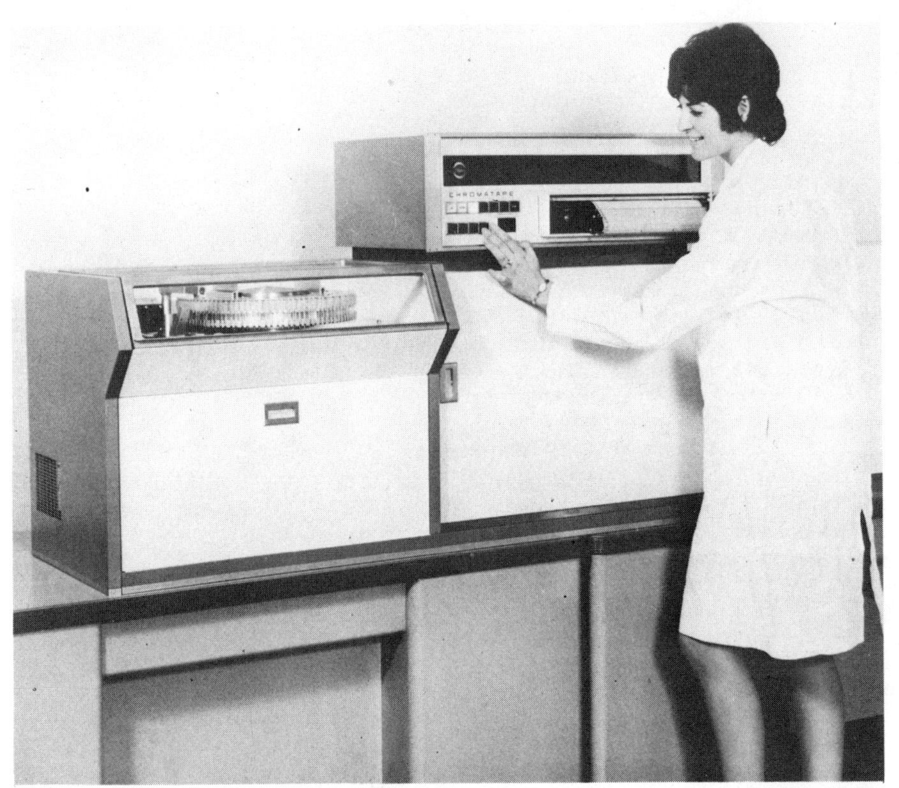

FIGURE 4.73 J. T. Baker CHROMATAPE™ system for automated thin-layer chromatography. Technician pushes button to start automated sequence of TLC steps. Samples are in micropipettes in circular tray at left. Readouts of results are traced out on dual chart in front of technician.

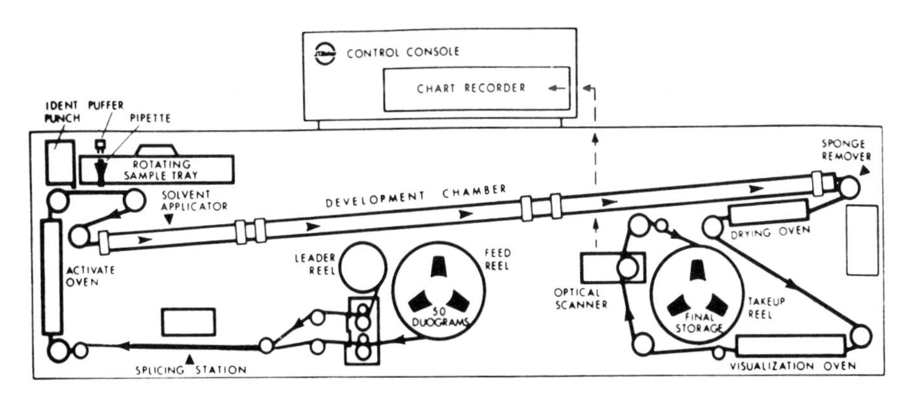

FIGURE 4.74 Flow diagram of CHROMATAPE™ processor.

development, the chromatograms are scanned optically by visible in situ reflectance spectroscopy, with the results being displayed on a dual-channel recorder. The output signal is also compatible with computer tape-punching equipment for data acquisition and analysis, and there are outlets provided on the instrument for addition of computer accessories.

REFERENCES

1. **Taylor, A. H.,** *J. Opt. Soc. Am.,* 4, 9 (1919).
2. **Taylor, A. H.,** *J. Opt. Soc. Am.,* 25, 51 (1935).
3. **Taylor, A. H.,** *J. Opt. Soc. Am.,* 21, 776 (1931).
4. **Benford, F.,** *J. Opt. Soc. Am.,* 24, 165 (1934).
5. **Wilcock, D. F. and Soller, W.,** *Ind. Eng. Chem.,* 32, 1446 (1940).
6. **Hoefert, H. J.,** *Z. Instrumentenk.,* 67, 118 (1959).
7. **Dearth, L. R., Shillcox, W. M., and van den Akker, J. A.,** *Tappi,* 43, 230A (1960).
8. **McLean, J. D.,** *Pulp Paper Mag. Can.,* 65, 434 (1964).
9. **Kortüm, G.,** *Reflexionsspecktroskopie,* Springer-Verlag, Berlin, 1969, 228-230. (English translation: Springer-Verlag, New York, 1969.)
10. **Van den Akker, J. A., Dearth, L. R., Olson, O. H., and Shillcox, W. M.,** *Tappi,* 35, 141A (1952).
11. **Hunter, R. S.,** National Paint, Varnish, Lacquer Assoc. Circ. No. 456, 69 (1934).
12. **Hunter, R. S.,** *J. Opt. Soc. Am.,* 30, 536 (1940).
13. **Hunter, R. S.,** *J. Res. Natl. Bur. Std.,* 25, 581 (1940).
14. **Wendlandt, W. W.,** *J. Chem. Educ.,* 45, A861, A947 (1968).
15. **Sanderson, J. A.,** *J. Opt. Soc. Am.,* 37, 771 (1947).
16. **Derksen, W. L. and Monahan, T. I.,** *J. Opt. Soc. Am.,* 42, 263 (1952).
17. **Jacquez, J. A., McKeehan, W., Huss, J., Dimitroff, J. M., and Kuppenheim, H. F.,** *J. Opt. Soc. Am.,* 45, 781 (1955).
18. **Tweet, A. G.,** *Rev. Sci. Instrum.,* 34, 1412 (1963).
19. **Michaelson, J. L.,** *J. Opt. Soc. Am.,* 28, 365 (1938).
20. **Hardy, A. C.,** *J. Opt. Soc. Am.,* 28, 360 (1938).
21. **Wendlandt, W. W. and Hecht, H. G.,** *Reflectance Spectroscopy,* John Wiley & Sons, New York, 1966.
22. **Hammond, H. K., III and Nimeroff, I.,** *J. Opt. Soc. Am.,* 42, 367 (1952).
23. **Anacreon, R. E. and Noble, R. H.,** *Appl. Spectrosc.,* 14, 29 (1960).
24. **McNicholas, H. J.,** *J. Res. Natl. Bur. Std.,* 1, 29 (1928).
25. **Helwig, B.,** *Licht,* 8, 242 (1938).
26. **Stenius, A. S.,** *J. Opt. Soc. Am.,* 45, 727 (1955).
27. **Dunn, S. T.,** Design and analysis of an ellipsoidal mirror reflectometer, Ph.D. thesis, Oklahoma State University, May, 1965.
28. **Dunn, S. T., Richmond, J. C., and Wiebelt, J. A.,** *J. Res. Natl. Bur. Std.,* 70C, 75 (1966).
29. **Kneissl, G. J.,** Paper presented at the 21st Pittsburgh Conference on Analytical Chemistry, Cleveland, Ohio, March, 1970.
30. **Judd, D. B. and Wyszecki, G.,** *Color in Business, Science and Industry,* 2nd ed., John Wiley & Sons, New York, 1963.
31. **Tonnquist, G.,** *J. Opt. Soc. Am.,* 45, 582 (1955).
32. **Barnes, L., Goya, H., and Zeitlin, H.,** *Rev. Sci. Instrum.,* 34, 292 (1963).
33. **Frei, R. W. and Frodyma, M. M.,** *Anal. Chim. Acta,* 32, 501 (1965).
34. **Frodyma, M. M., Lieu, V. T., and Frei, R. W.,** *J. Chromatogr.,* 18, 520 (1965).
35. **Frodyma, M. M. and Lieu, V. T.,** *Anal. Chem.,* 39, 814 (1967).
36. **Hatfield, W. E., Piper, T. S., and Klabunde, U.,** *Inorg. Chem.,* 2, 629 (1963).
37. **Asmussen, R. W. and Andersen, P.,** *Acta Chem. Scand.,* 12, 939 (1958).
38. **Wendlandt, W. W.,** *Thermal Methods of Analysis,* John Wiley & Sons, New York, 1965.
39. **Wendlandt, W. W., Franke, P. H., and Smith, J. P.,** *Anal. Chem.,* 35, 105 (1963).
40. **Wendlandt, W. W. and George, T. D.,** *Chemist-Analyst,* 53, 100 (1964).
41. **Wendlandt, W. W.,** in *Modern Aspects of Reflectance Spectroscopy,* Wendlandt, W. W., Ed., Plenum Pub., New York, 1968, 58-60.
42. **Wendlandt, W. W.,** *J. Chem. Educ.,* 40, 428 (1963).
43. **Wendlandt, W. W. and Dorsch, E. L.,** *Thermochim. Acta,* 1, 103 (1970).
44. **Symons, M. C. R. and Travalion, P. A.,** *Unicam Spectrovision,* 10, 8 (1961).
45. **Urbanyi, T., Swartz, C. J., and Lachman, L.,** *J. Am. Pharm. Assoc.,* 49, 163 (1960).
46. **Block, R. J., Durrum, E. L., and Zweig, G.,** *A Manual of Paper Chromatography and Paper Electrophoresis,* 2nd ed., Academic Press, New York, 1958.
47. Stahl, E., Ed., *Thin-layer Chromatography,* 2nd ed., Springer-Verlag, Berlin and Heidelberg, 1967 (German).
48. **Franglen, G.** (pp. 17-38), **Shellard, E. J.** (pp. 51-70), and **Jork, H.** (pp. 79-90), in *Quantitative Paper and Thin-layer Chromatography,* Shellard, E. J., Ed., Academic Press, London and New York, 1968.
49. **Seiler, N. and Möller, H.,** *Chromatographia,* 2, 319 (1969).
50. **Klaus, R.,** *J. Chromatogr.,* 16, 311 (1964).
51. **Klaus, R.,** *Pharm. Ztg.,* 112, 480 (1967).
52. **De Galan, L., van Leeuwen, J., and Camstra, K.,** *Anal. Chim. Acta,* 35, 395 (1966).
53. **Gordon, H. T.,** *J. Chromatogr.,* 22, 60 (1966).

54. **Hamman, B. L. and Martin, M. M.,** *Anal. Biochem.,* 15, 305 (1966).
55. **Beroza, M., Hill, K. R., and Norris, K. H.,** *Anal. Chem.,* 40, 1611 (1968).
56. **Goldman, J. and Goodall, R. R.,** *J. Chromatogr.,* 40, 345 (1969).
57. Zeiss Information No. 50-657/K-e, Carl Zeiss Inc., Oberkochen, Germany, 1968.
58. **Jork, H.,** *Z. Anal. Chem.,* 236, 310 (1968).
59. **Jork, H.,** *Cosmo Pharma,* 1, 33 (1967).
60. **Stahl, E. and Jork, H.,** *Zeiss Inform.,* 16 (No. 68), 52 (1968).
61. **Struck, H., Karg, H., and Jork, H.,** *J. Chromatogr.,* 36, 74 (1968).
62. **Jork, H.,** *J. Chromatogr.,* 33, 297 (1968).
63. **Jork, H.,** *Cosmo Pharma,* 4, 12 (1968).
64. **Jänchen, D. and Pataki, G.,** *J. Chromatogr.,* 33, 391 (1968).
65. **Goldman, J. and Goodall, R. R.,** *J. Chromatogr.,* 32, 24 (1968).
66. **Cravitt, S.,** Paper presented at the 4th Mid-Atlantic Regional Meeting of the American Chemical Society, Washington, D.C., February, 1969.
67. **Gurkin, M. and Kallet, E. A.,** *Am. Lab.,* October (1971).
68. **Lefar, M. S. and Lewis, A. D.,** *Anal. Chem.,* 42, 79A (1970).
69. **Lott, P. F. and Hurtubise, R. J.,** *J. Chem. Educ.,* 48, 481A (1971).

Chapter 5

APPLICATION TO COLOR MEASUREMENTS AND COLOR COMPARISON

1. Principle and Method of Color Measurements

A frequently encountered problem in many industrial and academic undertakings is the objective identification of a color in a form suitable for documentation. Since the color response of the human eye differs from person to person, or even in the same person when observations are made at different times, it is desirable to have a means of reporting colors in numerical form. Thus, the use of instrumental methods for the recording of colors seems logical, and since color manifests itself through a form of reflected light emerging from the object, it is not surprising that reflectance spectroscopy plays a major role in the field of color technology.

Of the color-order systems that have been developed,[1,2] only the CIE system (Commission Internationale de l'Eclairage) – also known as the ICI system (International Commission on Illumination) – will be considered in this discussion. Several authoritative literature sources are readily available in this field,[2-7] and the discussion in this chapter is therefore restricted to a basic working knowledge of the color-order system.

a. The CIE System

The most important color system used in connection with instrumental color measurements is the CIE system.[8-10] It is based on the concept of the three-dimensional nature of color, which permits additive color mixing of spectrum colors with the use of three light sources (primary lights) of widely different colors on a white screen. This concept has been known for a long time (Newton, 1730; Grassmann, 1853) and can be represented in a more concise form through Grassman's law: "The color matching functions or *Tristimulus Values* of the spectrum colors can be calculated for any specified set of primaries if they are known for one set of primaries."[11] In other words, color can be specified in terms of three numbers that represent the relative amounts of color added by the three primary lights. These amounts are the so-called tristimulus values. The limitations are, however, that no one set of three colors or primaries can reproduce every possible color by additive mixing. To overcome this problem, light from one of the primaries could be added to the color under investigation rather than being added to the two other primaries. This would really amount to a subtraction, and the test colors could then be described by the combined negative and positive values of primary colors. In this fashion any test light can be matched.

For the sake of convenience, the three tristimulus values are designated X, Y, and Z. The parameter Y is a measure of the lightness response and is also known as brightness or luminosity. The other parameters (X and Z) describe the aspect of color that allows it to be identified with various regions of the spectrum (hue) and that determines its excitation purity (saturation).

The tristimulus values can be presented as a unit trichromatic equation of the form

$$\text{Color (C)} = x(X) + y(Y) + z(Z), \tag{5.01}$$

where x, y, and z represent the chromaticity coordinates of C, obtained as a ratio of each individual tristimulus value over the sum total of all three values.

$$x = \frac{X}{X+Y+Z} \qquad y = \frac{Y}{X+Y+Z} \qquad z = \frac{Z}{X+Y+Z} \tag{5.02}$$

Since $x + y + z = 1$, only two of the three coordinates are needed to define the color, which permits the representation of colors in a so-called chromaticity diagram (Figure 5.01). The three-dimensional character of the tristimulus values is hence reduced to the two-dimensional type of the chromaticity coordinates x and y, which describe hue and saturation. The pair of x, y values is called chromaticity.

For a complete designation of the color of an object, a clear specification is required of the light source illuminating this object. In 1931 the CIE, in the process of setting up a standard observer and coordinate system, recommended the use of three standard illuminants: illuminant A, equivalent to a gas-filled incandescent lamp; illuminant B, representative of noon sunlight, and illuminant C, representative of daylight, such as that from an overcast sky. The same sources can be specified more accurately in terms of color temperatures by comparing their energy distribution in the visible range of the spectrum to that of a blackbody

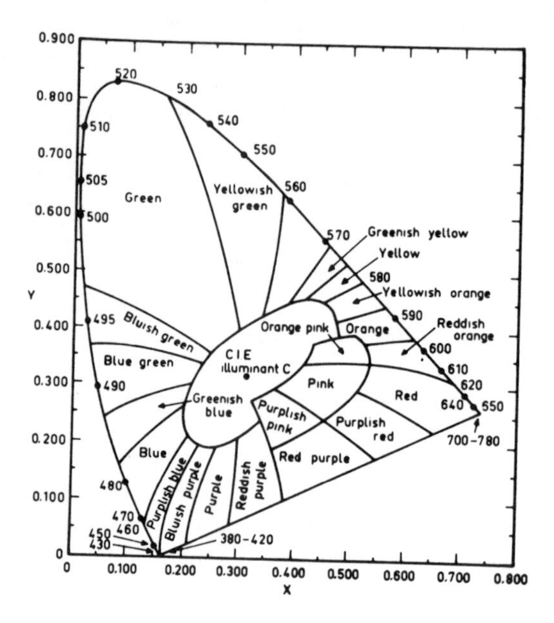

FIGURE 5.01 The (x, y)-chromaticity diagram showing the spectrum locus and the purple boundary (wavelength in nm).[20] (From Rodrigo, F. A., *Am. Heart J.*, 45, 809 (1953). With permission.)

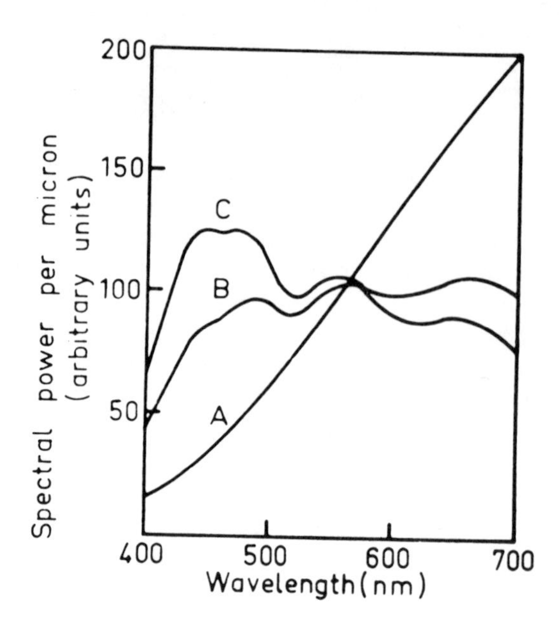

FIGURE 5.02 Spectral distributions of the three CIE standard sources A, B, and C.[5]

radiator.[12] Illuminant A would then consist of an incandescent lamp operated at a color temperature of 2854°K; illuminant B would make use of the same lamp in conjunction with a suitable filter[13] to give about 5000°K color temperature. Illuminant C corresponds in principle to A and C but uses a filter that brings its color temperature to about 6800°K. The relative spectral irradiance of the three CIE illuminants is represented in Figure 5.02.

On the basis of these concepts, one can measure a color instrumentally and specify a particular color with numerical values. There are essentially three types of instrumental methods for color measurements: the equicontrast method, the trichromatic method, and the spectrophotometric method.

b. The Equicontrast Method

This method uses the human eye as a detector of the radiation. Accuracy and reproducibility of this approach therefore depend on certain characteristics of the eye, such as defined for a "standard observer." These characteristics are normal color vision, limited size of the field of vision, limited lightness in this field (e.g., sample is without shape or texture) and a "rested" eye. The principle of operation is based on additive mixing of three primaries, e.g., red, green, and blue, and the matching of instrumentally measurable quantities of these three sources, first, to a standard reference source (white light) and then to the color that is to be measured. Such instruments have been described by several authors,[2,14-16] one of whom[16] has used as many as six different primary sources for the mixing of the light.

It is possible to eliminate the restrictions common to the human eye by replacing it with a suitable photodetector; this would then lead to the so-called trichromatic method.

c. The Trichromatic Method

In this method, light filters (tristimulus filters), positioned in the light path in front of the photocell, adapt the detector to the characteristics of the standard observer. This adaptation is usually provided in the red, green, and blue ranges of the spectrum and endows the photocell with an approximation of normal color vision. Ideally the photocell response should be proportional throughout the visible spectrum to some linear combination of the standard CIE distribution curves (see Figure 5.03). According to Grassmann's law, it would then be possible to test whether any two light beams have the same color, and consequently the direct measurement of tristimulus values X, Y, and Z could be carried out.[17] Photoelectric tristimulus colorimeters are used widely now and are discussed in detail in Chapter 4, Section 1. The accuracy of the measured tristimulus values depends to a large degree on how closely the photocell-tristimulus filter combination can be adapted to the characteristics of the CIE standard observer. Figure 5.03 shows an example of a widely used set of tristimulus filters[18] and of the kind of duplication one can expect with the CIE illuminant.

There are several ways in which the colorimeter readings can be interpreted. Perhaps the simplest way is that in which a white surface (e.g., freshly prepared magnesium oxide or barium sulfate) is used as a reference standard. An example of such a procedure as it could be used for a large number of commercially available instruments is briefly described in this section. The color is measured by illuminating it with CIE illuminant C and by taking the reference standard to read 100% reflectance when measured through each of the filters. The diffuse reflectance values R_x, R_y, and R_z are recorded and the tristimulus values are computed by multiplying the reflectance values with factors characteristic of the set of filters used. For the three-filter arrangement used in the Zeiss Elrepho Colormeter, for example, the following equations would be applied:[19]

$$X = 0.782\,R_X + 0.198\,R_Z$$

$$Y = R_y \qquad (5.03)$$

$$Z = 1.181\,R_Z\,.$$

These values can be converted to the chromaticity coordinates according to Equation 5.02 and plotted in the chromaticity diagram (Figure 5.01). The appropriate conversion factors can usually be found with the tristimulus filter specifications supplied by the manufacturer.

d. The Spectrophotometric Method

In cases where the highest possible accuracy is required, the investigator would turn to the spectrophotometric method. The determination of color from spectral reflectance curves is based on the measurement of tristimulus values for each individual spectral color. In principle, the method consists of measurement of the diffuse reflectance of each spectral color with a spectrophotometer. Any of the commercial reflectance attachments

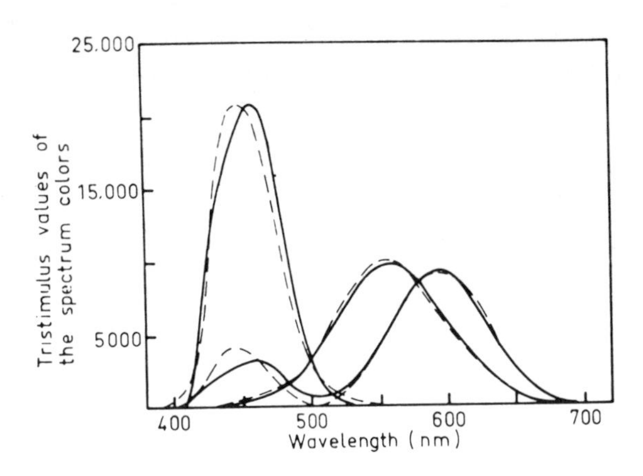

FIGURE 5.03 Curves showing the degree to which the Hunter tristimulus filters (———) combined with incandescent lamp and barrier-layer cell duplicate the CIE tristimulus values of the spectrum of illuminant C (- - - - -).[18]

discussed in Chapter 4, Section 2, can be used for this purpose. An excellent and detailed discussion and demonstration of this method on the basis of actual color samples has been given by Judd.[20] This chapter gives only a simplified account of the procedure, which should, however, enable the experimenter to carry out color measurements without having to resort to other literature sources.

The first step consists of recording the reflectance spectrum automatically with a double-beam instrument or manually at certain wavelength intervals with a single-beam instrument. The measurements would be made vs. a suitable standard such as MgO or $BaSO_4$. The second step concerns the conversion of the measured light-intensity values (% R) into tristimulus values with the use of suitable multiplying factors M_n (n could be x, y, or z) by which any spectral curve can be represented in terms of color coordinates. A mathematical representation of this concept would be

$$X = \int_{380}^{780} RM_X\, d\lambda, \qquad (5.04)$$

where R stands for percent reflectance, measured on the percent transmittance scale of the standard spectrophotometer. λ is the symbol for wavelength, and M is the multiplying factor. Similar equations can be written for the tristimulus values Y and Z.

Combining Equations 5.02 and 5.04 will result in the equations for the chromaticity coordinates:

$$x = \frac{\int_{380}^{780} RM_x \, d\lambda}{\sum_{n=x}^{Z} \int_{380}^{780} RM_n \, d\lambda} \quad (5.05)$$

$$y = \frac{\int_{380}^{780} RM_y \, d\lambda}{\sum_{n=y}^{Z} \int_{380}^{780} RM_n \, d\lambda} . \quad (5.06)$$

The multiplying factor M would have to be specified with regard to the type of standard CIE illuminant (A, B, or C) used. For the computation of the chromaticity coordinates, it is obvious that the use of Equations 5.05 and 5.06 would be rather tedious without a computer. However, one can adopt approximation methods that are based on integration at certain preselected wavelength intervals. Trichromatic coefficient computing forms based on this principle are usually provided by the instrument manufacturer and can be used for the determination of the color points. If such forms are not available, one can make readings at approximately 10-nm intervals throughout the visible region of the spectrum and multiply each reflectance value by a suitable multiplier MN_n (N:A, B, or C, depending on illuminant, n:x, y, or z). MN values as adopted in a modified form[21,22] are given in Table 5.01 for the spectral region 380 to 770 nm for CIE illuminants A, B, and C. The resulting values of X, Y, and Z are totaled to give the tristimulus values ΣX, ΣY, and ΣZ. The chromaticity coordinates are computed according to Equation 5.02 and plotted on the chromaticity diagram (Figure 5.01). Chromaticity coordinates of the spectrum colors taken at 5-nm intervals are presented in Table 5.02. With a wide number of modern instruments these computation steps are carried out automatically with a built-in computer, and the color coordinates are presented in digital or printed form in a matter of seconds (see Chapter 4, Section 2).

In order for the above method to be valid, the following conditions have to be fulfilled. Sharp spectrum peaks should be absent, and the sample should be completely free of fluorescence. Since reflectance spectra are usually broader than corresponding transmission spectra – due to scattering processes – the occurrence of sharp peaks is quite rare, and the first condition is easily fulfilled.

2. The Measurement of Multi-Component Pigment Systems

Mixing and matching procedures for pigment-containing materials are still largely an empirical task that is performed by professional color matchers with a wide experience in the visual examination of colors. Reflectance spectroscopy, however, does provide a convenient method for the analysis of materials of this type, particularly in cases where the colored pigments are mixed with an excess of white or light-yellow pigment; in practice, their absorption coefficients can be taken as zero. The latter situation means that the Kubelka-Munk theory is applicable and that at any given wavelength the coefficient of absorption of a pigmented material is a simple additive function of the absorption of the individual pigments present in the mixture. For the condition of infinite layer thickness, the Kubelka-Munk equation (see Chapter 2, Section 1) can be applied directly to express the reflectance R in a suitable form, particularly if dyed textiles, unvarnished papers, dry powders, concrete, etc. are investigated. In the case of glossy paint surfaces or materials such as linoleums, plastics, and printing inks, in which the pigment particles are bound together by means of a continuous medium, it is necessary to apply a correction for specular and internal reflection that occurs at the air/medium interface.[23,24] The equation

$$R'_\infty = 0.472\, R_\infty/(1 - 0.055\, R_\infty) \quad (5.07)$$

enables the conversion of the observed reflectance values R_∞ to the corrected values R'_∞ for glossy surfaces of the kind mentioned above, which contain pigments in a medium of refractive index ~ 1.5 (e.g., oil, oleo resins, or alkyd paints).

For the sake of convenience, Table 5.03 gives the corrected Kubelka-Munk values θ for observed reflectances. For a multipigment mixture adhering to the Kubelka-Munk theory, the scattering coefficient s remains the same for each component

TABLE 5.01

A Modified Table for the Computation of Chromaticity Coordinates[22]

Wave-length (nm)	Multiplier for illuminant A			Multiplier for illuminant B			Multiplier for illuminant C		
	MA_x	MA_y	MA_z	MB_x	MB_y	MB_z	MC_x	MC_y	MC_z
380	0.0010	0.0000	0.0048	0.0025	0.0000	0.0113	0.0036	0.0000	0.0164
390	0.0046	0.0001	0.0219	0.0123	0.0003	0.0585	0.0183	0.0004	0.0870
400	0.0193	0.0005	0.0916	0.0558	0.0014	0.2650	0.0841	0.0021	0.3992
410	0.0688	0.0019	0.3281	0.2091	0.0057	0.9970	0.3180	0.0087	1.5159
420	0.2666	0.0080	1.2811	0.8274	0.0248	3.9750	1.2623	0.0378	6.0646
430	0.6479	0.0265	3.1626	1.9793	0.0810	9.6617	2.9913	0.1225	14.6019
440	0.9263	0.0609	4.6469	2.6889	0.1768	13.4883	3.9741	0.2613	19.9357
450	1.0320	0.1167	5.4391	2.7460	0.3105	14.4729	3.9191	0.4432	20.6551
460	1.0207	0.2098	5.8584	2.4571	0.5050	14.1020	3.3668	0.6920	19.3235
470	0.7817	0.3624	5.1445	1.7297	0.8018	11.3825	2.2878	1.0605	15.0550
480	0.4242	0.6198	3.6207	0.8629	1.2609	7.3655	1.1038	1.6129	9.4220
490	0.1604	1.0398	2.3266	0.2960	1.9190	4.2939	0.3639	2.3591	5.2789
500	0.0269	1.7956	1.5132	0.0437	2.9133	2.4552	0.0511	3.4077	2.8717
510	0.0572	3.0849	0.9674	0.0810	4.3669	1.3694	0.0898	4.8412	1.5181
520	0.4247	4.7614	0.5271	0.5405	6.0602	0.6709	0.5752	6.4491	0.7140
530	1.2116	6.3230	0.3084	1.4555	7.5959	0.3705	1.5206	7.9357	0.3871
540	2.3142	7.5985	0.1625	2.6899	8.8322	0.1889	2.7858	9.1470	0.1956
550	3.7329	8.5707	0.0749	4.1838	9.6060	0.0840	4.2833	9.8343	0.0860
560	5.5086	9.2201	0.0357	5.8385	9.7722	0.0378	5.8782	9.8387	0.0381
570	7.5710	9.4574	0.0209	7.4723	9.3341	0.0206	7.3230	9.1476	0.0202
580	9.7157	9.2257	0.0170	8.8406	8.3947	0.0154	8.4141	7.9897	0.0147
590	11.5841	8.5430	0.0130	9.7329	7.1777	0.0109	8.9878	6.6283	0.0101
600	12.7103	7.5460	0.0096	9.9523	5.9086	0.0075	8.9536	5.3157	0.0067
610	12.6768	6.3599	0.0044	9.4425	4.7373	0.0033	8.3294	4.1788	0.0029
620	11.3577	5.0649	0.0020	8.1290	3.6251	0.0014	7.0604	3.1485	0.0012
630	8.9999	3.7122	0.0000	6.2135	2.5629	0.0000	5.3212	2.1948	0.0000
640	6.5487	2.5587	0.0000	4.3678	1.7066	0.0000	3.6882	1.4411	0.0000
650	4.3447	1.6389	0.0000	2.8202	1.0638	0.0000	2.3531	0.8876	0.0000
660	2.6234	0.9706	0.0000	1.6515	0.6110	0.0000	1.3589	0.5028	0.0000
670	1.4539	0.5327	0.0000	0.8796	0.3223	0.0000	0.7113	0.2606	0.0000
680	0.7966	0.2896	0.0000	0.4602	0.1673	0.0000	0.3657	0.1329	0.0000
690	0.4065	0.1467	0.0000	0.2218	0.0801	0.0000	0.1721	0.0621	0.0000
700	0.2067	0.0744	0.0000	0.1065	0.0384	0.0000	0.0806	0.0290	0.0000
710	0.1108	0.0398	0.0000	0.0538	0.0193	0.0000	0.0398	0.0143	0.0000
720	0.0556	0.0195	0.0000	0.0253	0.0089	0.0000	0.0183	0.0064	0.0000
730	0.0280	0.0100	0.0000	0.0120	0.0043	0.0000	0.0085	0.0030	0.0000
740	0.0144	0.0062	0.0000	0.0058	0.0025	0.0000	0.0040	0.0017	0.0000
750	0.0063	0.0021	0.0000	0.0024	0.0008	0.0000	0.0017	0.0006	0.0000
760	0.0032	0.0011	0.0000	0.0012	0.0004	0.0000	0.0008	0.0003	0.0000
770	0.0011	0.0000	0.0000	0.0004	0.0000	0.0000	0.0003	0.0000	0.0000

(From Smith, T., *Proc. Phys. Soc. (London)*, 46, 372 (1934). With permission.)

TABLE 5.02

Chromaticity Coordinates of the Spectrum Colors[20]

Wavelength (nm)	Chromaticity coordinates x	y	z
380	0.1741	0.0050	0.8209
385	0.1740	0.0050	0.8210
390	0.1738	0.0049	0.8213
395	0.1736	0.0049	0.8215
400	0.1733	0.0048	0.8219
405	0.1730	0.0048	0.8222
410	0.1726	0.0048	0.8226
415	0.1721	0.0048	0.8231
420	0.1714	0.0051	0.8235
425	0.1703	0.0058	0.8239
430	0.1689	0.0069	0.8242
435	0.1669	0.0086	0.8245
440	0.1644	0.0109	0.8247
445	0.1611	0.0138	0.8251
450	0.1566	0.0177	0.8257
455	0.1510	0.0227	0.8263
460	0.1440	0.0297	0.8263
465	0.1355	0.0399	0.8246
470	0.1241	0.0578	0.8181
475	0.1096	0.0868	0.8036
480	0.0913	0.1327	0.7760
485	0.0687	0.2007	0.7306
490	0.0454	0.2950	0.6596
495	0.0235	0.4127	0.5638
500	0.0082	0.5384	0.4534
505	0.0039	0.6548	0.3413
510	0.0139	0.7502	0.2359
515	0.0389	0.8120	0.1491
520	0.0743	0.8338	0.0919
525	0.1142	0.8262	0.0596
530	0.1547	0.8059	0.0394
535	0.1929	0.7816	0.0255
540	0.2296	0.7543	0.0161
545	0.2658	0.7243	0.0099
550	0.3016	0.6923	0.0061
555	0.3373	0.6589	0.0038
560	0.3731	0.6245	0.0024
565	0.4087	0.5896	0.0017
570	0.4441	0.5547	0.0012
575	0.4788	0.5202	0.0010
580	0.5125	0.4866	0.0009
585	0.5448	0.4544	0.0008
590	0.5752	0.4242	0.0006
595	0.6029	0.3965	0.0006
600	0.6270	0.3725	0.0005
605	0.6482	0.3514	0.0004
610	0.6658	0.3340	0.0002
615	0.6801	0.3197	0.0002
620	0.6915	0.3083	0.0002
625	0.7006	0.2993	0.0001
630	0.7079	0.2920	0.0001
635	0.7140	0.2859	0.0001
640	0.7190	0.2809	0.0001
645	0.7230	0.2770	0.0000
650	0.7260	0.2740	0.0000
655	0.7283	0.2717	0.0000
660	0.7300	0.2700	0.0000
665	0.7311	0.2689	0.0000
670	0.7320	0.2680	0.0000
675	0.7327	0.2673	0.0000
680	0.7334	0.2666	0.0000
685	0.7340	0.2660	0.0000
690	0.7344	0.2656	0.0000
695	0.7346	0.2654	0.0000
700	0.7347	0.2653	0.0000
705	0.7347	0.2653	0.0000
710	0.7347	0.2653	0.0000
715	0.7347	0.2653	0.0000
720	0.7347	0.2653	0.0000
725	0.7347	0.2653	0.0000
730	0.7347	0.2653	0.0000
735	0.7347	0.2653	0.0000
740	0.7347	0.2653	0.0000
745	0.7347	0.2653	0.0000
750	0.7347	0.2653	0.0000
755	0.7347	0.2653	0.0000
760	0.7347	0.2653	0.0000
765	0.7347	0.2653	0.0000
770	0.7347	0.2653	0.0000
775	0.7347	0.2653	0.0000
780	0.7347	0.2653	0.0000

(From Judd, D. B., in *Analytical Absorption Spectroscopy,* Mellon, M. G., Ed., John Wiley & Sons, New York, 1950. With permission.)

TABLE 5.03

Values of θ Corresponding to Observed Values of Reflectance*

	.0	.1	.2	.3	.4	.5	.6	.7	.8	.9
0	∞	213	105	71	53	42	34	29	26	23
1	20.4	18.5	16.9	15.6	14.5	13.5	12.5	11.8	11.1	10.5
2	9.9	9.4	9.0	8.6	8.2	7.8	7.5	7.2	6.9	6.7
3	6.4	6.2	6.0	5.8	5.6	5.4	5.25	5.1	4.95	4.8
4	4.7	4.6	4.4	4.3	4.2	4.1	4.0	3.9	3.8	3.7
5	3.6	3.5	3.4	3.35	3.29	3.23	3.16	3.10	3.04	2.98
6	2.92	2.86	2.80	2.74	2.69	2.64	2.59	2.54	2.49	2.45
7	2.41	2.36	2.32	2.28	2.25	2.21	2.17	2.14	2.10	2.07
8	2.04	2.00	1.97	1.94	1.91	1.88	1.85	1.82	1.80	1.77
9	1.75	1.72	1.69	1.67	1.65	1.63	1.61	1.58	1.56	1.54
10	1.52	1.50	1.48	1.46	1.44	1.42	1.40	1.38	1.36	1.34
11	1.33	1.32	1.30	1.28	1.27	1.25	1.23	1.22	1.20	1.19
12	1.18	1.16	1.15	1.14	1.12	1.11	1.10	1.09	1.08	1.06
13	1.05	1.04	1.03	1.02	1.01	1.00	0.99	0.98	0.97	0.96
14	0.95	0.94	0.93	0.92	0.91	0.90	0.89	0.88	0.87	0.86
15	0.85	0.84	0.83	0.82	0.81	0.805	0.80	0.79	0.78	0.775
16	0.77	0.76	0.75	0.745	0.74	0.73	0.72	0.713	0.707	0.701
17	0.695	0.689	0.682	0.676	0.670	0.664	0.658	0.652	0.646	0.640
18	0.634	0.628	0.622	0.616	0.610	0.604	0.598	0.593	0.588	0.583
19	0.577	0.574	0.569	0.564	0.559	0.554	0.549	0.544	0.540	0.535
20	0.531	0.527	0.522	0.518	0.514	0.510	0.505	0.501	0.497	0.493
21	0.489	0.485	0.481	0.477	0.473	0.469	0.465	0.461	0.457	0.454
22	0.450	0.446	0.442	0.439	0.435	0.431	0.427	0.424	0.420	0.417
23	0.413	0.410	0.407	0.403	0.400	0.397	0.394	0.391	0.387	0.383
24	0.380	0.377	0.374	0.370	0.367	0.364	0.361	0.358	0.356	0.353
25	0.351	0.348	0.345	0.343	0.341	0.338	0.336	0.333	0.331	0.328
26	0.326	0.324	0.322	0.319	0.317	0.314	0.312	0.310	0.307	0.305
27	0.303	0.301	0.299	0.296	0.294	0.292	0.290	0.288	0.286	0.284
28	0.282	0.279	0.277	0.275	0.273	0.271	0.269	0.267	0.265	0.263
29	0.261	0.259	0.257	0.256	0.254	0.252	0.250	0.248	0.246	0.244
30	0.242	0.240	0.239	0.237	0.235	0.234	0.232	0.230	0.228	0.226
31	0.225	0.223	0.222	0.220	0.218	0.216	0.215	0.213	0.211	0.210
32	0.208	0.207	0.205	0.204	0.203	0.201	0.200	0.199	0.197	0.196
33	0.195	0.193	0.192	0.190	0.189	0.188	0.187	0.186	0.184	0.183
34	0.181	0.180	0.179	0.178	0.176	0.175	0.174	0.173	0.172	0.171
35	0.170	0.169	0.167	0.166	0.165	0.164	0.163	0.162	0.161	0.160
36	0.159	0.158	0.156	0.155	0.154	0.153	0.152	0.151	0.150	0.149
37	0.148	0.147	0.146	0.145	0.144	0.143	0.142	0.141	0.140	0.139
38	0.138	0.137	0.136	0.135	0.134	0.133	0.132	0.131	0.130	0.129
39	0.128	0.127	0.1265	0.126	0.125	0.124	0.123	0.122	0.121	0.120
40	0.119	0.1185	0.118	0.117	0.116	0.115	0.114	0.1135	0.113	0.112
41	0.111	0.1105	0.110	0.1095	0.109	0.108	0.107	0.106	0.105	0.1045
42	0.104	0.103	0.1025	0.102	0.101	0.100	0.0995	0.099	0.0985	0.098
43	0.097	0.0965	0.096	0.0955	0.095	0.0945	0.0935	0.093	0.0925	0.092
44	0.091	0.090	0.0895	0.089	0.0885	0.0875	0.087	0.0865	0.086	0.0855
45	0.0845	0.084	0.0835	0.083	0.082	0.0815	0.081	0.0805	0.080	0.0795
46	0.079	0.0785	0.078	0.077	0.076	0.0755	0.075	0.0745	0.074	0.0735
47	0.073	0.0725	0.072	0.0715	0.071	0.0705	0.070	0.0695	0.069	0.0685

TABLE 5.03 (Continued)

48	0.068	0.0675	0.067	0.0665	0.066	0.0655	0.065	0.065	0.0645	0.064
49	0.0635	0.0635	0.063	0.0625	0.062	0.0615	0.061	0.0605	0.060	0.0595
50	0.0591	0.0587	0.0583	0.0579	0.0575	0.0571	0.0567	0.0563	0.0559	0.0555
51	0.0551	0.0547	0.0543	0.0539	0.0536	0.0532	0.0528	0.0524	0.0521	0.0517
52	0.0513	0.0509	0.0506	0.0502	0.0499	0.0496	0.0493	0.0489	0.0486	0.0482
53	0.0479	0.0475	0.0472	0.0469	0.0466	0.0462	0.0459	0.0456	0.0453	0.0449
54	0.0446	0.0443	0.0440	0.0436	0.0433	0.0430	0.0427	0.0423	0.0420	0.0417
55	0.0414	0.0411	0.0408	0.0404	0.0401	0.0398	0.0395	0.0391	0.0388	0.0385
56	0.0382	0.0379	0.0376	0.0373	0.0370	0.0367	0.0365	0.0362	0.0359	0.0356
57	0.0353	0.0350	0.0348	0.0345	0.0343	0.0340	0.0337	0.0334	0.0332	0.0329
58	0.0327	0.0324	0.0322	0.0319	0.0317	0.0315	0.0312	0.0309	0.0307	0.0304
59	0.0302	0.0300	0.0298	0.0296	0.0294	0.0291	0.0289	0.0286	0.0284	0.0282
60	0.0280	0.0278	0.0276	0.0273	0.0271	0.0269	0.0267	0.0264	0.0262	0.0260
61	0.0258	0.0256	0.0254	0.0252	0.0250	0.0248	0.0247	0.0245	0.0243	0.0241
62	0.0239	0.0237	0.0235	0.0233	0.0231	0.0229	0.0228	0.0226	0.0224	0.0222
63	0.0220	0.0218	0.0217	0.0215	0.0213	0.0211	0.0210	0.0208	0.0206	0.0205
64	0.0203	0.0201	0.0199	0.0197	0.0195	0.0194	0.0192	0.0190	0.0189	0.0187
65	0.0186	0.0184	0.0182	0.0181	0.0180	0.0178	0.0177	0.0175	0.0174	0.0172
66	0.0171	0.0169	0.0168	0.0166	0.0165	0.0163	0.0162	0.0160	0.0159	0.0158
67	0.0156	0.0155	0.0154	0.0152	0.0151	0.0150	0.0149	0.0147	0.0146	0.0145
68	0.0143	0.0142	0.0141	0.0139	0.0138	0.0137	0.0135	0.0133	0.0132	0.0131
69	0.0130	0.0129	0.0128	0.0127	0.0126	0.0125	0.0123	0.0122	0.0121	0.0120
70	0.0118	0.0116	0.0115	0.0114	0.0112	0.0111	0.0110	0.0109	0.0108	0.0106
71	0.0105	0.0104	0.0103	0.0102	0.0101	0.0100	0.0099	0.0098	0.0097	0.0096
72	0.0095	0.0094	0.0093	0.0092	0.0091	0.0090	0.0089	0.0088	0.0087	0.0086
73	0.0085	0.0084	0.0083	0.0083	0.0082	0.0081	0.0080	0.0079	0.0078	0.0078
74	0.0077	0.0076	0.0075	0.0075	0.0074	0.0073	0.0072	0.0071	0.0070	0.0070
75	0.0069	0.0068	0.0067	0.0067	0.0066	0.0065	0.0064	0.0063	0.00625	0.0062
76	0.0061	0.0060	0.00595	0.0059	0.0058	0.00575	0.0057	0.0056	0.00555	0.0055
77	0.0054	0.0053	0.00525	0.0052	0.00513	0.00506	0.00500	0.00495	0.00489	0.00483
78	0.00476	0.00469	0.00463	0.00457	0.00451	0.00446	0.00440	0.00433	0.00427	0.00420
79	0.00414	0.00408	0.00403	0.00397	0.00392	0.00387	0.00382	0.00377	0.00372	0.00367
80	0.00362	0.00357	0.00352	0.00347	0.00342	0.00337	0.00332	0.00327	0.00323	0.00318
81	0.00314	0.00310	0.00306	0.00301	0.00297	0.00293	0.00289	0.00285	0.00281	0.00276
82	0.00272	0.00268	0.00264	0.00260	0.00256	0.00252	0.00248	0.00244	0.00240	0.00236
83	0.00232	0.00227	0.00223	0.00219	0.00215	0.00211	0.00207	0.00203	0.00199	0.00195
84	0.00191	0.00188	0.00184	0.00181	0.00177	0.00174	0.00171	0.00168	0.00165	0.00162
85	0.00159	0.00156	0.00153	0.00150	0.00147	0.00144	0.00141	0.00138	0.00135	0.00132
86	0.00129	0.00126	0.00124	0.00121	0.00118	0.00115	0.00112	0.00110	0.00107	0.00104
87	0.00102	0.00099	0.00097	0.00095	0.00092	0.00090	0.00088	0.00085	0.00083	0.00081
88	0.00079	0.00077	0.00075	0.00073	0.00071	0.00069	0.00067	0.00065	0.00063	0.00061
89	0.00060	0.00058	0.00056	0.00054	0.00053	0.00051	0.00049	0.00048	0.00046	0.00045
90	0.00043	0.00042	0.00040	0.00039	0.00038	0.00036	0.00035	0.00033	0.00032	0.00031
91	0.00030	0.00028	0.00027	0.00026	0.00025	0.00024	0.00022	0.00021	0.00020	0.00019
92	0.00018	0.00018	0.00017	0.00016	0.00015	0.00014	0.00014	0.00013	0.00012	0.00011
93	0.00010	0.00010	0.00009	0.00008	0.00008	0.00007	0.00006	0.00006	0.00006	0.00005
94	0.00005	0.00004	0.00004	0.00003	0.00003	0.00003	0.00002	0.00002	0.00002	0.00002
95	0.00002	0.00001	0.00001	0.00001	0.00001	0.00001	0.00001	0.00000	0.00000	0.00000
96	0.00000	0.00000	0.00000	0.00000	0.00000	0.00000	0.00000	0.00000	0.00000	0.00000

*These figures incorporate the corrections applicable to gloss paints with media in dry film form of refractive index 1.50, illuminated normally and viewed at about 45°. Coefficient of external reflection of incident light = 0.040. Coefficient of internal reflection of light scattered back by the pigment = 0.555.[26]

(From Duncan, D. R., *J. Oil Colour Chem. Assoc.*, 45, 300 (1962). With permission.)

in the mixture. The Kubelka-Munk equation could be rewritten as follows

$$F(R'_\infty) = \frac{C_1K_1 + C_2K_2 + C_3K_3 + \ldots}{s} = \frac{\sum_1^n K_iC_i}{s}, \quad (5.08)$$

where K_i and C_i would correspond to the appropriate absorption coefficients and concentrations of the individual pigments. With small particle sizes, the scattering coefficient s becomes constant for various wavelengths and the following simple approximation can be used:

$$F(R'_\infty) \cong \sum_1^n K_iC_i. \quad (5.09)$$

This equation would be similar to the general expression developed earlier for multi-component analyses (see Chapter 2, Section 7) and, provided the white pigment is suitable as a reference standard, the same method can be used for a quantitative analysis of pigments. The absorption coefficient K would simply be replaced by the value τ, which is determined from the slope of the appropriate Kubelka-Munk plot. A different situation arises if the substrate itself is appreciably light-absorbing, in which case the reflectance values will have to be recalculated to absolute values or measured vs. a suitable standard such as MgO or $BaSO_4$. The new relationship would then be

$$F(R_\infty) - F_A(R_\infty) = \sum_1^n K_iC_i, \quad (5.10)$$

where $F_A(R_\infty)$ stands for the Kubelka-Munk function of the substrate itself. This does not matter, however, as long as the K values are still additive.

After having determined the optimal wavelength of analysis λ for a given set of pigments, one would derive a set of equations from Equations 5.09 and 5.10 and compute the relative concentrations of each pigment:

$$F(R_\infty)\lambda_1 - F_A(R_\infty)\lambda_1 = K_1(\lambda_1)C_1 + K_2(\lambda_1)C_2 + \ldots K_n(\lambda_1)C_n \quad (5.11)$$

$$F(R_\infty)\lambda_2 - F_A(R_\infty)\lambda_2 = K_1(\lambda_2)C_1 + K_2(\lambda_2)C_2 + \ldots K_n(\lambda_2)C_n. \quad (5.12)$$

At least as many equations have to be written as there are pigments to be analyzed.

Another exceptional case exists when the pigments of interest are present in very high concentrations that would result in deviations from the Kubelka-Munk theory. In such cases, the sample could be diluted with a neutral white powder; if this is not possible, the Kubelka-Munk function may be replaced by empirical functions which extend the linear function-to-concentration relationship.[25] Duncan[26-28] and other workers[29,30] have discussed many practical examples of pigment analysis based on the method discussed above. The time needed for the analysis of complex mixtures can be reduced by carrying out the computations of Equations 5.11 and 5.12 with a suitable "colorant mixture computer" such as the one designed specifically for this purpose by Davidson and Hemmendinger.[31] On a cathode ray screen this analogue computer displays the difference between the Kubelka-Munk functions (or other suitable functions) of the various components in a pigment mixture. With simple operations the pigments can be identified, and the proportions necessary to match a particular color are indicated. Details of this instrument have been discussed in several papers.[31-33]

Reflectance spectroscopy can be used also for the quantitative investigation of pigments and paints in the UV[34,35] and infrared[36,37] regions of the spectrum. Rupert[38] used diffuse reflectance methods to predict the colors obtainable by the mixing of pigments in paint manufacturing. The same technique was used to determine the hiding power, opacity, and aging effects of paints.[39-42]

The use of spectral reflectance for the identification of pigments in complex mixtures is another interesting aspect, which has been described in detail by Duncan.[26] An example is illustrated in Figure 5.04, which shows the reflectance spectra of two green paints (A and B). A represents copper phthalocyanine blue and B is Prussian blue. Under certain daylight conditions these two paints appear identical to the naked eye. From the spectrum it can be seen, however, that although both reflect substantially the same quantity of violet, blue, and blue-green light, paint A reflects more green and red and less yellow light than does paint B. Since green and red when mixed together produce a sensation of yellow, they will compensate for the lack of yellow in paint A, which in turn will lead to an apparently perfect match. One is therefore faced with a situation in which even an experienced color matcher could not distinguish one

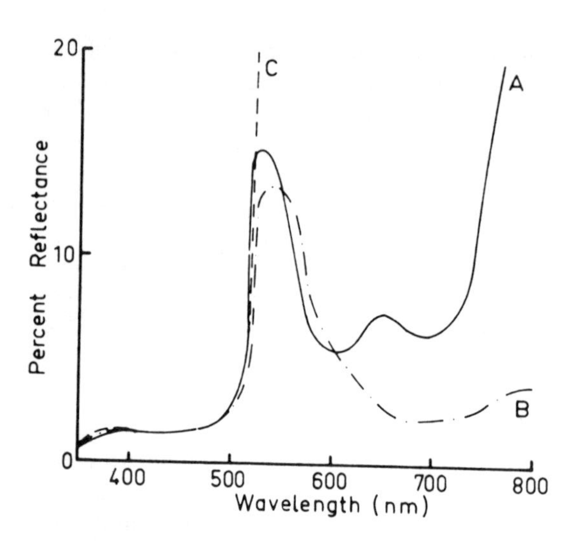

FIGURE 5.04 Reflectance curves of two metameric green paints. A, pigmented with copper phthalocyanine and middle chrome; B, pigmented with Prussian blue and middle chrome; C, pigmented with middle chrome only.[26] (From Duncan, D. R., *J. Oil Colour Chem. Assoc.*, 45, 300 (1962). With permission.)

paint from another under daylight conditions, but with the reflectance spectrophotometric curves, he would be on much firmer ground. He might be in a position to say that no green pigment, either alone or when mixed with white, would give a reflectance spectrum similar to that of either sample A or B, so that the spectra are probably mixtures of yellows and blues. Since blue pigments show little absorption below 520 nm, the shape of the reflectance spectrum in the region 400 to 520 nm should be typical of a yellow pigment, and, in fact, it almost coincides with that of paint C (see Figure 5.04) pigmented with pure middle lead chrome. Similar curves could be expected with organic yellows of similar color, but any different yellow, such as iron oxide yellow, would alter the spectrum. Above 520 nm curve A shows various maxima that are typical for copper phthalocyanine blue, whereas the B curve has the general, less irregular shape expected for the Prussian blue.

If, as in the case discussed above, only binary mixtures are involved, simple inspection of the spectra can yield considerable information. For more complicated mixtures, tabulated values for absorption coefficients K obtained at various wavelengths may be used for identification purposes. It is customary in the field of color technology to report the K values as ratios with another K value set for a particular wavelength. For example, Table 5.04 gives ratios of K/K_{425} for a group of typical yellow pigments. The reference K value, which serves to calculate the ratios, is chosen arbitrarily at wavelength 425, which for yellow pigments is in an area of the spectrum with relatively high absorption. If a yellow pigment in a pale yellow paint, plastic or the like, is to be identified with the use of this table, it will be sufficient to measure the reflectance at the marked wavelengths 400 to 550 nm. The readings are converted to corresponding Kubelka-Munk functions with the aid of Table 5.03. These values are divided by the Kubelka-Munk value obtained at 425 nm and the resulting data are compared with those in Table 5.04.

Similar qualitative reflectance spectroscopic studies have been carried out by Di Bernardo and Resnick[43] on green colors of copper

TABLE 5.04

Values of K/K_{425} for Typical Yellow Pigments[28]

Wavelength (nm)	400	425	450	475	500	525	550
S. African ocher	1.36	1.00	0.56	0.53	0.37	0.18	0.06
Marigold ocher	1.21	1.00	0.73	0.66	0.53	0.34	0.16
Synthetic yellow oxide	1.36	1.00	0.57	0.50	0.35	0.16	0.05
Middle chrome	0.93	1.00	0.95	0.88	0.70	0.07	0.01
Lemon chrome	1.05	1.00	0.82	0.63	0.20	0.01	0.00
Primrose chrome	1.00	1.00	0.82	0.41	0.11	0.01	0.00
Zinc chrome	1.39	1.00	0.58	0.15	0.02	0.00	0.00
Cadmium sulfide	0.97	1.00	0.01	0.67	0.03	0.00	0.00
C.1 Pigment Yellow 1	0.87	1.00	1.16	0.95	0.24	0.01	0.00
C.1 Pigment Yellow 3	0.87	1.00	0.97	0.32	0.01	0.00	0.00

(From Duncan, D. R., *Photoelectric Spectrometry Group Bulletin No. 16*, 483 (1965). With permission.)

phthalocyanine-benzidine yellow mixtures. Azo dyes were studied by Hannam and Patterson.[44] Clearly different reflectance spectra have been reported by Ulrich et al.[45] for the two pigments pumpkin and saffron, which ordinarily are very difficult to distinguish. Identification of white pigments can also be carried out by extending the reflectance techniques into the near-UV region of the spectrum.[29,46] Weber[47] and Jacobsen[48] have reported on such investigations.

Investigations such as those discussed in this section usually precede final color-matching procedures. They are still rather crude approximations, and the second step would involve the accurate measurement and computation of the color coordinates x, y, and z according to the previously discussed procedure (Section 1d).

3. Applications to Various Systems

a. Biological Systems

Many of the systems discussed, for example under "Foodstuff" (Section 3c) or in the chapter on chromatographic applications (Chapter 7), could also be classified under this heading. A typical application of diffuse reflectance spectroscopy to biological systems is the determination of oxygen saturation in blood samples. The method was first introduced by Brinkman and Zijlstra[49] as "reflection oximetry" and was further developed by Rodrigo.[50] Rodrigo's work was carried out at about 650 nm, and he proposed a relationship between oxygen saturation OS and the reflected radiation of the form

$$OS = A + [B/I_r(\lambda)] \tag{5.13}$$

for cases of blood-layer thickness ⩾ 3mm. In this equation, A and B are constants, which depend on the instrumental geometry, wavelength and intensity of irradiated light. $I_r(\lambda)$ is the diffusely reflected light intensity at wavelength λ. Since in this method constants A and B depend on instrumental factors, Polanyi and Hehir[51] built a reflection oximeter that permitted the absolute measurement of OS. Their instrument uses polychromatic light from a tungsten lamp at a 0° angle of incidence. The sample is viewed at a 45° angle by two separate detector and filter sets at two different wavelengths (λ_1, 660 nm; λ_2, 805 nm). The ratio of resistance of the two photocells is measured by a wheatstone bridge. The function of OS based on this mode of operation was given by Polanyi and Hehir[51] as

$$OS = A_r + B_r\,[I_r(\lambda_2)/I_r(\lambda_1)\,], \tag{5.14}$$

where $A_r = A_t$ and $B_r = H.B_t$ (A_t and B_t come from the equation $OS = A_t + B_t[OD(\lambda_1/OD\,\lambda_2)]$,[51] which is used in the transmission mode, and they are therefore related to extinction coefficients of oxyhemoglobin and reduced hemoglobin at λ_1 and λ_2 respectively). H is a positive function of the two wavelengths λ_1 and λ_2 and is close to unity. For the concrete case mentioned above, with $\lambda_1 = 660$ nm and $\lambda_2 = 805$ nm, Equation 5.14 becomes

$$OS = 1.13 - 0.28\,[I_r(\lambda_{805})/I_r(\lambda_{660})]. \tag{5.15}$$

This relationship of oxygen saturation OS and light reflected at 805 nm and 660 nm, respectively, was found to be linear.

Some applications of a rather unusual nature have been reported. Lubnow[52] used this technique to measure the color of hair, plumage and other parts of birds. He studied the classification possibilities of two strains of lark, using melanin as coloring matter in their plumage. Other workers[53,54] have reported on the spectral reflectance of hummingbird feathers and obtained similar results for certain types of feathers from one bird to another.

Derksen and Monahan[55] described the construction of a reflectometer for the measurement of diffuse reflectance in the visible and infrared regions of the spectrum. The instrument was used for an investigation of human skin. Figure 5.05 shows the spectra from the skin of a medium-dark

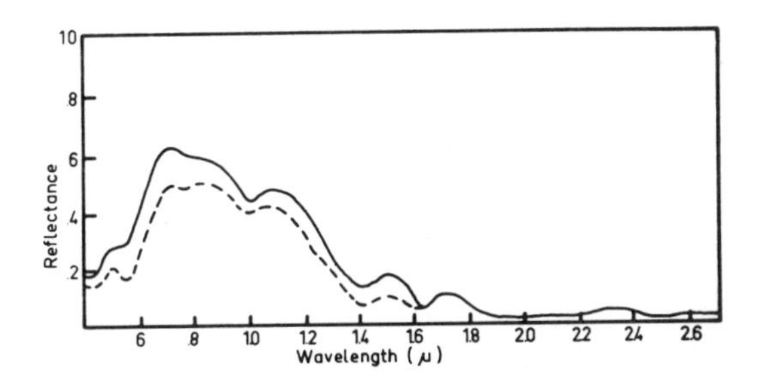

FIGURE 5.05 Reflectance of human skin: fair Caucasian skin (———) and medium-dark Negro skin (- - - - -).[55] (From Derksen, W. L. and Monahan, T. I., *J. Opt. Soc. Am.*, 42, 263 (1952). With permission.)

Negro and a fair, untanned Caucasian. The highest reflectance values are observed in the near-infrared region 750 to 1,200 nm. The instrument uses a hemispherical mirror for the collection of diffusely reflected light. The angle of incident radiation is 9° to the sample surface. Luckiesh et al.[56] used reflectance techniques to investigate the tanning of human skin as a function of various UV sources and the time of irradiation. Studies were carried out on a blonde, an intermediate, and a brunette subject.

Reflectance studies of leaves, foliage, and algae have been reported by a number of workers.[57,58] Here the method served to control the chlorophyll production in sea plants or to study the influence of chemicals on the appearance of foliage.

The possibilities of using diffuse reflectance techniques for investigations of systems, such as the ones described in this section, are endless, and it remains up to the individual investigator to see the value of such an approach and to use his ingenuity for devising suitable experiments.

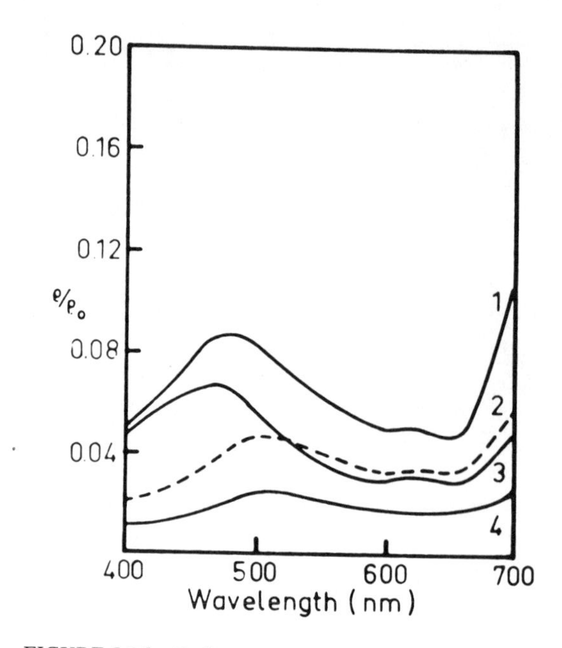

FIGURE 5.06 Reflectance spectra of carpets: 1, blue; 2, green; 3, blue; 4, dark green.[59] (From Moon, P., *J. Opt. Soc. Am.*, 32, 238 (1942). With permission.)

b. Building Materials

Spectral reflectance has found wide acceptance in the investigation of building materials. In some cases it has been used in quality control or in solving matching problems. The technique has also been valuable to illumination engineers for the study of lighting conditions with different interior materials. Moon[59-63] has measured a wide range of samples; one of his studies dealt with various kinds of floor coverings,[59] including linoleums, oak and cork flooring, asphalt tiles, and carpets of several colors and shades. Some typical reflectance spectra for linoleums and carpeting fabrics are given in Figures 5.06 and 5.07. The measurements were carried out with a Hardy Reflectance Spectrophotometer; reflection factors (sum of diffuse and specular reflectance) and trichromatic coefficients were calculated from reflectance spectra by the previously mentioned method (Section 1d).

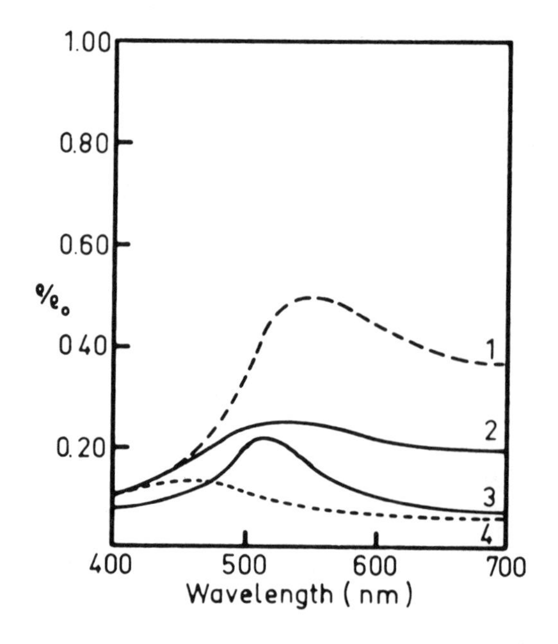

FIGURE 5.07 Reflectance spectra of linoleums: 1, light green; 2, green; 3, dark green; 4, dark blue.[59] (From Moon, P., *J. Opt. Soc. Am.*, 32, 238 (1942). With permission.)

Similar studies have been carried out with some materials used in schools, such as chalk boards, wall paints, window shades and metal finishes. Table 5.05 gives true reflection factors (corrected to absolute values) and trichromatic coefficients for illuminants A and C for a number of chalk boards conventionally used in schools. The blackboards showed nonselective reflectance spectra.

The data needed for the lighting and decoration of rooms include reflection factors of ceramic tiling material used in bathrooms and kitchens,[61] acoustic materials[62] for ceilings and walls, and wall-coating materials,[63] such as wallpaper, latex

TABLE 5.05

Reflection Factors ρ and Trichromatic Coefficients of Chalk Boards for Illuminants A and C[60]

Name	Approx. color	A			C		
		x	y	ρ	x	y	ρ
Vitrolite,® white	White	0.4478	0.4093	0.906	0.3111	0.3196	0.899
Nucite,® black	Black	0.4463	0.4073	0.0739	0.3085	0.3152	0.0743
Vitrolite, black	Black	0.4429	0.4050	0.0660	0.3042	0.3089	0.0666
Slate	Black	0.4424	0.4067	0.0627	0.3046	0.3119	0.0633
Vitrolite, cream	Cream	0.4620	0.4118	0.768	0.3284	0.3338	0.757
Nucite, cream	Cream	0.4697	0.4141	0.633	0.3369	0.3426	0.621
Vitrolite, green	Light green	0.4322	0.4428	0.540	0.3120	0.3622	0.554
Carrara, jade	Light green	0.4289	0.4469	0.481	0.3097	0.3659	0.495
Nucite, green	Green	0.4111	0.4348	0.133	0.2827	0.3478	0.142

(From Moon, P., *J. Opt. Soc. Am.*, 32, 243, (1942). With permission.)

paints, etc. In the case of ceramic tiles, both gloss and color may be of importance, and the influence of specular reflectance as an additional component to diffuse reflectance can be a decisive factor. Usually the illumination engineer is interested in total reflection properties, which include both reflection modes. Figure 5.08 presents reflectance spectra of a group of green ceramic tiles and the corresponding color coordinates and true reflection factors ρ are tabulated in Table 5.06.

Taylor[34] and other workers[55] (see also bibliography in Reference 34) have reported on the reflectance measurements of metal surfaces, wood, white plaster, and other building materials. Reflection factors for the visible and UV regions of the spectrum were presented. Other spectral reflectance data have been reported for various ceramic materials,[64-67] sanitation ware,[68] and chrome magnesite brick materials.[69]

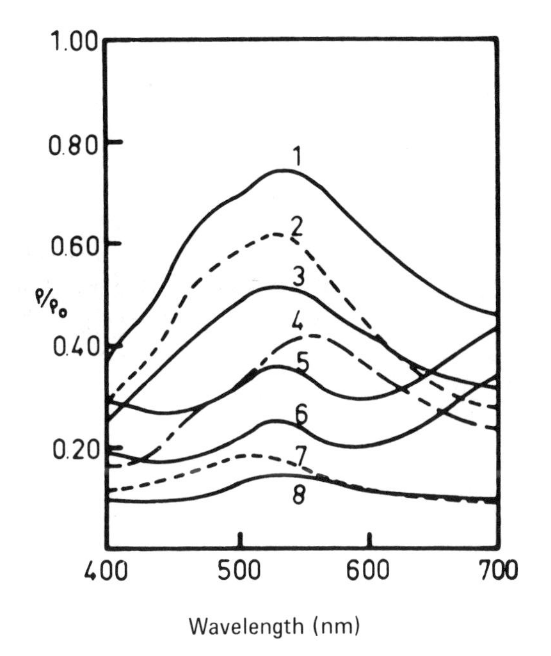

FIGURE 5.08 Reflectance spectra of some green tiles.[61] (For an explanation of the numbers see Table 5.06.) (From Moon, P., *J. Opt. Soc. Am.*, 31, 482 (1941). With permission.)

c. Foodstuff

In recent years attempts have been made to eliminate the personal element in visual comparison of color changes and color differences in foods; as a consequence, spectral reflectance techniques have been adopted in many food laboratories and food-processing industries. A few examples for the application of diffuse reflectance spectroscopy in this field are given here.

A General Electric Recording Reflectometer was used to correlate changes in reflectance with changes in the ascorbic acid content of certain foods, such as asparagus, beans, broccoli, cauliflower, corn, spinach, peaches and strawberries during deep-freeze storage.[70] The foods were studied at temperatures of +10, 0, and −20°F over a 12-month period. Even at these low temperatures the differences in temperature resulted in definite changes in the visible reflectance spectra of the samples. Changes of reflectance also seemed to be related closely to changes in the ascorbic acid content (see Table 5.07 and Figures 5.09 and 5.10).

TABLE 5.06

Reflection Factors ρ and Trichromatic Coefficients for Some Ceramic Tiles Illuminated with Illuminants A and C[61]

No.	Approx. color	A x	A y	A ρ	C x	C y	C ρ
1	Light green	0.4346	0.4298	0.652	0.3085	0.3490	0.667
2	Light green	0.4146	0.4376	0.494	0.2921	0.3495	0.518
3	Light green	0.4328	0.4315	0.441	0.3072	0.3499	0.451
4	Light jade green	0.4513	0.4426	0.361	0.3336	0.3810	0.363
5	Green	0.4484	0.4192	0.312	0.3144	0.3387	0.313
6	Green	0.4527	0.4214	0.214	0.3195	0.3462	0.214
7	Cossack dark green	0.4289	0.4382	0.120	0.3055	0.3545	0.123
8	Dark green	0.4063	0.4284	0.133	0.2806	0.3326	0.142

(From Moon, P., *J. Opt. Soc. Am.*, 31, 482 (1941). With permission.)

TABLE 5.07

Reduced Ascorbic Acid Content of Foods[70]

Food	Ascorbic acid content after freezing (mg/100 g)	Ascorbic acid retained* after 12 months of storage, % +10°F	0°F	-20°F
Asparagus	40	10	100	100
Beans, green	14	5	70	100
Beans, wax	22	15	75	100
Broccoli	78	15	75	90
Cauliflower	78	5	35	80
Corn	9	5	35	80
Spinach	31	10	45	90
Peaches	6	25	75	75
Strawberries	38	25	85	100

*To nearest 5%.
(From Guerrant, N. B., *J. Agric. Food Chem.*, 5, 207 (1957). With permission.)

Kent-Jones et al.[71,72] have discussed the design of a simple reflectometer that they used for the measurement of various grades of wheat flour. Naughton et al.[73,74] have investigated the heme pigments in tuna fish by spectral reflectance in order to study the undesirable "greening" process, which is usually observed on precooking yellow-fin tuna. By looking at the various heme pigments, it was possible to trace the "greening" to an anomalous heme protein oxidation (see Figure 5.11). Kraft and Ayres[75] and Pirko and Ayres[76] have carried out similar studies on heme pigments in beef. They succeeded in indicating the relative proportions of myoglobin, metmyoglobin and oxymyoglobin on the basis of the reflectance peaks.

The changes associated with the irradiation of meat and meat extracts have been studied by Ginger et al.[77] with the use of reflectance techniques. Spectral changes due to cooking and pickling of meats were reported by Tappel.[78]

Several reflectance spectroscopic instruments have been compared and rated as to their suitability for the measurement of canned tomato juice.[79] The Beckman DU Spectrophotometer equipped with standard reflectance attachment, the Hunter Colormeter, and the Photovolt Reflection Meter were some of the instruments tested. A scale has been developed by Yeatman et al.[80] for the reflectance spectroscopic measurement of tomato puree colors with a tomato-reflectance colorimeter especially developed for this purpose by Hunter and Yeatman.[81] This technique has also been used to study sugars,[82] the browning of milk upon heating,[83] the darkening of peanut butter as a result of roasting procedures,[84] and many other procedures.

Reflectance techniques have been found to be suitable also for the investigation of food dyes. Yamaguchi et al.[85] measured food colorants adsorbed on filter paper. A logical extension of this work was the investigation of food dyes by spectral reflectance after separation on thin-layer

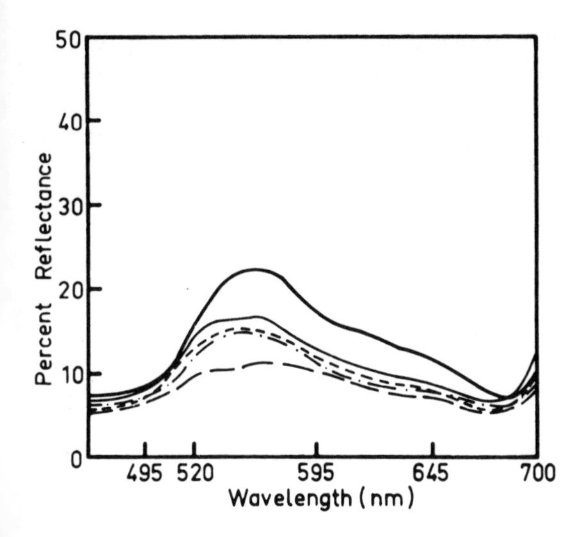

FIGURE 5.09 Reflectance spectra of spinach.[70] ——— unblanched. Blanched for 1.5 min: ·–·–·– before freezing; ——— stored for 12 months at –20°F; - - - - - - stored for 12 months at 0°F; – – – stored for 12 months at +10°F. (From Guerrant, N. B., *J. Agric. Food Chem.*, 5, 207 (1957). With permission.)

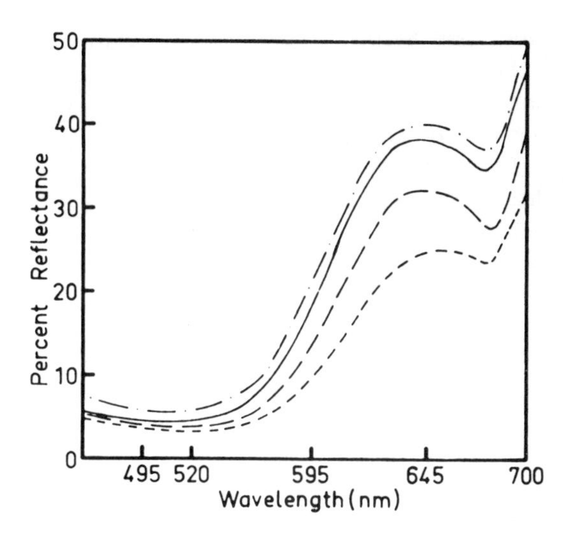

FIGURE 5.10 Reflectance spectra of strawberries.[70] –·–· before freezing; ——— stored for 12 months at –20°F; - - - - - - stored for 12 months at 0°F; – – – stored for 12 months at +10°F. (From Guerrant, N. B., *J. Agric. Food Chem.*, 5, 207 (1957). With permission.)

chromatograms.[86] Some factors affecting the reflectance spectra of food dyes adsorbed on an alumina matrix were studied also.[87]

d. Geological Systems

Many reflectance measurements in the geological field are carried out by the specular reflectance mode, but diffuse reflectance spectroscopy or the measurement of total reflectance has found its place in this area also. An excellent discussion on the measurement of color in connection with microscopic ore examination has been given by Piller.[88] Color is considered to be one of the most important properties in the identification and classification of minerals, but before the adoption of reflectance techniques its description depended solely on the individual observer's color response. Various reasons can be given for the fact that quantitative instrumental specification of mineral samples observed under the microscope was not adopted earlier. These are the lack of familiarity of the microscopist with the handling of colorimetric quantities and methods, the complexity and variety of color systems, and the lack of reflectance spectroscopic instrumentation suitable for combination with a microscope. The last reason is no longer valid since suitable microscope reflectometers now are manufactured commercially. The work by Piller, for example, was carried

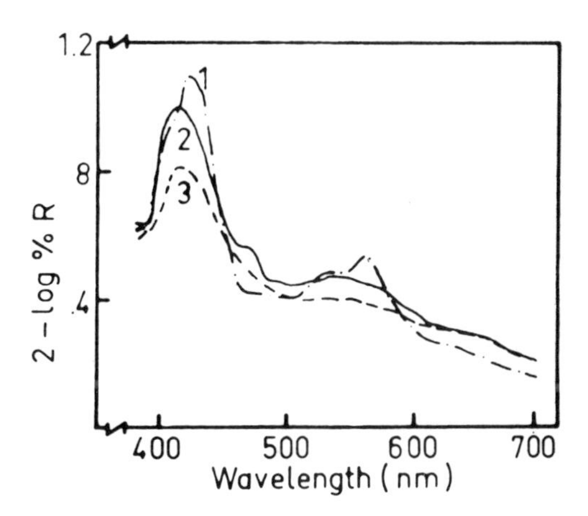

FIGURE 5.11 Reflectance curves showing pigments of cooked tuna flesh. 1, Denatured globin hemochrome of reduced, cooked tuna flesh; 2, 3, denatured globin hemochrome of cooked tuna flesh.[74] (From Naughton, J. J., Zeitlin, H., and Frodyma, M. M., *J. Agric. Food Chem.*, 6, 933 (1958). With permission.)

out with Zeiss equipment that was originally designed by him. Joyce, Loebl is marketing a microdensitometer which has been described in Chapter 4, Section 1. Instrumental problems encountered in earlier periods have been discussed by several workers.[3,89-92] For the actual color classification of minerals, Piller[88] proposed the use of the CIE color system and the computation of tristimulus values from spectral reflectance curves (see Section 1d). Table 5.08 shows the trichromatic coefficients and Helmholtz units for a large number of minerals as compiled from several literature sources.

The specification of color according to Helmholtz was often deemed more desirable for such systems, since they correspond more closely to hue and saturation[3] (also known as chroma). The Helmholtz quantities (dominant wavelength λ_f and spectral chromaticity coefficient p_e) indicate that the chromaticity is defined by a mixture of one pure spectral color and the achromatic color. λ_f can be obtained from the chromaticity diagram (Figure 5.01) at the spectrum color line intersection s of a straight line going from the illuminant point C through the color point G (G is defined by coordinates x, y) of the system of interest. The excitation purity p_e is defined as the ratio GC/SC. Helmholtz units can also be computed by a set of diagrams published for example by Hardy.[3] The Helmholtz units can be supplemented by the value Y (= R_{vis}) of the brightness.

The errors encountered in microscopic reflectance measurements have been discussed by Bowie and Henry,[89] Ehrenberg,[90] v. Gehlen and Piller,[91] Leow,[92] and Piller.[96] Experimental errors, such as incorrect microscope adjustments, glare of the lenses, fluctuations in the electronic equipment, insufficient monochromaticity of the irradiated light, or poor sample surfaces, can largely be eliminated by proper techniques and suitable instrumentation. Systematic relative errors, however, were believed not to exceed 1%.

Color values from three minerals, as obtained by several workers, as well as the difference of these values due to the difference in observer, are compiled in Table 5.09. The results indicate that instrumental color measurements of ore minerals can be as sensitive as the visual color perceptions, but they are definitely more reliable and reproducible. The relatively large difference for the covellite may be due to larger differences in surface quality of the samples and to the fact that this mineral has a strongly inflected spectral reflectance curve.

Other workers have discussed diffuse reflectance spectroscopic investigations of geological materials.[98,99] Ashburn and Weldon[99] have reported on an interesting investigation of desert surfaces by this same technique. For this purpose an instrument had to be constructed that enabled measurements of diffuse reflectance of an extended surface in the spectral range 390 to 650 nm. The principal components of this apparatus are an integrating sphere with an entrance aperture and an opal glass window placed in front of a photomultiplier. The diffuse radiation was rendered monochromatic with a wedge interference filter placed between the glass window and the detector. Wavelength selection was done with a rack and pinion assembly. The half-width of the band pass was about 90 Å, which makes it possible to measure reflectance over 2π steradians but still restricts the flux through the filter to a small fixed angle. Some spectra (albedo plotted vs. wavelength) are shown in Figure 5.12. The term "albedo" is defined as the ratio of the radiation reflected from a body to the total incident radiation and is often used in connection with meteorological applications or in the investigation of planets ("planetary albedo"). Albedo values for various types of surfaces measured at wavelength intervals from 0.4 to 0.65 μ are listed in Table 5.10 for various illumination conditions. The measurements were made from heights between 1.5 and 2 m above the surface, except for measurements of the salt bed, which were made from a helicopter at 200 m altitude. The data revealed that, for all types of desert terrain investigated, the diffuse reflectance in the blue portion of the spectrum is significantly less than in the red. No significant change in data could be detected when the height of observation was varied between 2 and 300 m.

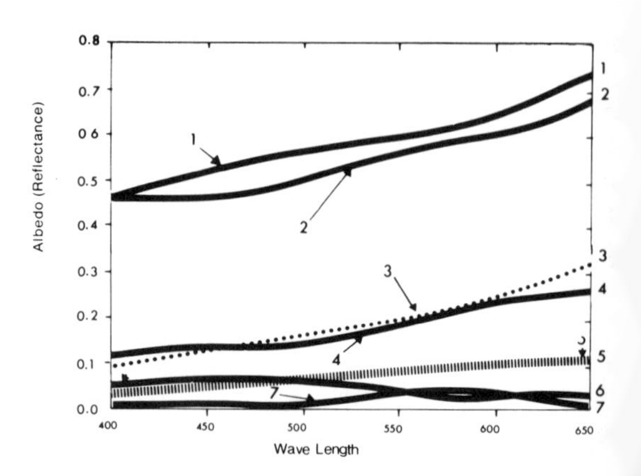

FIGURE 5.12 Diffuse reflectance of representative types of desert surfaces. 1, Salt bed; 2, non-saline dry playa; 3, granite pediment; 4, brush and wind-blown sand; 5, lag surface of red scoria on sand; 6, basaltic lava; 7, alfalfa.[99] (From Ashburn, E. V. and Weldon, R. G., *J. Opt. Soc. Am.*, 46, 442 (1956). With permission.)

TABLE 5.08

Trichromatic Coefficients and Helmholtz Units of Measurement of Some Ore Minerals[88]

Mineral	Immersion medium and direction of vibration	Brightness Y or visual reflectivity R_{vis}%	CIE trichromatic coefficients (referred to the equal energy illuminant E) x	y	Helmholtz units (referred to the CIE illuminant C) Dominant wavelength λ_f	Excitation purity p_e	Number of measured values Rλ	Reference
Antimonite	air ‖a	41.35	0.3018	0.3088	479	0.040	9	93
	air ‖b	30.81	0.3064	0.3167	494	0.013	9	
	air ‖c	47.44	0.2965	0.3067	481	0.060	9	
	air ‖a	41.22	0.2988	0.3057	479	0.052	35	88
	air ‖c	47.42	0.2930	0.3020	480	0.078	35	
	oil ‖a	25.72	0.2954	0.3026	479	0.068	35	
	oil ‖c	31.74	0.2843	0.2942	480	0.119	35	
Arsenopyrite	air ‖a (‖c)	52.82	0.3089	0.3171	501	0.003	9	93
	air ‖b (‖a)	53.30	0.3146	0.3226	574	0.030	9	
	air ‖c (‖b)	52.73	0.3167	0.3257	574	0.041	9	
Galena	air	44.24	0.2987	0.3009	470	0.060	35	88
	oil	28.85	0.2949	0.2957	469	0.082	35	88
Bournonite	air ‖a	46.29	0.3029	0.3118	482	0.030	9	93
	air ‖b	34.28	0.3016	0.3090	480	0.040	9	
	air ‖c	35.96	0.2989	0.3056	478	0.053	9	
Chalcophanite	air ω	26.15	0.2868	0.2916	476	0.146	37	94
	air ε	10.59	0.3005	0.3068	478	0.046	37	
	oil ω	11.97	0.2709	0.2760	477	0.188	37	
	oil ε	1.84	0.2934	0.2998	478	0.081	37	
Chalcocite (rhombic)	air ‖a	32.84	0.2958	0.3044	479.5	0.064	9	93
	air ‖b	32.58	0.2957	0.3039	479.5	0.064	9	
	air ‖c	32.61	0.2967	0.3040	479	0.064	9	
Covellite	air ω	6.79	0.2288	0.2424	479	0.377	14	91
	air ε	23.87	0.2706	0.2796	479	0.185	4	
	oil ω	1.60	0.3073	0.2066	553	0.436	14	
	oil ε	10.21	0.2382	0.2759	484	0.304	4	
	air ω	6.70	0.2202	0.2276	475.5	0.388	9	93
	air ε	22.54	0.2838	0.2867	474.5	0.129	9	
Hematite	air ω	29.59	0.2991	0.3093	484	0.046	6	95
	air ε	25.89	0.2977	0.3073	482	0.055	6	
	oil ω	14.90	0.2916	0.3012	481	0.084	6	
	oil ε	11.83	0.2899	0.2987	480	0.093	6	
Ilmenite	air ω	20.19	0.3083	0.3121	−567	0.013	6	95
	air ε	17.24	0.3138	0.3153	−495	0.010	6	
	oil ω	7.55	0.3089	0.3107	−561	0.020	6	
	oil ε	7.57	0.3173	0.3131	−498	0.028	6	
Pyrrhotite	air ω	34.47	0.3272	0.3300	581	0.081	6	95
	air ε	39.54	0.3232	0.3284	580	0.068	6	
	oil ω	22.02	0.3388	0.3423	579	0.148	6	
	oil ε	22.46	0.3320	0.3372	578	0.114	6	
	air ω	34.96	0.3271	0.3306	580	0.083	9	93
	air ε	39.66	0.3238	0.3296	578	0.072	9	

TABLE 5.08 (Continued)

Pyrite (Elba)	air	53.36	0.3310	0.3430	573.5	0.130	12	88
	oil	41.10	0.3382	0.3469	576.5	0.158	12	
π-Hematite	air ω	28.01	0.2996	0.3089	482	0.047	6	95
	air ϵ	24.34	0.2986	0.3080	482	0.050	6	
	oil ω	13.58	0.2909	0.3007	481	0.088	6	
	oil ϵ	10.56	0.2885	0.2985	481	0.098	6	
Umangite	air ω	13.69	0.2954	0.2717	-564	0.150	37	94
	air ϵ	16.45	0.2867	0.2966	486	0.257	37	
	oil ω	5.21	0.2957	0.2828	420	0.100	37	
	oil ϵ	5.60	0.2855	0.2793	466	0.137	37	

(From Piller, H., *Mineralium Deposita,* 1, 175 (1966). With permission.)

TABLE 5.09

Differences Δ x and Δ y and Δ R_{vis}/R_{vis} of Trichromatic Coefficients x and y and of Brightness R_{vis} Evaluated from Results Measured by Different Authors[88]

Reference	Mineral	Direction of vibration	Differences of the trichromatic coefficients		Relative differences of the brightness or visual reflectivity
			Δ x	Δ y	$\frac{\Delta Y}{Y} = \frac{\Delta R_{vis}\%}{R_{vis}}$
88	Antimonite in air	‖a	+0.0030	+0.0031	≈+0.3
		‖c	+0.0035	+0.0047	≈+0.04
93, 96, 99	Pyrrhotite in air	ω	−0.0001	+0.0006	≈+1.4
		ϵ	+0.0006	+0.0012	≈+0.3
91, 93	Covellite in air	ω	−0.0086	+0.0148	≈−1.3
		ϵ	−0.0132	+0.0071	≈−5.8

(From Piller, H., *Mineralium Deposita,* 1, 175 (1966). With permission.)

TABLE 5.10

Albedo of Various Desert Surfaces[99]

Surface	Time	Sky	Albedo 0.40	0.45	0.50	0.55	0.60	0.65
1. Salt bed	0730 PST	2/10 clouds	0.46	0.52	0.57	0.60	0.66	0.74
Pink salt	1100 PST	Clear	0.23	0.17	0.17	0.20	0.28	0.28
2. Nonsaline playa	1400	Clear	0.47	0.47	0.51	0.57	0.60	0.68
3. Red volcanic lag surface	1400	Clear Forest fire smoke west	0.036	0.058	0.060	0.080	0.112	0.122
4. Brush and windblown sand	1200	Clear	0.115	0.142	0.152	0.194	0.242	0.268
5. Granite pediment	1100	Clear	0.113	0.135	0.164	0.201	0.235	0.328
6. Alfalfa	1300	Clear	0.012	0.014	0.017	0.046	0.034	0.020
7. Sand near granite pediment	1030	Clear	0.114	0.127	0.137	0.178	0.217	0.232
8. Basaltic lava	1430	Clear Forest fire smoke	0.058	0.061	0.071	0.042	0.037	0.031

(From Ashburn, E. V. and Weldon, R. G., *J. Opt. Soc. Am.*, 46, 442 (1956). With permission.)

e. Paper and Pulp Material

Reflectance spectroscopy methods are widely used in this field for the measurement of color and for the quantitization of such factors as whiteness, brightness and gloss of papers. Since the literature in this field is quite extensive, only a few examples dealing specifically with diffuse reflectance are discussed in this section. Investigations of cellulose layers and papers in connection with chromatographic procedures are discussed in Chapter 7.

In one of the earlier investigations in this area, Taylor[34] studied the reflectance factors of commercial newspaper grades. Different grades of colored paper were measured by Anacreon and Noble[100] and by Stenius.[101] Luner and Chen[102] have discussed the use of spectral reflectance to control whiteness and brightness of the wood pulp during the paper-manufacturing process.

For the final product, the paper, the white color is an important criterion of its quality. The colorimetric measurement of numerous grades of paper shows that differences in whiteness are not merely due to changes in lightness but also to real color differences involving a variety of hues. Obviously these colors would be of a very low saturation. A comprehensive study[103,104] has shown that at the same lightness a touch of blue suggests a higher quality of white, whereas any other hue would suggest a lower quality. Only recently[105] has a system been proposed by which the tristimulus values of white colors obtained by spectral reflectance can be arranged according to a scale of such values. The system is designed so that certain differences between observers in this "aesthetic scale" can be taken into account. Other critical and extensive treatments of parameters and techniques for the measurement of brightness and whiteness of paper have been given by Stenius,[106,107] particularly in connection with measurement techniques based on the Zeiss Elrepho Colormeter. Figure 5.13 shows a diagram of the interrelationship of several parameters which are customarily measured with the Elrepho in the quality control of papers and other cellulose materials.

f. Pharmaceutical Products

Diffuse reflectance spectroscopy is a well-established technique in pharmaceutical science and industries. Its major use is in the investigation of aging and illumination effects, temperature stability, and the effect of humidity on tablets, powders, creams and liquid pharmaceutical products such as emulsions. The technique also has an important place in production control.

Reflectance measurements of creams, powders, granules, and coated tablets were reported by McKeehan and Christian[112] on a more qualitative

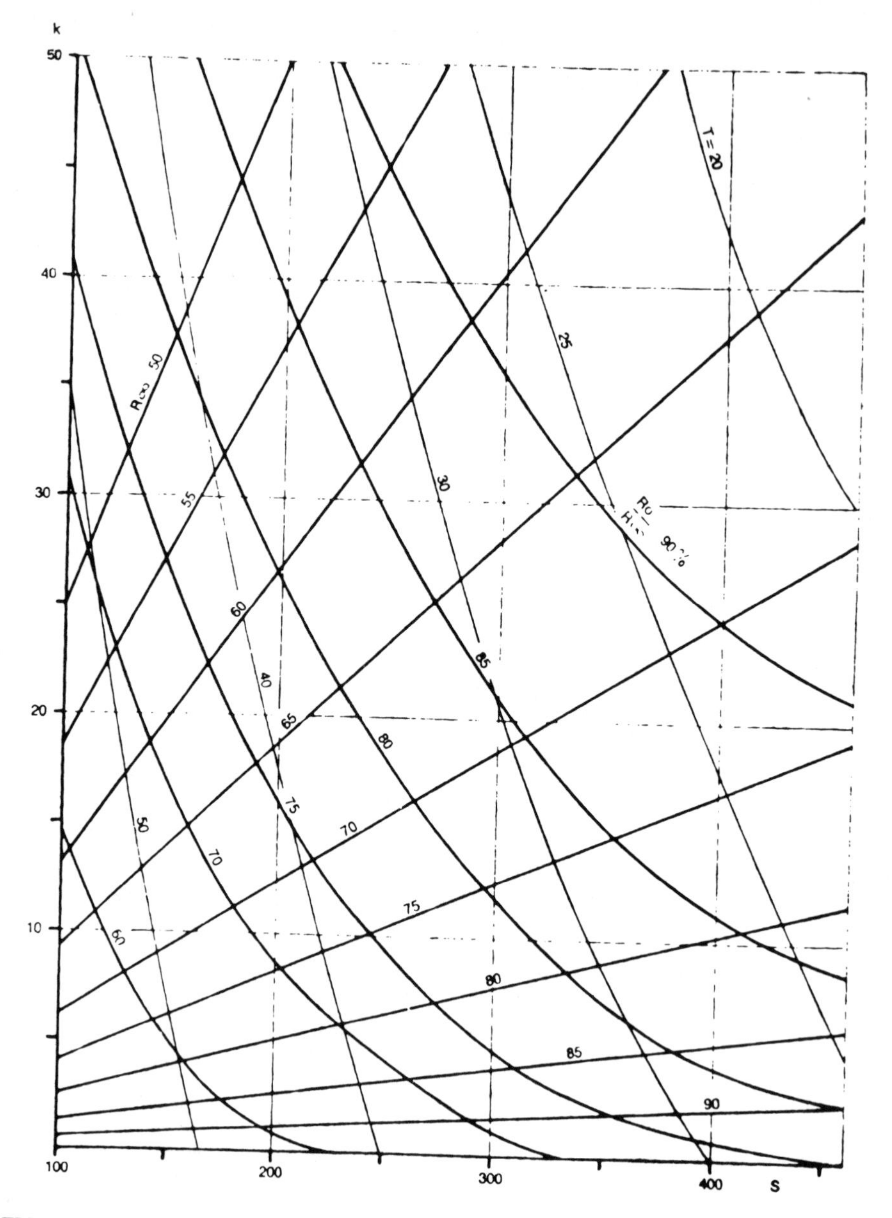

FIGURE 5.13 Relationship between absorption coefficient k, light-scattering coefficient s, reflectance of a layer of sheets of infinite thickness R_∞, opacity R_0/R_∞ and transparency T for a certain grade of paper with a weight of 60 g/m².[107] (From Stenius, A. S., *Zeiss Inform.*, 16, 21 (1968). With permission.)

basis. Lachman, Urbanyi, Swartz and other workers[113-118] have investigated the usefulness of reflectance techniques for a quantitative approach to problems such as the ones mentioned above. A holder that permits the measurement of tablets in the reflectance attachment of the Beckman DU Spectrophotometer has been described[114] (see Chapter 4, Section 3d). A precision of 2% and better was reported. This sample holder was then used for an investigation of the light fastness of several soluble dyes permitted for the coloring of tablets.[114-117] Normal and excessive illumination conditions of a similar spectral range were examined for periods of up to 84 days (see Figures 5.14 and 5.15). Table 5.11 lists some rate constants for the fading process of a number of these dyes in compressed tablets. The same group[118] reported on a spectral reflectance study of the influence of various protective color glasses on the fading of tablets.

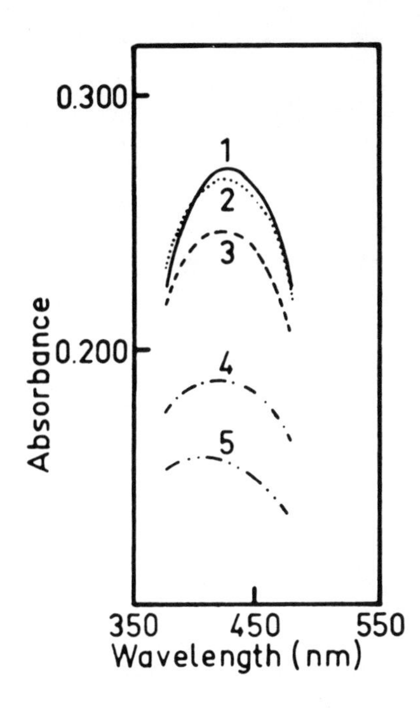

FIGURE 5.14 Plots of the visible absorption spectra of FD&C Yellow No. 5 after different intervals of storage under normal and exaggerated illumination. 1, Initial curve; 2, 42 days normal illumination; 3, 84 days normal illumination; 4, 42 days exaggerated illumination; 5, 84 days exaggerated illumination.[115] (From Lachman, L., Swartz, C. J., and Cooper, J., *J. Am. Pharm. Assoc.*, 49, 213 (1960). © 1960 by the Am. Pharm. Assoc., published in *J. Am. Pharm. Assoc.*, 49, 165 (1960). With permission.)

The fading of a certified red dye (FD&C No. 3) in tablets as a function of concentration, time, and light intensity has been investigated by Everhard and Goodhart.[119] A linear relationship was found between the usual concentration of the dye in the tablet and the Kubelka-Munk function F(R), calculated from relative reflectance values. A first-order-rate equation for the fading process was proposed:

$$\ln F(R_t) = -ktI + \ln F(R'_t). \qquad (5.16)$$

In this equation $F(R_t)$ is the Kubelka-Munk

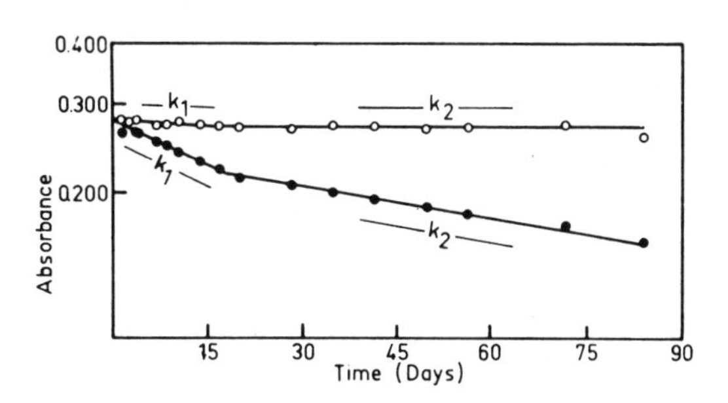

FIGURE 5.15 Influence of light intensity on the fading of the surface of tablets colored with FD&C Yellow No. 5.[115] (From Lachman, L., Swartz, C. J., and Cooper, J., *J. Am. Pharm. Assoc.*, 49, 213 (1960). © 1960 by the Am. Pharm. Assoc., published in *J. Am. Pharm. Assoc.*, 49, 165 (1960). With permission.)

TABLE 5.11

Rate Constants for the Fading of Colors in Compressed Tablets in Days^{-1} [115]

	Ordinary lighting			Accelerated lighting		
Color	$k_1 \times 10^3$	$k_2 \times 10^3$	$k_3 \times 10^3$	$k_1 \times 10^3$	$k_2 \times 10^3$	$k_3 \times 10^3$
FD&C Red No. 1	15.8	5.11	1.09	69.0	19.5	4.4
FD&C Red No. 3	71.2	22.2	5.9	264.0	25.5	9.2
FD&C Blue No. 1	5.48	1.63	–	50.6	9.4	–
FD&C Blue No. 2	3.45	–	–	5.75	0.15	–
FD&C Green No. 3	4.55	0.75	–	29.2	10.9	4.98
D&C Green No. 5	0.25	–	–	59.8	20.4	1.88
FD&C Yellow No. 5	2.04	0.38	–	16.3	4.78	–
D&C Yellow No. 10	16.4	8.70	1.22	72.1	22.1	3.29
FD&C Violet No. 1	14.6	6.13	2.26	78.0	35.6	8.36
D&C Orange No. 3	0.25	–	–	20.7	2.88	–

(From:Copyright 1960 by the American Pharmaceutical Association, Originally published in the *Journal of the American Pharmaceutical Association,* 49, 165 (1960). (From Lachman, L., Swartz, C. J., Urbanyi, T., and Cooper, J., *J. Am. Pharm. Assoc.,* 49, 165 (1960). With permission of American Pharmaceutical Assoc.)

TABLE 5.12

Time Required for Objectionable Fading at Low Light Intensities as Calculated from Accelerated Light Conditions[119]

Concentration of dye (% w/w)	Hours at 540 fc	Calcd. hours at 50 fc	Calcd. hours at 10 fc
0.060	17	180	920
0.030	7	75	380
0.015	2	20	110

 Originally published in the Journal of the American Pharmaceutical Association, *52*, 281 (1963). (From Everhard, M. E. and Goodhart, F. W., *J. Pharm. Sci.*, 52, 281 (1963). With permission of American Pharmaceutical Assoc.)

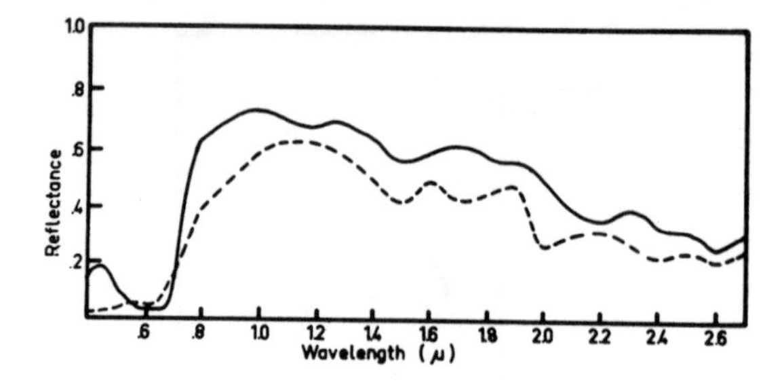

FIGURE 5.16 Reflectance spectra of wool serge (- - - - -) and cotton twill (———).[55] (From Derksen, W. L. and Monahan, T. I., *J. Opt. Soc. Am.*, 42, 263 (1952). With permission.)

function calculated at time t, and $F(R'_t)$ is the Kubelka-Munk function for t = 0. I stands for light intensity and k is the kinetic rate constant, calculated to be $6.1 \pm 0.3 \times 10^{-5}$ hr^{-1} foot-candles^{-1}. With this equation it is possible to calculate the fading that occurs under normal light conditions from the fading which has been measured at high light intensities. The advantage of this method is clear from Table 5.12.

Earlier attempts to correlate the fading of tablets at different intensities of the incident light were not successful. Each intensity had to be considered as a separate case with a different kinetic constant associated with it.[115,117]

Everhard et al.[120] have discussed the use of reflectance techniques for matching the color of solid dosage forms in pharmaceutical science. The chromaticity coordinates and brightness of close to 50 colorants, which are customary and permitted in this field, have been reported. The coordinates x, y, and z were calculated from the reflectance spectra of the tablets by a method similar to the one described earlier in Section 1d. The principles of multi-component analysis, as explained in Chapter 2, Section 7, were applied to calculate the concentrations of colorants required for a match. Similar techniques have been discussed by Duncan[26-28] in connection with the investigation of pigment-containing materials.

g. Textiles

Textiles are among the major fields of application for reflectance spectroscopic techniques. Since many of the samples are highly structured, the frequently adopted geometry of irradiation and viewing of the sample is d^R (any suitable angle). The diffuse illumination of the sample surface can be effected with the use of an integrating sphere, semisphere or ellipsoidal mirror, in order to avoid shadow effects. Just a few examples from the voluminous literature in this field will be mentioned in this section in order to convey an idea of the potential use of this technique for problems in the textile industry.

One of the earlier, more sophisticated applications of diffuse reflectance spectroscopy to fabrics was reported by Hardy,[121] who developed the first recording spectrophotometer for reflectance work and actually used a piece of green silk to illustrate the reproducibility of his instrument. Since then, a large number of investigators have reported on color measurements of dyed fabrics.

Moon and Cettei[122] analyzed 229 different samples of cloth used for men's clothing with regard to reflection factors. The study was undertaken in order to establish guide lines for the choice of materials which, when worn by pedestrians, contrast well with dark roads. The Hardy Color Analyzer (Hardy Spectrophotometer),[121] which represents a typical application to illumination engineering, was used for this study. Derksen and Monahan[55] used a custom-made reflectometer with a semisphere design for the measurement of some fabrics in the visible and near-infrared regions of the spectrum (see Figure 5.16).

The reflectance technique can be useful for the investigation of dyeing mechanisms. This was demonstrated by Giles et al.[123] for the dyeing of viscose rayon and other fabrics. Spectral reflectance serves conveniently also for the study of factors such as bleaching due to the influence of light,[124] and the effect of temperature, humidity, and aging[125] on various fabrics. Fourt and Sookne[126] applied diffuse reflectance methods to cotton yarns. As would be expected from previous discussions, they found that the angle of incident radiation was critical for the reproduction of spectra. Another critical factor was the position of the yarns in regard to the plane of light (see Table 5.12). Laundering processes have been studied with this technique,[127-129] enabling the evaluation of the efficiency of soaps and detergents.

Another area of use of spectral reflectance in the textile industry is the evaluation of optical brighteners, which are often used in modern detergents. Vaeck[103] has discussed a new method for the determination of whiteness as a result of such chemicals. This method is independent of fluorescent components also present in the system. Vaeck achieved this with a set of new color filters (Zeiss FMX/L) which can reproduce the color at the existing sample-surface illumination.

A comprehensive treatment of the use of spectral reflectance techniques in the textile field has been given by Müller-Gerber;[125] it includes many of the previously mentioned applications.

The use of differential reflectance techniques was first advocated by Lermond and Rogers[130]

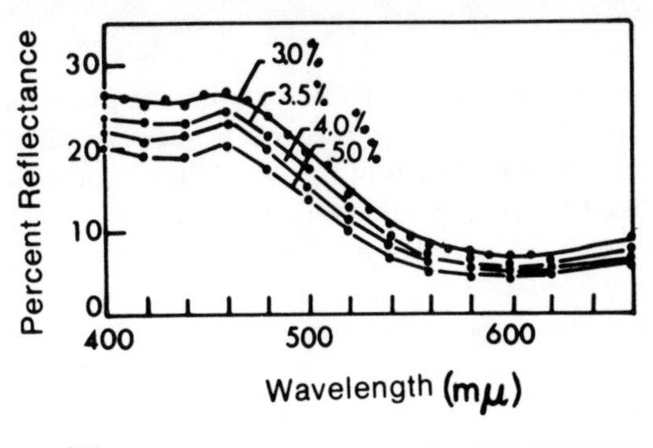

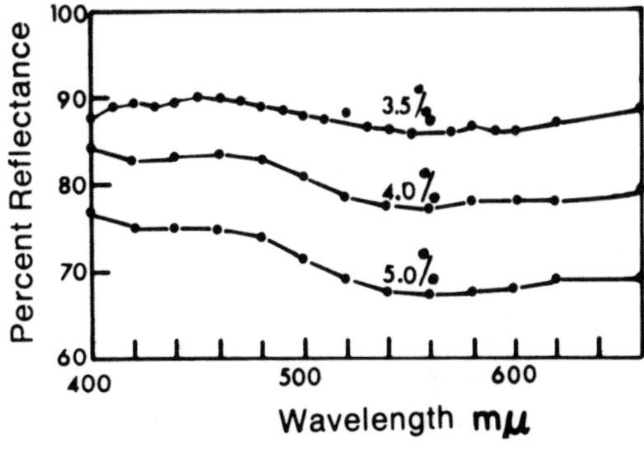

FIGURE 5.17 *a.* Reflectance, relative to magnesium carbonate, of cotton samples printed with Diagen Blue MSG solution. *b.* Differential reflectance of Diagen Blue MSG samples relative to the 3% sample as standard at 100% reflectance.[130] (From Lermond, C. A. and Rogers, L. B., *Anal. Chem.*, 27, 340 (1955). With permission.)

TABLE 5.13

Studies on Yarns Dyed with Ponsol Brown, Showing Reproducibility of Measurements and Effects on Reflectance of Pressing Surface of Mounted Sample and of Changing Direction of Yarn[130]*

Wavelength (nm)	Removal and replacement of given sample, % R ± SD** (4 measurements)	Pressed surface, (Av. % R of 2 measurements)	Direction of threads in sample changed 90°, % R ± SD (4 measurements)
400	7.67 ± 0.06	8.25	7.20 ± 0.10
420	5.93 ± 0.10	6.27	5.63 ± 0.07
440	4.79 ± 0.03	5.13	4.63 ± 0.04
460	4.68 ± 0.01	5.10	4.50 ± 0.06
480	5.62 ± 0.03	6.19	5.39 ± 0.15
500	7.76 ± 0.07	8.44	7.55 ± 0.10
540	11.50 ± 0.10	12.20	10.90 ± 0.10
580	14.90 ± 0.10	15.80	13.90 ± 0.10
620	32.50 ± 0.20	33.80	31.40 ± 0.40
660	54.70 ± 0.10	55.80	53.00 ± 0.20
700	62.10 ± 0.20	62.60	51.00 ± 0.30

*Samples were measured against magnesium carbonate.
**SD (standard deviation) =

(From Lermond, C. A. and Rogers, L. B., *Anal. Chem.*, 27, 340 (1955). With permission.)

and has proven to be particularly advantageous for strongly dyed textiles with low reflectance where obviously a dilution of the sample is not possible (see Chapter 2, Section 6b). The advantage of the differential reflectance technique is clearly shown in Figures 5.17a and b, which depict the reflectance spectra of cotton samples printed with different amounts of reflectance spectra of Diagen Blue MSG. When measured vs. an $MgCO_3$ standard, the lines of the spectra are crowded together in the spectral region of about 600 nm. However, using a standard with a 3% dye concentration as a reference material, which would be set at 100% R on the spectrophotometer scale, will result in a significant scale expansion effect (see Figure 5.03). At this concentration level, small differences in absolute reflectance actually represent large concentration differences, which are more clearly presented in the differential curves. Similar results have been obtained with other samples, the least concentrated of which had a reflectance of only 2% measured vs. $MgCO_3$.

Other systems investigated by Lermond and Rogers include yarns dyed with Ponsol Brown (see Table 5.13), nylon samples dyed with Acetamine Yellow CG, cotton prints of Ponsol Golden Orange 4R, etc. Several methods have been proposed by the same workers for the expression of concentration of dyes on fabrics. Plotting the Kubelka-Munk function vs. percent concentration resulted in smoothly bent curves suitable for analytical purposes, but obviously these systems did not quite adhere to the Kubelka-Munk theory, except at very low concentrations. Plotting the Kubelka-Munk function vs. the logarithm of concentration, however, usually resulted in an extended linear relationship.

REFERENCES

1. **Wyszecki, G.,** *Farbsysteme,* Musterschmidt Verlag, Berlin, 1960.
2. **Judd, D. B. and Wyszecki, G.,** *Color in Business, Science and Industry,* 2nd ed., John Wiley & Sons, New York, 1963, 264.
3. **Hardy, A. C.,** Ed., *Handbook of Colorimetry*, M.I.T. Press, Cambridge, 1936.
4. **Evans, R. M.,** *An Introduction to Color,* John Wiley & Sons, New York, 1948.
5. Optical Society of America, Committee on Colorimetry, *The Science of Color,* Thomas Y Crowell Co., New York, 1953.
6. **Wright, W. D.** *The Measurement of Color,* 3rd ed., Van Nostrand, Reinhold, New York, 1964.
7. **Billmeyer, F. W., Jr. and Saltzman, M.,** *Principles of Color Technology*, John Wiley & Sons, New York, 1966.
8. *Proceedings of the 8th Session of the Commission Internationale de l'Eclairage, Cambridge, 1931,* Cambridge University Press, 1932.
9. **Judd, J. B.,** *J. Opt. Soc. Am.,* 23, 359 (1933).
10. **Judd, J. B.,** *Natl. Bur. Std. Circular 478,* 1950.
11. **Grassmann, H.,** *Poggendorf's Ann. Physik Chemie,* 89, 69 (1853).
12. **Planck, M.,** *Ann. Physik,* 4, 553 (1901).
13. **Davis, R. and Gibson, K. S.,** *Natl. Bur. Std. Circular 114,* 1931.
14. **McNicholas, H. J.,** *J. Res. Natl. Bur. Std.,* 1, 793 (1928).
15. **Priest, I. G.,** *J. Res. Natl. Bur. Std.,* 15, 529 (1935).
16. **Donaldson, R.,** *Proc. Phys. Soc. (London),* 59, 554 (1947).
17. **Guild, J.,** *J. Sci. Instrum.,* 11, 69 (1939).
18. **Hunter, R. S.,** *Natl. Bur. Std. Circular 429,* 1942.
19. Zeiss Information No S50-679-e, Carl Zeiss Inc., Oberkochen, Germany, 1968.
20. **Judd, D. B.,** in *Analytical Absorption Spectroscopy,* Mellon, M. G., Ed., John Wiley & Sons, New York, 1950.
21. **Smith,T. and Guild, J.,** *Trans. Opt. Soc. (London),* 33, 73 (1931).
22. **Smith, T.,** *Proc. Phys. Soc. (London),* 46, 372 (1934).
23. **Saunderson, J. L.,** *J. Opt. Soc. Am.,* 32, 727 (1942).
24. **Duncan, D. R.,** *J. Oil Colour Chem. Assoc.,* 32, 296 (1949).
25. **Frei, R. W., Ryan, D. E., and Lieu, V. T.,** *Can. J. Chem.,* 44, 1945 (1966).
26. **Duncan, D. R.,** *J. Oil Colour Chem. Assoc.,* 45, 300 (1962).
27. **Duncan, D. R. and Mesner, H.,** *Paint Research Station, Technical Paper No. 177,* 1952.
28. **Duncan, D. R.,** *Photoelectric Spectrometry Group Bulletin No. 16,* 483 (1965).
29. **Pratt, L. S.,** *Chemistry and Physics of Organic Pigments,* Chapman and Hall, London, 1947, 301.
30. **Stearns, E. I.,** *Am. Dyestuff Rep.,* 40, 562 (1951).
31. **Davidson, H. R.,** *Am. Paint J.,* 46, 9 (1962).

32. Davidson, H. R. and Hemmendinger, H., *J. Opt. Soc. Am.*, 48, 281 (1958).
33. Davidson, H. R., Hemmendinger, H., and Landry, I. L. R., *J. Soc. Dyers Colour.*, 79, 577 (1963).
34. Taylor, A. H., *J. Opt. Soc. Am.*, 24, 192 (1934).
35. Wilcock, D. F. and Soller, W., *Ind. Eng. Chem.*, 32, 1446 (1940).
36. Saunderson, J. A., *J. Opt. Soc. Am.*, 37, 771 (1947).
37. Vesse, V. C., *Offic. Dig. Fed. Paint Varnish Prod. Clubs, No. 227*, 217 (1943).
38. Rupert, F. F., *J. Opt. Soc. Am.*, 20, 661 (1930).
39. Parker, D. H., *Paint Ind. Mag.*, 72, 18 (1957).
40. Tough, D., *J. Oil. Colour Chem. Assoc.*, 39, 169 (1956).
41. Bruins, A. W., *Chem. Weekbl.*, 46, 282 (1950).
42. Meacham, J. A., *Am. Paint J.*, 23, 23 (1938).
43. Di Bernardo, A. and Resnick, P., *J. Opt. Soc. Am.*, 49, 480 (1959).
44. Hannam, A. R. and Patterson, D., *J. Soc. Dyers Colour.*, 79, 192 (1963).
45. Ulrich, W. F., Kelly, F., and Nelson, D.C., *Beckman Reprint R-6134,* Beckman Instruments Inc., Fullerton, California, 1959.
46. Tilleard, D. L., *Paint Research Station Memorandum, No. 197* (1952).
47. Weber, H. H., *Farbe Lack,* 63, 586 (1957).
48. Jacobsen, A. E., *J. Opt. Soc. Am.*, 38, 442 (1948).
49. Brinkman, I. and Zijlstra, W. G., *Arch. Chir. Neerl.*, 1, 177 (1949).
50. Rodrigo, F. A., *Am. Heart J.*, 45, 809 (1953).
51. Polanyi, M. L. and Hehir, R. M., *Rev. Sci. Instrum.*, 31, 401 (1960).
52. Lubnow, L., *Zeiss Inform.*, 51, 10 (1964).
53. Dorst, J., *Mem. Museum Natl. Hist. Nat. (Paris), Ser. A: Zool.*, 1, 125 (1951).
54. Greenewalt, C. H., Brandt, W., and Friel, D. D., *J. Opt. Soc. Am.*, 50, 1005 (1960).
55. Derksen, W. L. and Monahan, T. I., *J. Opt. Soc. Am.*, 42, 263 (1952).
56. Luckiesh, M., Holladay, L. L., and Taylor, A. H., *J. Opt. Soc. Am.*, 20, 423 (1930).
57. Wong, C. L. and Blevin, W. R., *Aust. J. Biol. Sci.*, 20, 501 (1967).
58. Shibata, K., *J. Biochem. (Tokyo),* 45, 599 (1958).
59. Moon, P., *J. Opt. Soc. Am.*, 32, 238 (1942).
60. Moon, P., *J. Opt. Soc. Am.*, 32, 243 (1942).
61. Moon, P., *J. Opt. Soc. Am.*, 31, 482 (1941).
62. Moon, P., *J. Opt. Soc. Am.*, 31, 317 (1941).
63. Moon, P., *J. Opt. Soc. Am.*, 31, 723 (1941).
64. Andrews, A. I. and Zwerman, C. H., *J. Am. Ceram. Soc.*, 22, 65 (1939).
65. Pask, J. A., *Bull. Am. Ceram. Soc.*, 20, 50 (1941).
66. Emery, F. H., *Bull. Am. Ceram. Soc.*, 20, 381 (1941).
67. Peskin, W. L., *Bull. Am. Ceram. Soc.*, 20, 402 (1941).
68. Merry, W. H., *Bull. Am. Ceram. Soc.*, 36, 236 (1956).
69. Bautsch, H. J., *Silikat. Tech.*, 9, 552 (1958).
70. Guerrant, N. B., *J. Agric. Food Chem.*, 5, 207 (1957).
71. Kent-Jones, D. W. and Martin, W., *Analyst,* 75, 127 (1950).
72. Kent-Jones, D. W., Amos, A. J., and Martin, W., *Analyst,* 75, 133 (1950).
73. Naughton, J. J., Frodyma, M. M., and Zeitlin, H., *Science,* 125, 121 (1957).
74. Naughton, J. J., Zeitlin, H., and Frodyma, M. M., *J. Agric. Food Chem.*, 6, 933 (1958).
75. Kraft, A. A. and Ayres, J. C., *Food Technol.*, 8, 290 (1954).
76. Pirko, P. C. and Ayres, J. C., *Food Technol.*, 11, 461 (1957).
77. Ginger, I. D., Lewis, V. T., and Schweigert, B. S., *J. Agric. Food Chem.*, 3, 156 (1955).
78. Tappel, A. L., *Food Res.*, 22, 479 (1957).
79. Robinson, W. B., Wishnetsky, T., Ransford, J. R., Clark, W. L., and Hand, D. B., *Food Technol.*, 6, 269 (1952).
80. Yeatman, J. N., Sidwell, A. P., and Norris, K. N., *Food Technol.*, 14, 16 (1960).
81. Hunter, R. S. and Yeatman, J. N., *J. Opt. Soc. Am.*, 51, 552 (1961).
82. Gilett, T. R. and Holven, A. L., *Ind. Eng. Chem.*, 35, 210 (1943).
83. Burton, H. J., *Dairy Res.*, 21, 194 (1954).
84. Morris, N. J., Lohmann, I. W., O'Connor, R. T., and Freeman, A. F., *Food Technol.*, 7, 393 (1953)
85. Yamaguchi, K., Fujii, S., Tobata, T., and Kato, S., *J. Pharm. Soc. Jap.*, 74, 1322 (1954).
86. Frodyma, M. M., Frei, R. W., and Williams, D. J., *J. Chromatogr.*, 13, 61 (1964).
87. Frei, R. W. and Zeitlin, H., *Anal. Chim. Acta,* 32, 32 (1965).
88. Piller, H., *Mineralium Deposita,* 1, 175 (1966).
89. Bowie, S. H. U. and Henry, N. F. M., *Trans. Inst. Mining Met.*, 73, 467 (1963/64).
90. Ehrenberg, H., *Z. Wiss. Mikrosk. Tech.*, 66, 32 (1964).
91. v. Gehlen, K. and Piller, H., *Petrography,* 10, 94 (1964).
92. Leow, J. H., *Econ. Geol.*, 61, 598 (1966).

93. **Simpson, P. R.,** *Can. Mineralogist,* 8, 673 (1964/66).
94. **v. Gehlen, K.,** Unpublished work.
95. **v. Gehlen, K. and Piller, H.,** *Mineral. Mag.,* 35, 335 (1965).
96. **Piller, H.,** *Mineral. Mag.,* 36, 242 (1967).
97. **v. Gehlen, K. and Piller, H.,** *Neues Jahrb. Mineral. Monatsh.,* 4, 97 (1965).
98. **Rösch, S.,** *Opt. Acta,* 11, 267 (1964).
99. **Ashburn, E. V. and Weldon, R. G.,** *J. Opt. Soc. Am.,* 46, 442 (1956).
100. **Anacreon, R. E. and Noble, R. H.,** *Appl. Spectrosc.,* 14, 29 (1960).
101. **Stenius, A. S.,** *J. Opt. Soc. Am.,* 45, 727 (1955).
102. **Luner, P. and Chen, D.,** *Tappi,* 46, 98 (1963).
103. **Vaeck, S. V.,** *Ann. Sci. Textiles Belges,* 1, 95 (1966).
104. **Vaeck, S. V.,** *Textilveredelung,* 1, 658 (1966).
105. **Loof, H.,** *Papier,* 21, 297 (1967).
106. **Stenius, A. S.,** *Tappi,* 46, 183A (1963).
107. **Stenius, A. S.,** *Zeiss Inform.,* 16, 21 (1968).
108. **Budde, W. and Chapman, S. M.,** *Pulp Paper Mag. Can.,* 67, T206 (1968).
109. **Schmidt, K. G.,** *Papier,* 12, 141 (1959).
110. **Hisey, R. W. and Cobb, H. W.,** *Tappi,* 42, 122 (1959).
111. **Braun, W. and Kortüm, G.,** *Zeiss Inform.,* 16, 27 (1968).
112. **McKeehan, C. W. and Christian, J. E.,** *J. Pharm. Sci.,* 46, 631 (1957).
113. **Lachman, L. and Cooper, J.,** *J. Am. Pharm. Assoc.,* 48, 226 (1959).
114. **Urbanyi, T., Swartz, C. J., and Lachman, L.,** *J. Am. Pharm. Assoc.,* 49, 163 (1960).
115. **Lachman, L., Swartz, C. J., Urbanyi, T., and Cooper, J.,** *J. Am. Pharm. Assoc.,* 49, 165 (1960).
116. **Lachman, L., Swartz, C. J., and Cooper, J.,** *J. Am. Pharm. Assoc.,* 49, 213 (1960).
117. **Lachman, L., Weinstein, S., Swartz, C. J., Urbanyi, T., and Cooper, J.,** *J. Am. Pharm. Assoc.,* 50, 141 (1961).
118. **Swartz, C. J., Lachman, L., Urbanyi, T., and Cooper, J.,** *J. Am. Pharm. Assoc.,* 50, 145 (1961).
119. **Everhard, M. E. and Goodhart, F. W.,** *J. Pharm. Sci.,* 52, 281 (1963).
120. **Everhard, M. E., Dickcius, D. A., and Goodhart, F. W.,** *J. Pharm. Sci.,* 53, 173 (1964).
121. **Hardy, A. C.,** *J. Opt. Soc. Am.,* 18, 96 (1929).
122. **Moon, P. and Cettei, M. S.,** *J. Opt. Soc. Am.,* 28, 277 (1938).
123. **Giles, C. H., Rahman, S. M. K., and Smith, D.,** *Text. Res. J.,* 31, 679 (1961).
124. **Lanner, H. T.,** *Text. Res. J.,* 33, 351 (1963).
125. **Müller-Gerber, L.,** *Spinner Weben Textilveredlung,* 78, 2 (1960).
126. **Fourt, L. and Sookne, A. M.,** *Text. Res. J.,* 21, 469 (1951).
127. **Hurwitz, M. L.,** *Am. Dyestuff Reptr.,* 35, 83 (1946).
128. **Barker, G. E. and Kern, C. R.,** *J. Am. Oil Chem. Soc.,* 27, 113 (1950).
129. **Kling, A.,** *Fette, Seifen, Anstrichm.,* 65, 285 (1963).
130. **Lermond, C. A. and Rogers, L. B.,** *Anal. Chem.,* 27, 340 (1955).

Chapter 6

APPLICATIONS TO PROBLEMS IN PHYSICAL, INORGANIC, AND ORGANIC CHEMISTRY

1. Surface Phenomena

The use of diffuse reflectance spectroscopy in the area of surface studies is one of the most elegant applications of this technique. The only alternate method for the investigation of species adsorbed on fine powders would be the use of transmission spectroscopy through thin layers of the powder material. This has been done with relatively good success in the infrared region of the spectrum, e.g., in the investigation of chemisorption phenomena of ethylene on alumina[1] or the study of mixed alumina-silica gel surfaces.[1,2] In the infrared, light scattering occurs at a relatively low level, but, in the near-infrared and particularly in the visible and UV regions of the spectrum, these scattering processes become serious enough to make quantitative investigation quite impossible. The difficulty in reproducing layer thickness on these thin layers of adsorbent is another drawback of the transmission method.

In view of these circumstances, it is not surprising that diffuse reflectance spectroscopy has become the standard method in this field. The following sections indicate the possible uses of the technique for problems of this nature.

a. Sample Preparation

In order to obtain meaningful data for the physical-chemical studies of adsorption phenomena, the preparation of samples is extremely important. Since the influence of particle size of the adsorbent or diluent has been found to affect the shape of reflectance spectra,[3,4] standardization of grinding and sifting procedures is recommended. Kortüm[5] has suggested grinding the powders in a ball mill or mortar. Since the particle size depends on the length of grinding time, this and the grinding conditions had to be standardized. For the grinding process it is usually recommended to use porcelain ball-mill containers of 50 to 250 cm^3 volume, operating with 2 large or 6 small balls made of the same material. Agate containers and balls are occasionally indicated for harder powder materials. Within one dilution series, the same set of grinding equipment should be used; Kortüm suggested at least six hours of grinding for a homogeneous mixing of standard and sample material. Considering the overall reliability of the reflectance technique, abrasion from the grinding equipment during prolonged use had a negligible effect on the measurements.

Various workers have reported satisfactory reproducibility of grain size by first grinding the materials in an agate mortar for only 15 min and then sifting them through fine mesh screens (e.g., 200 mesh). The structure of the sample surface is another parameter that should be kept in mind. The surface should be smooth and free from gloss, particularly if directional irradiation (at, for example, 45°) is employed. Shadow effects can seriously alter reflectance values from one sample to another. These surface conditions are fulfilled quite satisfactorily when a sample cell such as the one described by Barnes et al.[6] is used for packing and preparation of the sample. A quartz glass cover serves as the sample window, and gloss effects are negligible when a zero irradiation angle is used. If one does not use a cover glass, a metal or glass tamp should be used to compress the sample slightly to obtain a reasonably smooth surface. In this fashion, the sample could simply be placed in an open planchet. If gloss becomes a problem, a tiny layer of fine powder can be sifted over the compressed sample surface.

The pressure exerted on the sample during the packing process should also be standardized as much as possible, since reflectance spectra show a slight dependency on packing density. The distance from the sample surface to the walls of the integrating sphere should not vary; this condition is fulfilled nicely when using an appropriate sample holder, such as the one described by Barnes et al.[6] For loosely packed powders, a layer thickness of between 2 and 5 mm will be sufficient for the condition of infinite layer thickness, i.e., absolutely no light will penetrate the sample layer. This condition is desirable if the results are to be interpreted on the basis of the well-known Kubelka-Munk theory and if interferences from the background are to be eliminated. The use of reference standards has been discussed in Chapter 3, Section 2. Often one chooses the same reference material that is being used as an adsorbent or neutral diluent in the sample to be investigated. When choosing a proper diluent, one has to keep in mind that powders such as carbonates and a large number of oxides, e.g., SiO_2 or Al_2O_3, exhibit rigorous adsorption properties, particularly in a freshly regenerated state. Materials such as LiF or starch are therefore often preferred as diluents since they are relatively neutral.

The presence of atmospheric humidity is generally considered an undesirable factor in investigations of this kind. The reason for this is the highly polar nature of the water molecules, which then can compete successfully for the active adsorption sites on the surface of the powders. In most cases, sample preparation should, therefore, be carried out under careful exclusion of water vapor, i.e., under controlled atmospheric conditions. The weighing, mixing, grinding, and sample-packing procedures are preferably carried out in a glove box. A plexiglass drybox with dynamic air-drying through a column of molecular sieves or silica gel could be used. The transfer of the sample from the oven to the drybox atmosphere and from there to the instrument can be done in a desiccator over phosphorus pentoxide or other suitable drying agents. The degree of dryness in the drybox can be monitored with a small amount of P_2O_5 powder, which cakes up in an insufficiently dry environment. The sample cell[6] can be sealed tightly with silicone grease at the edges of the planchet to exclude moisture while the sample is being transferred and recorded.

The actual adsorption process poses no particular problems. With solid adsorbates such as nitrophenols or nitroanilines, the substances are simply added to the adsorbent in controlled amounts. The mixing and grinding procedures ensure a homogeneous mixing if done rigorously and for a sufficient length of time. Sufficient time should also be allotted to the sample in order to establish an adsorption equilibrium. Adsorption from the gas phase is usually carried out in a closed system, which is then heated to achieve a high vapor pressure of the adsorbate. This approach has the advantage of excellent control of temperature concentration and atmospheric conditions. It has, however, the disadvantage of slow equilibration of the system and is not suitable for adsorbates with low vapor pressure. Control of experimental conditions poses considerable problems when the adsorbate is to be applied to the adsorbent in the form of a solution. The effect of the solvent has to be taken into account, and removal of the solvent from the adsorbent-adsorbate system is important. The solvent can be withdrawn by a vacuum stripping procedure while the sample is kept in a controlled atmosphere, or, if sufficiently volatile and harmless and present in minute amounts, it can be left to evaporate into the atmosphere of the drybox. The advantage of adsorption from solutions lies in the fact that the tedious grinding procedures needed to achieve a homogeneous mixture can be eliminated. It need not be specified that before exposing the gaseous or dissolved adsorbates to the adsorbent, the powders have to be pretreated to obtain a uniform particle size and controlled state of regeneration.

For analytical work the sample preparation procedures need not be as rigorous as described in this section.

b. Adsorbent-adsorbate Interaction

Early observations of color changes of many organic systems upon adsorption on active surfaces have been reported by Weitz et al.[7,8] and by de Boer and Houben.[9] This has stimulated investigations of this field with the aid of diffuse reflectance spectroscopy. Several interpretations have been given for the phenomena observed. Polarization, Lewis acid-base interaction, formation of electron donor-acceptor complexes are but a few of the reasons given, and there is still a certain amount of controversy about this subject. Basically these interactions can all be classified under the term chemisorption.

Zeitlin et al.[10] have studied the bathochromic and hypsochromic shifts of reflectance spectra observed upon adsorption of Hg(II) compounds on various active adsorbents. Similar shifts are known to occur in solution and have been interpreted on the basis of solvent-solute interactions with the polarity of the solvent or solute playing a decisive role.[11,12] Zeitlin et al.[10] have therefore assigned solvent properties to the adsorbent and solute properties to the adsorbate (Hg salts), and they have used a similar approach for the explanation of the data presented in Table 6.01.[10] The final λ_{max} values were measured 60 to 90 days after mixing the samples; it was assumed that the reaction was complete after this time. For a general interpretation of the results in Table 6.01, the data on the HgI_2-alumina system had to be disregarded because of the anomalous behavior of this system.[13-15] As can be seen from the table, the enhanced polarizability of the anions with a decrease in size results in a bathochromic shift in the order Br $>$ Cl and S $>$ O. The higher polarizing

TABLE 6.01

Wavelength (nm) of the Initial and Final λ_{max} of Hg(II) Compounds Mixed with Alumina, Silica Gel, and Sodium Fluoride[10]

Compounds	λ_{max} (initial)		λ_{max} (final)	
Alumina				
HgI_2	575	290	–	290
$HgBr_2$		248		255
$HgCl_2$		249–230		233–234
HgS	570		570	
HgO	501		501	
Silica gel				
HgI_2	568	269–272	–	317–320
$HgBr_2$		240–243		263–264
$HgCl_2$		–		220–225
HgS	563–565		568–569	
HgO	518		517–520	
Sodium fluoride				
HgI_2	573	293	573	293
$HgBr_2$		248–249		248–249
$HgCl_2$		233		233
HgS	560–566		560–566	
HgO	520–528		520–528	

(From Zeitlin, H., Goya, H., and Waugh, L. T., *Nature,* 198, 178 (1963). With permission.)

power observed for alumina and silica gel in comparison to sodium fluoride is also evident on the basis of the larger shifts of λ_{max} observed for the first two adsorbents. The same group later attempted to extend the polarization theory to organic adsorbates. Mononitrophenols were chosen as the first systems for such an investigation,[16-19] since it was observed visually that they readily underwent a color change upon adsorption on suitable substrates. Starch, talcum, lithium, sodium, and potassium carbonates were used as adsorbents in this study. A large bathochromic shift was observed for the reflectance spectra of *o*-, *p*-, and *m*-nitrophenol with solid carbonates as adsorbents (Table 6.02). With variation of the size of the cation in the carbonate, the shift decreased in the order K > Na > Li. It was suggested[16] that the mononitrophenols exist in an ionized form on the adsorbent surface, but the data were insufficient to fully support this theory. An extension of this investigation was carried out by Zeitlin, Frei, and McCarter.[17] Only *o*-nitrophenol was studied, and a group of alkaline earth oxides served as adsorbents. The role of air humidity and moisture contamination of the adsorbents was closely investigated.

Examination of the reflectance maxima for the *o*-nitrophenol adsorbed on the alkaline earth oxides (Figure 6.01, Table 6.03) revealed bathochromic shifts similar to the ones observed in the previously discussed paper.[16] The order is as expected: BaO > SrO > CaO > MgO > BeO. The beryllium oxide followed this order only in the air-dry state and otherwise behaved in an unpredictable fashion. This unexpected behavior of the beryllium can be explained on the basis of x-ray crystallographic studies. BeO is known to crystallize in the zinc-blende structure, whereas all the other alkaline earth oxides crystallize in the rock salt configuration. For the regularly observed effect of the cations on the mononitrophenols, the following interpretation was advanced: when the

TABLE 6.02

Wavelength Maxima (nm) of the Reflectance Spectra of Mononitrophenols Adsorbed on Alkaline Carbonates[16]

Carbonate	*o*-Nitrophenol	*m*-Nitrophenol	*p*-Nitrophenol
Li_2CO_3	425	331	325
Na_2CO_3	448	347	392
K_2CO_3	488	387	408

(From Zeitlin, H., Kondo, N., and Jordan, W., *Phys. Chem. Solids,* 25, 641 (1964). With permission of Pergamon Press.)

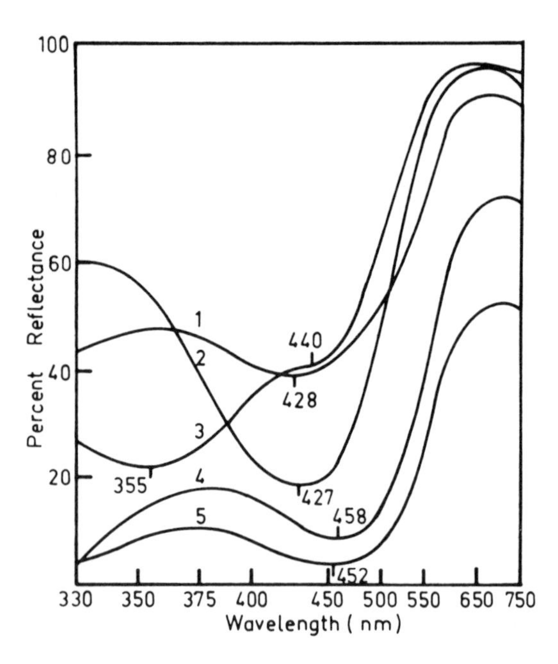

FIGURE 6.01 The reflectance spectra of *o*-nitrophenol adsorbed on alkaline earth oxides preheated to 600°C: $\overset{3}{A}$, on BeO; $\overset{2}{B}$, on MgO; $\overset{1}{C}$, on CaO; $\overset{5}{D}$, on SrO; $\overset{4}{E}$, on BaO.[17] (From Zeitlin, H., Frei, R. W., and McCarter, M., *J. Catal.,* 4, 77 (1965.)

TABLE 6.03

Wavelength Maxima (nm) of the Reflectance Spectra of *o*-Nitrophenol Adsorbed on Alkaline Earth Oxides[17]

		Regeneration temperature		
Oxide	Air-dry	200°C	600°C	1000°C
BeO	416 ± 2	434 ± 3	440 ± 3	444 ± 3
MgO	420 ± 2	420 ± 2	427 ± 2	428 ± 3
CaO	424 ± 2	426 ± 2	428 ± 3	430 ± 3
SrO	434 ± 2	440 ± 2	452 ± 2	ca. 454
BaO	440 ± 2	447 ± 2	458 ± 3	

(From Zeitlin, H., Frei, R. W., and McCarter, M., *J. Catal.,* 4, 77 (1965). With permission of Academic Press Inc.)

alkaline earth metals are arranged in the order of increasing size, e.g., Be to Ba, the effective nuclear charge decreases in this order due to increased screening by additional electron levels. The cation in the crystal lattice of the adsorbent exerts a polarizing action on the phenol as a result of the adsorbent-adsorbate interaction. This polarizing effect of the cations on the nonlocalized bond system of the phenol should increase in the order $Ba^{2+} < Sr^{2+} < Ca^{2+} < Mg^{2+} < Be^{2+}$. Such a polarization should result in a partial freezing or localization of the electron cloud of the phenol, rendering electronic excitation more difficult and thereby accounting for the regular shift of the adsorption peaks toward the more energetic shorter wavelength.

Table 6.03 shows the effect of co-adsorbed moisture and provides some insight into its role. The color change, which can be followed by the displacement of reflectance spectra, depends upon humidity. A regular bathochromic shift takes place in the reflectance spectrum of the phenol-alkaline earth oxide system with increased regeneration temperatures of the adsorbent. These results are in agreement with those of Kortüm et al.[20-22] It is believed that competition for the active adsorption sites takes place between the species undergoing adsorption and the water molecule. The heat treatment of the adsorbent before mixing with the phenol brings about a gradual and progressive elimination of co-adsorbed water. This in turn results in a more effective and direct exposition of the phenol to the polarizing action of the adsorbent cations. The changes are reversible, because the spectra change back to the appearance of air-dry samples upon exposure of the samples to air humidity. Kortüm et al.[20,21] felt that this hypsochromic shift due to the influence of air humidity could be attributed to a transition from chemisorption (taking place in the first monomolecular layer between the adsorbent and the adsorbed species) to physical adsorption (van der Waals' forces). The data in Table 6.03 also reveal a relationship between moisture effect and the size of the cation. The effect is most pronounced with the large cations Ba^{2+} and Sr^{2+}. The behavior of beryllium oxide is again anomalous.

The effect of preheating or regeneration of adsorbent on the position of the adsorption maxima was also investigated with food dyes adsorbed on chromatographic adsorbents.[18] Again, a small bathochromic shift was observed as a result of increasing regeneration temperatures. Zeitlin and Lieu[19] have further extended the scope of this investigation by including di- and trinitrophenols as well as the sodium salts of all the nitrophenols. They attempted to shed some light on the nature of the adsorbed phenolic species and to find more evidence for the earlier hypothesis that the species that actually undergoes adsorption is the phenoxide anion.[16,17] The results are summarized in Table 6.04. A hypsochromic shift is observed in the series from K_2CO_3 to Na_2CO_3 to Li_2CO_3; this agrees with previous observations.[16] For the phenolic salts, however, either no shifts or only very small ones (5 to 6 nm) were found. The following interpretation was advanced for this phenomenon: since the spectra of the salts are those of the phenoxide anions in the adsorbed state, little or no influence on the part of the effective nuclear charges of the carbonate cations can be exerted, which results in a lack of significant shifts. On the other hand, with the mononitrophenols, which are relatively weak acids, it is assumed that the equilibrium condition is in favor of the nonionized phenols in the adsorbed state. However, this equilibrium is still influenced by the effective nuclear charge of the adsorbent cations; consequently, the degree of ionization increases in the order $K^+ < Na^+ < Li^+$.

In conclusion, one can say that the maxima observed for the mononitrophenols on carbonates and, for that matter, also on alkaline earth oxides and other similar adsorbents are believed to be characteristic of an equilibrium state of the free phenol and its ionized form. In the case of the di- and trinitrophenols – which, according to their pK values (see Table 6.04), are much stronger acids – the equilibrium state would be shifted primarily toward the ionized modification, and little differential polarizing effect would be expected from the different adsorbent cations. This is supported again by experimental data (Table 6.04), where shifts are only 16 and 6 nm for the dinitrophenols, and negligible for the trinitrophenol. On the basis of these data, however, it was not possible to establish whether the ionization actually involved a separation of the proton from the phenol molecule or whether stretching of the oxygen-hydrogen bond took place.

In order to broaden the base of these investigations, nitroanilines were substituted for the nitrophenols in another study.[23] As can be seen from Table 6.05 and Figure 6.02, similar shifts were observed as with the previously studied nitrophenols, except that the shifts were less pronounced. No attempt was made to explain the nature of the adsorbed species. Small batho-

TABLE 6.04

Absorption Maxima (nm), pK Values and Total Peak Shifts of Various Phenols Adsorbed on Alkaline Carbonates[19]

Phenols	$LiCO_3$	Na_2CO_3	K_2CO_3	Total shift	pK
o-Nitrophenol	425	448	488	63	7.23
p-Nitrophenol	325	383	408	83	7.14
m-Nitrophenol	331	347	387	56	8.35
2,4-Dinitrophenol	408	418	424	16	4.11
4,6-Dinitro-*o*-cresol	424	422	430	6	4.42
2,4,6-Trinitrophenol	418	418	421	3	0.84

(From Zeitlin, H. and Lieu, V. T., *J. Catal.*, 4, 546 (1965). With permission of Academic Press Inc.)

TABLE 6.05

Absorption Maxima (nm ± 3 nm) of the Reflectance Spectra of *o*-Nitroaniline Adsorbed on Alkaline Earth Oxides[23]

Oxide	Air-dry	Regeneration temperature 200°C	Regeneration temperature 600°C
BeO	409	412	415
MgO	412	416	421
CaO	416	420	424
SrO	419	424	430
BaO	425	434	–
Total shift	16	22	

(From Frei, R. W., Zeitlin, H., and Fujie, G., *Can. J. Chem.*, 44, 3051 (1966). With permission of the National Research Council of Canada.)

chromic shifts were observed as a result of increased regeneration temperatures also; this phenomenon was explained earlier.

In contrast to previous studies with nitrophenols,[16-19] which were carried out on a qualitative basis only, it was attempted in this more recent work[23] to ascertain whether a correlation existed between intensity of the absorption maxima of the mononitroanilines and the adsorbent used. Figure 6.02 shows the relative peak heights for equal mole ratios of *p*-nitroaniline with alkaline earth oxides of 200-mesh particle size and measured under the same conditions (i.e., 300°C regeneration) vs. MgO as a reference standard. The observed trend shows a decrease in color intensity in the order MgO > CaO > SrO > BaO for *p*-nitroaniline, with BeO being the only exception to the rule. Observation of such a trend is not new. Kortüm and Vogel[21] reported an increase in intensity of the absorption maxima of malachite green-*o*-carbonic acid lactone adsorbed on alkaline metal chlorides and sulfates of the same grain size in the order LiCl > NaCl > KCl > RbCl > CsCl and $MgSO_4 > CaSO_4 > BaSO_4$.

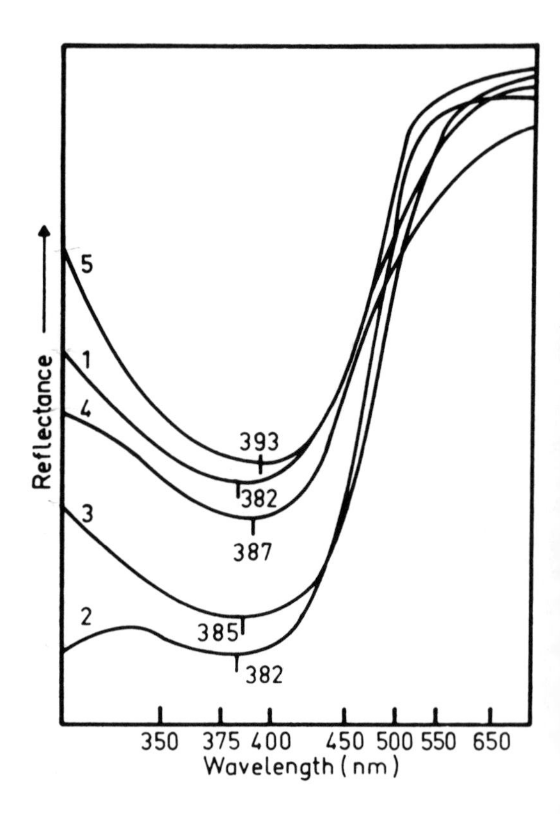

FIGURE 6.02 Reflectance spectra of 0.04 g *p*-nitroaniline adsorbed on 0.05 mole of (1) BeO, (2) MgO, (3) CaO, (4) SrO, (5) BaO of 200-mesh particle size and regenerated at 300°C.[23] (From Frei, R. W., Zeitlin, H., and Fujie, G., *Can. J. Chem.*, 44, 3051 (1966). With permission of the National Research Council of Canada.)

An explanation for the trends exhibited is provided by the polarization theory, whereby the intensity of the color formation is a function of the differentiating nuclear charge exerted by the various cations of the adsorbents. If the adsorption model that was used to explain the influence of moisture on changes in reflectance spectra is valid, it is reasonable to assume that the specific active surface of the adsorbent should exert an influence. The fact that BeO does not fit into the scheme supports such an assumption, since it is know that the alkaline earth oxides, with the exceptio of BeO, crystallize with the rock salt structure. In contrast, BeO has the 4.4 coordinated zinc-blende structure.

Further support is lent to this view by the spectrum shown in Figure 6.03, which exhibits a decrease in peak height with increasing size of the *anion* in the order $MgO > MgCO_3 > MgSO_4$ for *p*-nitroaniline.

Other diffuse reflectance spectroscopic investigations of adsorbed organic systems carried out by Zeitlin and his group included acetone,[3,24] 2,4-dinitrophenylhydrazone,[3,24] 4-methyl-2-pentanone,[3] Michler's ketone[25,26] (4,4′-bisdimethylaminobenzophenone), benzophenone,[26] and dimethylaminobenzaldehyde.[26] The studies were carried out in the UV region of the spectrum, and various commercial adsorbents such as acidic, neutral, and basic alumina as well as silica gel were investigated. Of particular interest were shifts of adsorption maxima as a function of the time of adsorption. These hypsochromic shifts (up to 13 nm),[26] which occurred as time progressed, were attributed to the molecules occupying somewhat lower energy states in the crystal lattice immediately after mixing the sample. Later, when a full state of dispersion has been reached on the active adsorbent surface, a higher energy state can be expected.

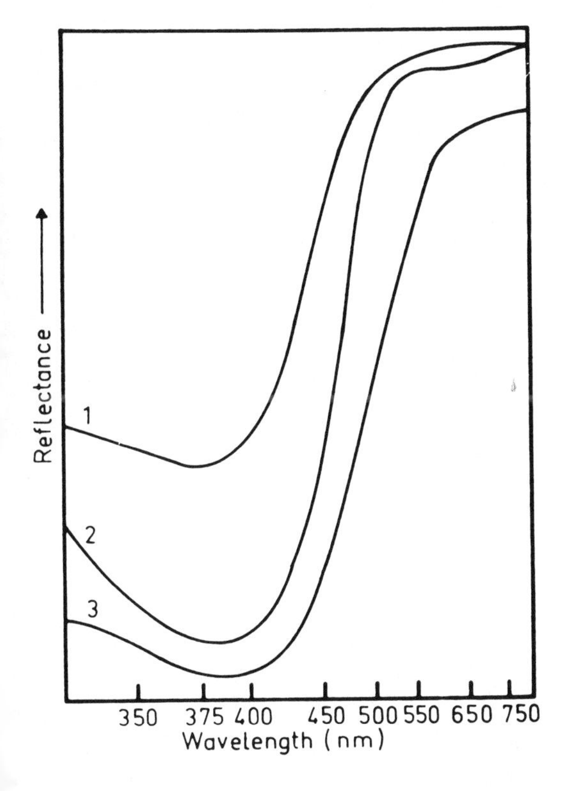

FIGURE 6.03 Reflectance spectra of 0.04 g *p*-nitroaniline adsorbed on 0.05 mole of (1) $MgSO_4$, (2) $MgCO_3$, (3) MgO, of 200-mesh particle size and regenerated at 200°C.[23] (From Frei, R. W., Zeitlin, H., and Fujie, G., *Can. J. Chem.*, 44, 3051 (1966). With permission of the National Research Council of Canada.)

Other attempts to correlate reflectance spectral changes with the degree of adsorbent adsorbate interaction have been made by Schwab et al.[27,28] A group of basic dyes of the benzene-derivative type (e.g., dimethylaminoazobenzene, aminoazobenzene, nitroaniline, nitrodimethylaniline, and others) have been investigated. Bathochromic shifts were obtained for each dye with increasing acidity of the adsorbent (see Figure 6.04) in the order quartz powder < Al_2O_3 air-dried < SiO_2 air-dried < Al_2O_3 dried at 200° < SiO_2 dried at 200° < Al_2O_3 dried at 900° < SiO_2 washed with HCl < pure solid adsorbent < Al_2O_3 washed with HCL < α-Al_2O_3 < bentonite. It was not possible to establish a similar order with the basic adsorbates, and the authors declined to commit themselves to a firm explanation of these phenomena. The varying acidity could be explained on the basis of active hydroxy groups or a Lewis-type acidity in connection with incomplete coordination; on the other hand, the data could also be interpreted on the basis of the polarization theory as discussed by Zeitlin et al.[16-19]

Griffiths et al.[29] have shown some uses of diffuse reflectance spectroscopy in the UV region of the spectrum for a group of inorganic compounds. They used lithium fluoride as a neutral diluent of low activity but noticed that surface phenomena seriously interfered in applications of a strictly inorganic nature.

There is no doubt that G. Kortüm can be credited with having made the major contribution toward the use of diffuse reflectance in the investigation of adsorption phenomena. An excellent discussion of his many investigations in this field is given in his book *Reflexionsspektroskopie*.[5] An English translation of this book has been published recently. Therefore, the discussion of his work is kept as brief as possible.

Transitions from chemisorption to physical adsorption have been investigated by Kortüm et al.[20] with the system *p*-dimethylaminoazobenzene (DMAB) on regenerated $BaSO_4$ and $CaSO_4$ salts.

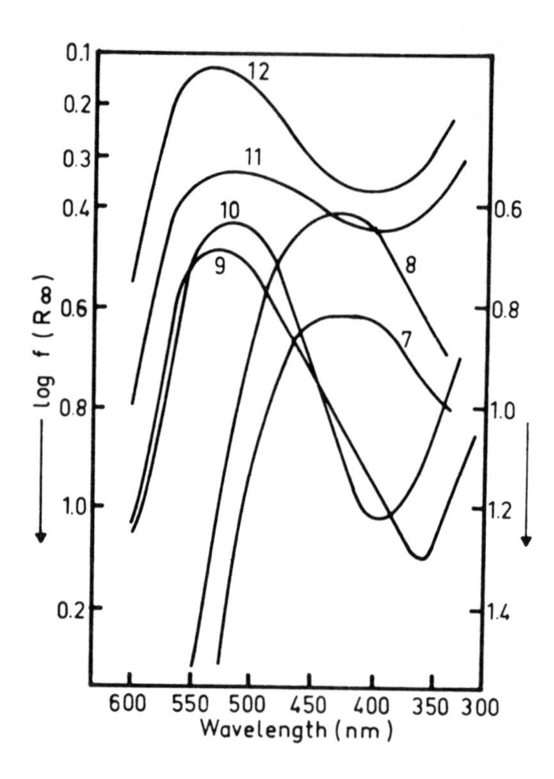

FIGURE 6.04 (7) 0.25 mg *p*-dimethylaminoazobenzene on 1 g Al_2O_3 (air-dry); (8) 0.42 mg *p*-dimethylaminoazobenzene on 1 g Al_2O_3, 200°C; (9) 0.15 mg *p*-dimethylaminoazobenzene on 1 g α-Al_2O_3; (10) 0.15 mg *p*-dimethylaminoazobenzene on 1 g α-Al_2O_3 + HCl; (11) 0.2 mg *p*-dimethylaminoazobenzene on 1 g γ-Al_2O_3, 900°C; (12) 0.2 mg *p*-dimethylaminoazobenzene on 1 g γ-Al_2O_3 HCl acidic.[27] (From Schwab, G. M. and Schneck, E., *Z. Physik. Chem.* (Frankfurt am Main), 18, 206 (1958). With permission.)

On an adsorbent that was heated to sufficiently high temperatures so that all the co-adsorbed water was eliminated, an instant color change of the yellow DMAB to a red form was observed upon adsorption. In Figure 6.05, this red band can be seen between 18,000 and 21,000 cm^{-1}; it is attributed to adsorbent-adsorbate interaction in the first monomolecular layer (chemisorption). With increasing adsorbate concentration, some of the DMAB is adsorbed in the second and subsequent layers by physical adsorption (van der Waals' forces), and as a result one observes the formation and gradual predominance of a second band that appears at 24,000 cm^{-1} and is due to the yellow modification of the DMAB. Kortüm explained this chemisorption process on the basis of a Lewis acid-base interaction, with the adsorbent acting as the acid. This interaction is, of

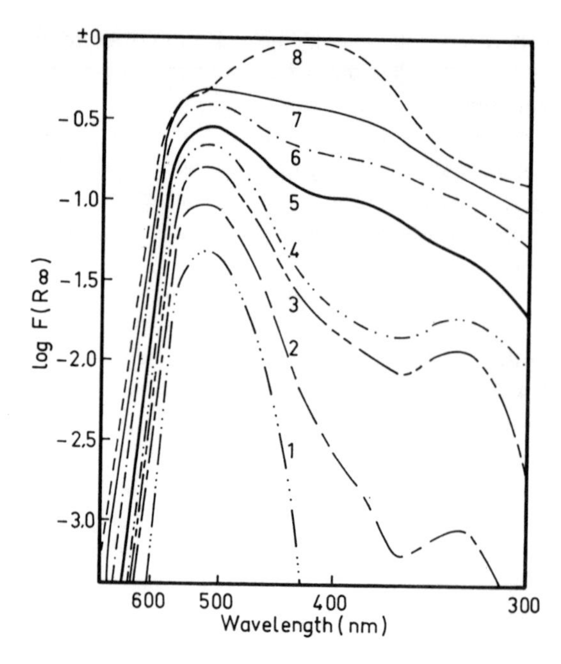

FIGURE 6.05 Reflectance spectra of *p*-dimethylaminoazobenzene adsorbed on dry $BaSO_4$ as a function of concentration. (1) x = 1.40 x 10^{-4}; (2) x = 2.42 x 10^{-4}; (3) x = 4.07 x 10^{-4}; (4) x = 6.69 x 10^{-4}; (5) x = 1.035 x 10^{-3}; (6) x = 1.572 x 10^{-3}; (7) x = 2.531 x 10^{-3}; (8) x = 4.412 x 10^{-3}.[20] (From Kortüm, G., Vogel, J., and Braun, W., *Angew. Chem.*, 70, 651 (1958). With permission of Verlag Chemie.)

course, only possible in the first monomolecular layer.

According to Kortüm,[5] the adsorption of mercury halides on active adsorbents, such as carried out by Zeitlin et al.,[10] can also be classified as a Lewis acid-base reaction, except that in this case the adsorbent acts as a base and the mercuric salt as an acid. A mercury salt that frequently has been investigated in the adsorbed state is HgI_2. For some time, visual color changes have been known to occur with the adsorption of red HgI_2 on various active adsorbents.[7,8] Weitz interpreted the observations with the formation of individual Hg^{2+} and I^- ions due to cleavage of the covalent bond. Zeitlin et al.[13,14] have studied the HgI_2-alumina system in detail with the aid of diffuse reflectance spectroscopy and microscopic, as well as x-ray, diffraction studies. Reflectance spectra of polymorphic forms of HgI_2 on alumina are shown in Figure 6.06. The mercuric iodide mixed with alumina shows a gradual increase in absorbance of the band at 290 nm, whereas the maximum at 575 nm decreases simultaneously until it disappears completely. This change did not

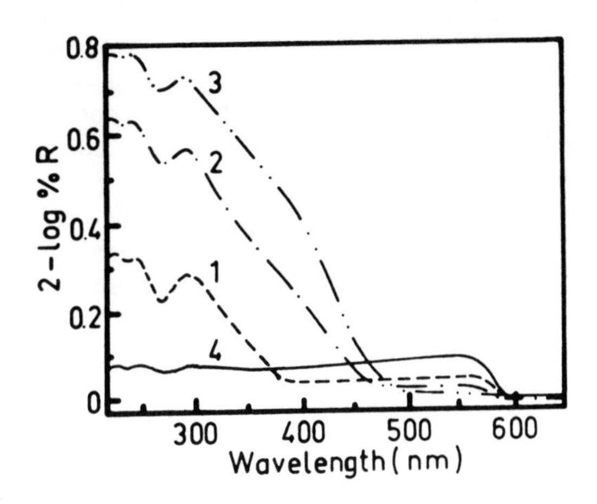

FIGURE 6.06 Comparison of the reflectance spectra of polymorphic forms of mercuric iodide: 1, red mercuric iodide on alumina; 2, red-yellow mercuric iodide on alumina; 3, yellow mercuric iodide on alumina; 4, red mercuric iodide on sodium fluoride.[13] (From Zeitlin, H. and Goya, H., *Nature,* 183, 1041 (1959). With permission.)

take place when HgI_2 was adsorbed on lithium fluoride.[14,30] Upon exposure of the HgI_2-alumina sample to water vapor, a reverse reaction took place, and the band at 575 nm reappeared.[14] On the basis of the combined experimental evidence (including x-ray data), it was concluded that an otherwise metastable (at room temperature) yellow modification of the HgI_2 was being formed upon adsorption of the stable red form. The energy required for this transformation was believed to have been released on adsorption of the mercuric iodide on the alumina, and the yellow form remained remarkably stable in the adsorbed state. This postulation was further supported by the fact that no such change was observed on the relatively inactive lithium fluoride surface. Kortüm has investigated this phenomenon reflectance spectroscopically with a HgI_2-MgO system.[15] His conclusion was that neither of the postulations made by Weitz and Zeitlin et al., respectively, was valid. The similarity of spectra of HgI_2 in a KI-solution and in the adsorbed state, plus the complete absence of Debye-Scherrer HgI_2 lines in an x-ray study of this HgI_2-MgO system prompted him to suggest the existence of a tetrahedral mercury complex on the surface of the adsorbent. This complex was believed to be of a nature similar to the one existing in solution, and it was deemed possible that active sites on the adsorbent could partially displace the I^--ligands on the mercury complex.

Participation of the adsorbent in an actual complexation process has also been suggested by Frei et al.[31] in a spectral reflectance study of some metal chelates in the adsorbed state. The bathochromic shifts observed for the metal chelates of pyridine-2-aldehyde-2-quinolylhydrazone (PAQH) adsorbed on alkaline earth and alkaline carbonates were of the order observed previously:[16-19] $BaCO_3 > SrCO_3 > CaCO_3 > MgCO_3$, and $K_2CO_3 > Na_2CO_3 > Li_2CO_3$, and they were interpreted on the basis of the polarization theory. The complexes investigated were of the following structure:

$$[Co(III)\,(PAQH)_2]^+$$
$$Ni(II)\,(PAQH)_2$$
$$Cu(II)\,(PAQH)_2,$$

all octahedral. (The structure of the iron(III) complex was unknown.) As can be seen from Table 6.06, surprisingly large bathochromic shifts are observed on cellulose and starch. At equilibrium conditions with atmospheric humidity,

TABLE 6.06

Reflectance and Transmittance Maxima (nm ± 2 nm) of PAQH Complexes in Various Media[31]

PAQH complex	EtOH	SiO_2	Al_2O_3	Cellulose	Starch
Fe complex	470	475	490	475	475
Ni complex	475	479	494	518	520
Cu complex	478	483	498	509	508
Co complex	494	500	522	530	530

(From Frei, R. W. and Zeitlin, H., *Can. J. Chem.*, 47, 3902 (1969). With permission of National Research Council of Canada.)

these two chromatographic adsorbents would be classified as relatively weak, and one would expect the maxima to appear near or below those observed for silica gel. Other reflectance spectroscopic studies[32] (see Figure 7.30) have revealed that the copper and nickel species observed on cellulose are probably identical to the 1:1 metal-PAQH chelates reported in solution by Heit and Ryan.[33] When the brownish bis-chelates were applied to a cellulose or starch layer, a change to the 1:1 modification was observed visually to occur within a few seconds. All these observations suggested a more unusual adsorbent-adsorbate interaction than one would ordinarily expect on the basis of polarization. It was therefore postulated that the cellulose- or starch-water complex may be blocking some of the coordination sites on the metal, hence permitting only the formation of a 1:1 species. As a result, the formation of a mixed ligand species, e.g., of the formula

$$[Ni\,(PAQH) - Cellulose{:}(H_2O)_X]$$

with unknown charge, was proposed. Cobalt behaved quite differently because of the 3^+ oxidation state and the charged nature of the complex.

Some of the systems investigated by Kortüm et al.[5,22] were classified as electron donor-acceptor complexes, such as defined by Briegleb.[34] They involved adsorption of iodine and bromine on various adsorbents, e.g., alkaline halides,[5,35] alkaline earth, and aluminum oxide.[36] In the chemisorption process that occurs upon adsorption of the halogens on these powders, the adsorbent acts as electron donor by virtue of its negative ions, and the adsorbed halogen acts (as expected) as an electron acceptor. The formation of complexes of the structure X^-I_2 and X^-Br_2 has been proposed, where X^- could be I^-, Br^-, Cl^-, or OH^-. The reflectance spectra have been found to be in good agreement with similar spectra in solution. For adsorption of iodine on dry Aerosil®,[37] no change in spectrum was observed from that measured for iodine vapor in the non-adsorbed state. This suggests that in the latter case the iodine undergoes only physical adsorption. It was essential to heat the adsorbents in order to eliminate co-adsorbed water, since the presence of water molecules leads to additional bands at higher wavelength, and the spectra become time-dependent. The reflectance spectra of bromine adsorbed on silica gel and Aerosil®, dried at different temperatures, have also been measured.[38] The earlier results were confirmed, and more precise data about the adsorption process were obtained.

A few less common surface phenomena involving redox, ring cleavage, and photochemical reactions have also been reported by Kortüm and co-workers.[21,39-44] The adsorption of benzene on an SiO_2-Al_2O_3 mixed catalyst surface[39] has been studied under vacuum by making use of diffuse reflectance spectroscopy in the infrared region. The resulting spectra gave evidence for the existence of a $C_6H_6H^+$ charge-transfer complex, but in addition a $C_6H_6^{\,+}$ radical was detected, whose formation was attributed to the high electron affinity of the Al^{3+}-ion in the mixed catalyst, which causes an actual redox reaction between the adsorbate and the adsorbent. The coexistence of these two species in turn was attributed to an equilibrium condition of the nature

$$C_6H_6H^+ + \text{catalyst} \rightleftharpoons C_6H_6^{\,+} + \text{catalyst} + H^+,$$

which under suitable conditions permitted both species to be present. Analogous reactions were observed for many other aromatic hydrocarbons. The behavior of triphenylchloromethylene has been investigated on adsorbents such as SiO_2, CaF_2, and $MgSO_4$; as a result of the redox surface interaction, the triphenylmethyl cation was formed.[40] On adsorbents of higher activity, such as Al_2O_3, BeO, MgO, and BaO, the oxidation-reduction process is believed to proceed even further.

Another consequence of an extremely strong adsorbent-adsorbate interaction is the occurrence of a reversible ring cleavage of ring-shaped aromatic compounds. An example of this is provided by the adsorption of malachite green-*o*-carboxylic acid lactone on suitable adsorbents such as alkaline halides. With this system, which has been investigated in detail by Kortüm and Vogel,[21] a blue color had developed upon adsorption of the colorless lactone on a preheated, water-free, adsorbent surface. The blue species was identified by comparison of the reflectance spectrum to solution spectra of the malachite green, in which the lactone ring had been cleaved. The following cleavage reaction is suspected to occur on the adsorbent surface:

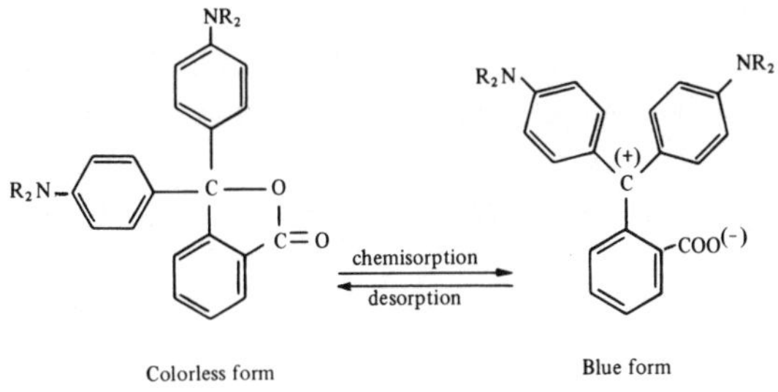

Colorless form ⇌ (chemisorption / desorption) Blue form

(6.01)

The blue form is believed to exist in the adsorbed state as a zwitterion, as evidenced by infrared studies that confirmed the presence of a COO^--group.[41] The desorption occurs at a very slow rate upon subjecting the sample to air saturated with water vapor. The water molecules then compete for the adsorption sites and gradually displace the organic species. The energy required for this ring cleavage is very high and the reaction does not occur readily in solution. It is therefore surprising that upon adsorption the same reaction can take place instantaneously and even at room temperature. In an explanation of this phenomenon, it is believed that the adsorbent exerts a catalytic influence on the system, which results in a substantial decrease of the activation energy, hence permitting the reaction to proceed at room temperature.[21]

Spiropyrans are another group of compounds that undergo reversible ring cleavage at room temperature when adsorbed on a suitable adsorbent. Systems of this nature adsorbed on MgO and NaCl, for example, have been studied by Kortüm and Bayer.[42]

The changes of activation energies, as evident from the previously discussed ring-cleavage reaction, can also significantly influence the occurrence of photochemical processes.[43,44] A blue quinoid-type structure is formed by 2-(2′,4′-dinitrobenzyl)-pyridine, brought about by a tautomeric shift from the methylene bridge to a nitro group when adsorbed on a suitable adsorbent and irradiated. The reaction reverses in the dark to the original colorless form. This reaction and the kinetics associated with it have been investigated by means of reflectance spectroscopy.[43]

Another diffuse reflectance spectroscopic investigation of a similar nature was carried out by Kortüm and Braun.[44] It involved the oxidation of anthracene in the adsorbed state under UV radiation in the presence of oxygen. If adsorbed on alumina, the resulting product, anthraquinone, undergoes a further oxidation process, whose path depends on the properties of the alumina used. Compounds such as chinizarin (1,2-dioxyanthraquinone), alizarine, or chryazin are some of the end products that have been reported.[5,44] The same reactions can occur on silica gel, but with a much slower reaction rate, and on KCl the anthracene is only oxidized to the anthraquinone.

Other studies by Kortüm and his group include the diffuse reflectance spectroscopic investigation of thermochromic and piezochromic properties of compounds such as ethylene, bi-flavylene or dimethyldiacridine adsorbed on MgO and Al_2O_3.[45,46] Occasionally the near-infrared region of the spectrum was employed in these studies. A recent example of the use of diffuse reflectance spectroscopy in this region by Kortüm's group is a study of the reactions of gaseous NO and NOCl with the powdered surfaces of alumina, silica gel, silica-alumina, MgO, and CaO.[47] NO, which has electron-donor-acceptor properties, proved a useful adsorbate for investigating the surface states of the various oxides. The knowledge of the spectra of NO on the various oxide adsorbents also aided in the interpretation of the spectra of adsorbed NOCl. All oxides showed bands at 2,250 cm^{-1} with NO. These bands were of medium intensity and in the cases of silica, alumina, and silica-alumina could be attributed to the interaction of NO with the strained oxygen bridges of the surface by formation of $Me\text{-}O^-$ and $Me \ldots N{\equiv}O^+$. In the case of MgO and CaO, which have cubic structures, the observed bands could be due to an interaction with $Me^{2+}O^{2-}$-groups of the surface. In the alumina and silica-alumina samples, additional bonds at longer wavelength attributable to bonding between NO and Al-atoms (Lewis acid) were also observed.

Ishii et al.[48] have investigated the analytical application of diffuse reflectance spectroscopy in the infrared region of the spectrum to analytical chemistry. A diffuse reflectance attachment was added to an infrared spectrophotometer, and the spectra of several inorganic salts were measured. A linear relationship was found between the reflective absorbance of the compounds studied and the logarithm of their concentrations within the working range studied. The relationship was not influenced by specular reflectance.

The use of spectral reflectance techniques for the investigation of surface phenomena can be of considerable interest to the analytical chemist.

Determination of compound stability, position, and intensity of reflectance spectra as a function of temperature of the environment,[49] regeneration temperature of the adsorbents,[18,31] drying temperatures for the adsorbent-adsorbate systems,[32] time study of the effect of storage,[49] influence of light, etc. are but a few examples of studies that can aid the analyst in finding out the optimum analysis conditions for systems of interest. Other parameters investigated in a similar fashion are pH dependence[32] and the influence of spray-reagent composition[32] on the conditions of analytical systems in the adsorbed state. These types of investigation have been found to be particularly useful in connection with chromatographic systems for in situ evaluation of the separated compounds. They are discussed in Chapter 7.

Let us now consider a practical example where diffuse reflectance spectroscopy has been used to advantage in a study of problems arising in the drug industry. Lach and co-workers have studied the interactions between the various components in a number of medicinal formulations, and their results have led them to make some very interesting suggestions with regard to screening programs which could be implemented in the preparation of new formulations.[50–56]

Lach's interest in the problem arose from the consideration that solid-solid interactions, such as chemisorption, between the active ingredients in medicinal formulations and the inert adjuvants might affect the action of the preparation. Diffuse reflectance spectroscopy was used to study the interactions between oxytetracycline, phenothiazine, anthracene, and salicyclic acid and various common adsorbents, such as magnesium trisilicate, magnesium oxide, and alumina.[50] Spectral evidence for interactions believed to be charge-transfer complex formation was obtained for the various combinations of drug and adjuvant tested. Typical of the methodology is the following example. From 15 to 100 mg of the active ingredient, such as oxytetracycline, was added to 2 g of the inert material, such as magnesium trisilicate. The powders were placed in a 50-ml vial with 25 ml of water to act as a dispersing medium. The bottle was capped and the sample was equilibrated for 24 hr. The suspension was then filtered; the powder was dried in a vacuum oven, and the diffuse reflectance spectrum of the powder sample was measured. The spectra were measured using a Beckman DU spectrophotometer with diffuse reflectance sphere attachment. The samples were held in two plastic, black-coated cells that had a center circular hole 1 in. diam. and 1/8 in. deep. Magnesium carbonate was used as a reference standard for all measurements. As a control, 2 g of the inert material was physically mixed with the indicated amount of active material.

The diffuse reflectance spectra obtained for the oxytetracycline-magnesium silicate mixture showed a significant difference between the spectrum of the equilibrated mixture and the physically mixed sample and also from those of the individual components. The equilibrated mixture was straw yellow in color, while the physical mixture was white or faintly yellow. The diffuse reflectance spectra revealed a bathochromic shift, with a new shoulder for the equilibrated mixture as compared to the control in the region 400 to 700 nm, as well as decreased reflectance. The reflectance in the UV-region was also markedly decreased, with a new band appearing at 315 nm. This band may have been due to an interaction between oxytetracycline and the adsorbent, with the adsorbent clarifying an already existing peak in this region, or it may be the reflectance spectrum of a film of the drug adsorbed on the adjuvant. Similar results were obtained for other systems studied. The samples were also subjected to compression to simulate the pill-making process, and similar results were obtained. Thus, there was clear evidence of interaction between the active ingredients and the adjuvants in the systems studied.

Lach and Bornstein next investigated the interactions between anthracene, prednisone, and hydrochlorothiazole and a number of metallic adjuvants, such as alumina, and nonmetallic adjuvants, such as stearic acid.[51] Visual color changes were again noted, with the diffuse reflectance spectra revealing bathochromic and hyperchromic shifts in the equilibrated samples as compared to the controls. The spectral changes were attributed to charge-transfer chelation. The interactions of oxytetracycline and oxytetracycline HCl with a number of metallic and nonmetallic adjuvants were also investigated using the previously outlined methodology.[52] Equilibration, compression, and moisture techniques were again applied, and the results indicated that a metal ion or polyfunctional adsorbent molecule is necessary for interactions to occur.

The results of these studies indicated to the investigators that a preliminary screening program to discover possible interactions between the active ingredients and inert fillers in drug preparations might be desirable. Data were not available on rates of adsorption and dissolution, but the authors suggested that some thought should be given to the problem. When the dosage contains a small amount of drug in relation to the amount of adjuvant, the chemisorption interactions between the two could significantly alter the dose response obtained. This could be important in various formulations, including pills, capsules, suspensions, and ointments. An alternate possibility, which must also be considered, is that the active ingredient may itself become coated with a layer of adjuvant, thus further complicating the problem.

Lach et al. next investigated the possibility of interactions between some common dyes used in drug preparations and a variety of adjuvants.[53] It had been noted that the colors of some preparations tended to change or fade. Reflectance studies again showed that interactions could occur when a metallic or polyfunctional adjuvant was present. The general order of metallic adjuvant interaction with the dyes was found to be Mg(II) > Ca(II) > Zn(II) ~ Al(III). The reflectance data were supported by visible changes in color, elution studies, and accelerated fading techniques.

Studies were also performed on ferrous sulfate equilibrated in both aqueous and nonaqueous systems with a variety of metallic adjuvants, and significant spectral variations were again observed.[54] The conclusion was that iron salts in dosage forms may undergo chemisorption reactions with the adjuvants, so that the concentration of iron found by analysis may not actually be available for a therapeutic response. Diffuse reflectance spectra of systems including tetracycline and its derivatives, bis-hydroxycoumarin, and methantheline bromide with various metal adjuvants revealed spectral changes that varied in their intensity.[55] While some systems showed relatively minor changes, others revealed large bathochromic and hyperchromic changes and new band formation. In a further study, the metallic chelates of bis-hydroxycoumarin and furosemide were prepared and isolated, and their diffuse reflectance spectra proved to be comparable to those of the equilibrated drug-adjuvant systems.[56] Thus, the results indicated that the interactions observed between the active ingredients and adjuvants in drug formulations proceed by a mechanism similar to chelation.

The studies of Lach and co-workers illustrate the analytical usefulness of diffuse reflectance spectroscopy in studying surface interactions. In this case, the investigations have revealed that a knowledge of the possible interactions between the various components of medicinal formulations may result in a better product and could conceivably result in a considerable saving to the manufacturer (not to mention the consumer) if more efficient and stable preparations could be achieved.

Another recent example where surface interactions have been studied by diffuse reflectance spectroscopy was reported by Hakusui et al.[57] The spectra of colored clays were measured in the presence of water, methanol, and benzene in an attempt to identify the species responsible for the coloration of acid clay upon the addition of benzidine and some related aromatic diamines. The color appeared to be due to the formation of mono- and di-valent cations through the oxidation of the diamines, the mode of interaction of the cation depending on the kind and amount of liquid present. The lightening of the color which was observed upon drying the samples was attributed to the reaction

$$D^{2+} + DH^{+} \rightleftharpoons 2D^{+} + H^{+}.$$

Some applications of a different nature are briefly discussed in the following three sections.

c. Determination of Active Surface Areas of Powders

Some of the observations discussed in the previous section make it evident that chemisorption can only occur in the first monomolecular layer and that the adsorption forces in the second and subsequent molecular layers could be attributed to relatively weak van der Waals' forces (physical adsorption). This shift from one adsorption state to another frequently results in drastic color changes. Typical examples are the systems *p*-dimethylaminoazobenzene:$BaSO_4$,[20] which was interpreted as a Lewis acid-base interaction and malachite green *o*-carboxylic acid lactone (MGL):NaCl,[21] where the adsorbent-adsorbate interaction is believed to be a reversible ring cleavage. Kortüm and Oelkrug[58] have put this phenomenon to practical use for the determination of active surface areas of powder adsorbents. Quantitative spectral reflectance investigations of this system (Figure 6.07), followed by plotting of the Kubelka-Munk function $F(R_\infty)$ vs. adsorbent/adsorbate mole ratios over a large concentration range, result in the calibration curves as depicted in Figure 6.08. This isothermal plot can be interpreted as follows. Up to about 5×10^{-4} mole fraction the Kubelka-Munk function retains its validity. With a further increase in adsorbate concentration, a deviation from linearity is observed due to adsorption of some molecules in the second and subsequent layers (physical adsorption). The curve finally reaches a limiting value, at which point the first monomolecular layer is believed to be completely covered. Extrapolation of the two straight branches of the plot enables the determination of this saturation concentration, which provides a relative measure of the surface area of the adsorbent. Kortüm and Oelkrug[58] measured the malachite green lactone on dry NaCl of different particle size. For the larger particle size a saturation concentration x_{mA} 3.85×10^{-4} and for the smaller particle size (larger surface area) a saturation concentration of x_{mB} 4.33×10^{-4} were reported. This corresponds to a relative ratio of surfaces $x_{mB}/x_{mA} = 1.12$. The same surfaces were measured by the BET method, using N_2 gas as an adsorbate. The relative ratio obtained by this method was 1.11, which is in excellent agreement with the diffuse reflectance spectroscopic method. The measurement of absolute values is of course impossible because it is not known how the complex organic molecules adhere to the surface of the adsorbent. Nevertheless, it is believed that this method has good potential, e.g., for the determination of relative surface areas of chromatographic adsorbents. The large organic molecules that would be used in such a study are similar to the majority of compounds in chromatographic separations, and the resultant surface-area measurements would be more realistic

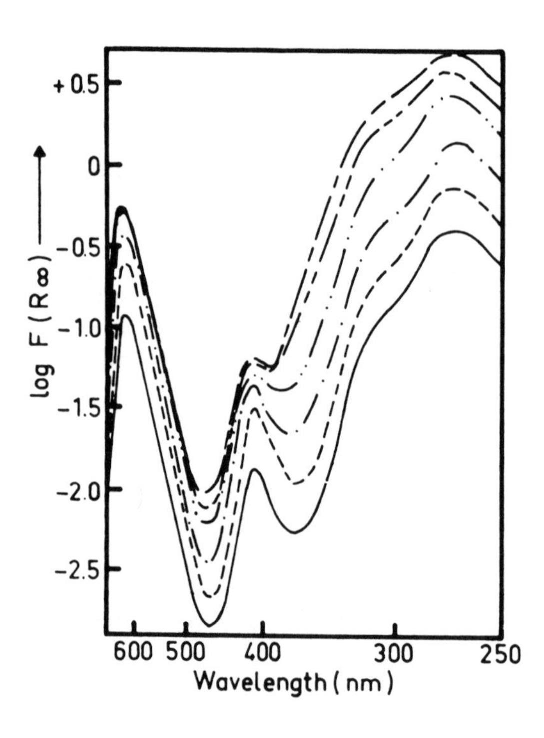

FIGURE 6.07 The dependence of Kubelka-Munk functions of several bands of MGL on concentrations of MGL adsorbed on dry NaCl. Concentration in mole fractions: $x = 1.80 \cdot 10^{-4}$ $x = 3.80 \cdot 10^{-4}$ $x = 7.39 \cdot 10^{-4}$ $x = 14.74 \cdot 10^{-4}$ $x = 23.86 \cdot 10^{-4}$ $x = 44.79 \cdot 10^{-4}$

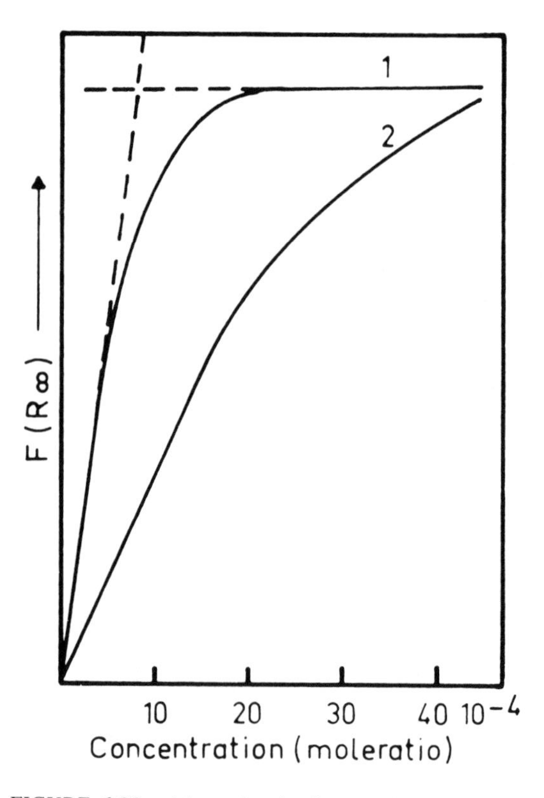

FIGURE 6.08 Adsorption isotherms of MGL adsorbed on dry NaCl.[21]
(1) Chemisorption (2) Physical adsorption. (From Kortüm, G. and Vogel, J., *Chem. Ber.*, 93, 706 (1960). With permission of Verlag Chemie GMBH.)

than if measured by the BET method with the use of nitrogen as an adsorbate. Also, nitrogen tends to penetrate into cavities that are not readily accessible to larger molecules; thus, unrealistically high surface values may be obtained. Knowledge of reliable data on relative surface areas of a wide range of chromatographic adsorbents would facilitate the choice of the right adsorbent and the prediction of loading capacities, thus preventing overloading of adsorption chromatographic systems.

d. Kinetic Studies on Surfaces

Many reactions occur at such fast rates that kinetic studies have to be carried out at low temperatures or with rather involved flash techniques. When the same systems are adsorbed on suitable powder surfaces, the same reaction may occur, but at a much slower rate. Kinetic processes of this nature can therefore conveniently be studied with diffuse reflectance spectroscopic techniques. In fact, hardly any other method is available for the investigation of kinetic processes in the solid or adsorbed state. A specific example has been studied by Kortüm et al.,[43] with the reversible photochemical reaction of 2-(2′,4′-dinitrobenzyl)-pyridine adsorbed on substrates such as NaCl, LiF, and SiO_2. The reverse reaction, going from blue to colorless upon storing the system in the dark (see Equation 6.02), was investigated in this study.

Blue form ⇌ ($h\nu$) Colorless form

(6.02)

The procedure involved sealing off the samples in quartz cells; after irradiation they were kept in thermostated containers. The reflectance was measured at 600 nm at certain time intervals, and pure adsorbent (e.g., silica gel) was used as a reference standard. The Kubelka-Munk functions were obtained from absolute reflectance measurements as derived from the following equation:

$$R'_{\infty sample} = \frac{R_{\infty sample}}{R_{\infty SiO}} \times \frac{R_{\infty SiO_2}}{R_{\infty MgO}} \times R'_{\infty MgO}. \qquad (6.03)$$

R'_∞ stands for absolute values of MgO and sample, and R_∞ stands for relative experimental values for the sample measured vs. SiO_2 and SiO_2 measured vs. MgO. Plotting the logarithm for the Kubelka-Munk function $F(R'_\infty)$ vs. time for four different temperatures resulted in the kinetic plots presented in Figure 6.09. Two first-order reactions were obtained for each temperature. One can be attributed to the reaction shown in Equation 6.02 and the other to the formation of a colorless cation formed on account of the OH group on the silica gel (see structure 6.04).

(6.04)

For adsorbents such as LiF and NaCl only one straight line plot, which is attributable to the first reaction (Equation 6.02), was obtained. The kinetic constant k was then calculated in the usual manner by taking the slopes of the kinetic plots presented in Figure 6.09. Activation energies were computed from the slopes of another set of curves obtained by plotting the logarithm of k vs. 1/T (T, absolute temperature). The equation for the computation of the energy term is seen below.

$$E = -\frac{2.30\ R\Delta \log k}{\Delta\ (1/T)} \qquad (6.05)$$

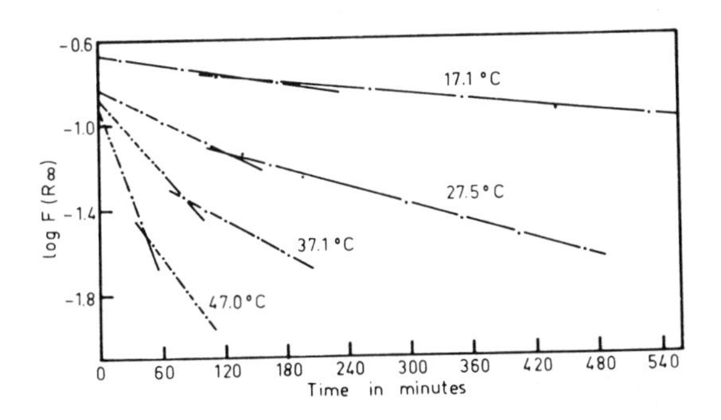

FIGURE 6.09 Fading reaction of 2-(2′,4′-dinitrobenzyl)-pyridine as a function of temperature: plot of log $F(R_\infty)$ versus time in minutes at temperatures of 17.1, 27.5, 37.1 and 47.0°C.[43] (From Kortüm, G., Kortüm-Seiler, M., and Bailey, S. D., *J. Phys. Chem.*, 66, 2439 (1962). With permission of the American Chemical Society.)

For the reactions of 2-(2′,4′-dinitrobenzyl)-pyridine adsorbed on silica gel, the activation energy for one reaction was determined as 15 kcal/mole and for the other as 17 kcal/mole. These values are about three times larger than the corresponding activation energies measured in solution.

Kinetic methods of analysis for systems in solution have become quite well known.[59] It is conceivable that the elegant technique discussed above could be adapted by the analytical chemist to similar catalytic phenomena occurring in the solid state, e.g., on chromatographic adsorbents.

e. Determination of Equilibrium Constants

The knowledge of equilibrium constants for association or dissociation equilibria of compounds in the adsorbed state could be useful for general analytical problems related to such systems or to problems of catalysis or kinetics, for example.

Kortüm and Braun[60–63] have carried out several experiments to study the feasibility of diffuse reflectance spectroscopy as a means of measuring equilibrium constants in the solid state. The systems under investigation were anthracene-*s*-trinitrobenzene and pyrene-*s*-trinitrobenzene adsorbed on NaCl and SiO_2.

For similar problems in solution, deviations from the Bouger-Lambert-Beer law can be used in certain systems for the computation of equilibrium constants. Kortüm and Braun used deviations from the Kubelka-Munk theory at low concentration of adsorbate to calculate analogous data in the adsorbed state. A suitable maximum (charge transfer bands) had to be chosen. An iteration procedure[60] was used to compute the constants (for details see Reference 5), which in general were found to be several orders of magnitude smaller than corresponding constants measured in solution (e.g., chloroform). Attempts to measure dissociation constants of the same systems at elevated temperatures were not very successful[61] because of difficulties in the control of moisture content of the samples. In addition, an irreversible oxidation of the aromatic molecules was observed on highly regenerated silica gel, which rendered these systems unsuitable for examinations of this nature. On the other hand, if perfectly dry sodium chloride was used as an adsorbent, the dissociation equilibrium was found to be independent of the temperature. Re-examination of these systems[62] revealed that the deviation from linearity on Kubelka-Munk plots at low concentration of pyrene-*s*-trinitrobenzene on NaCl was not attributable to the formation of a molecular complex (as predicted earlier[60]) but was merely a result of self-absorption of the NaCl. Once the relative reflectance values were corrected to absolute values (see Equation 6.03), no such deviation was found. This discounts completely the possibility of a dissociation occurring on the surface of the sodium chloride. For silica gel, however, evidence that such a dissociation could occur was still strong, and re-examination of the silica gel:hexamethylbenzene-*s*-trinitrobenzene system carried out by Braun and Kortüm in a more recent project[63] gave further support to this postulation. The mathematical procedures employed in this work were similar to those mentioned previously.[60] Analogous procedures have been used for the determination of equilibrium constants for adsorbed iodine molecules (I_4).[35]

In conclusion, it can be said that the dissociation phenomenon is believed to be strongly dependent on co-adsorbed water. This is a condition needed for the establishment of a true equilibrium in the two-dimensional boundary phase. In other words, the molecules should be able to move freely on the surface, which is impossible if by chemisorption they are rigidly bonded to the crystal lattice of the adsorbent. This view is supported by the observation that no dissociation occurs on completely dry sodium chloride surfaces, for example.[62]

2. Spectral Reflectance at High and Low Temperatures

Most of the work in diffuse reflectance spectroscopy has been carried out at ambient temperatures. Even in cases where samples were heated in order to investigate certain temperature effects on kinetic processes, removal of co-adsorbed moisture, irreversible decomposition, and other chemical changes, the samples will have cooled down to ambient temperatures by the time measurements are carried out. The discussion in this section is, therefore, mainly concerned with diffuse reflectance measurements made while the sample is maintained at higher or lower temperatures with the use of suitable cells, heating blocks, and cooling devices. With these methods, valuable additional information can often be obtained on a particular system.

a. High-temperature and Dynamic Reflectance Spectroscopy

Spectral reflectance work at higher temperatures is far better known than studies at reduced temperature levels. The major investigators in this field are Wendlandt and co-workers,[64–75] and several reviews[64–67] are now available on this topic.

Two approaches to high-temperature reflectance spectroscopy have been reported: (a) measurement of reflectance spectra at fixed temperature (isothermal operation), which will generally be referred to as high-temperature reflectance spectroscopy (HTRS),[68] and (b) a dynamic method of continuous reflectance measurement at a fixed wavelength over a certain temperature range, which is referred to as dynamic reflectance spectroscopy (DRS).[69] The working principle of these two modes of operation is depicted in Figures 6.10 and 6.11. Figure 6.10 represents HTRS curves for a system that is being heated to temperatures increasing in the order T_1, T_2, T_3, T_4. . . . T_n, and at each temperature interval a spectrum is recorded. As one can see, the maximum at wavelength λ_1 increases with an increase in temperature at the expense of the other maximum at λ_2. Although these spectra already can provide valuable information about thermal transition phenomena of the sample, it is far more practical and accurate to determine minimum and maximum thermal transition temperatures with the aid of dynamic reflectance spectroscopy. The DRS curves corresponding to Figure 6.10 are shown in Figure 6.11a and b, measured at fixed wavelengths λ_1 and λ_2, respectively. These "isolambdic" curves (% reflectance vs. temperature) show the exact temperature for the start and the end of a thermal transition. In other words, DRS yields data similar to those obtained with other established thermal analysis techniques, such as differential thermal analysis DTA, thermogravimetric analysis TGA, and others. Since it does not interfere with thermodynamic and gravimetric effects, it can be regarded as a valuable complementary method for the study of thermal stability and the structural changes as a function of temperature changes.

The construction of sample holders suitable for DRS and HTRS work, as well as a brief description of the instrumental setup, has been given in Chapter 4, Section 3c. Sample preparation pro-

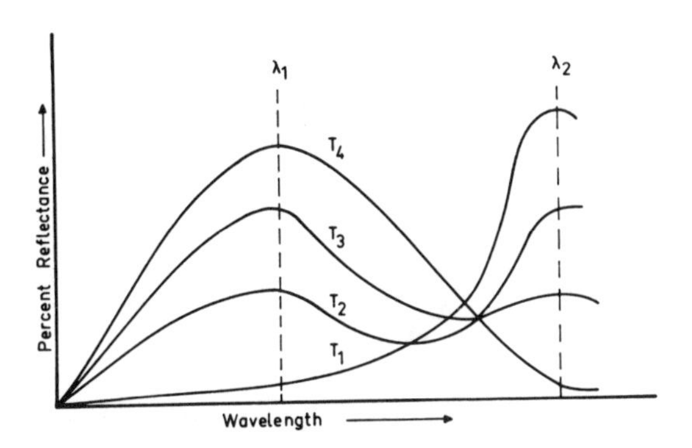

FIGURE 6.10 High-temperature reflectance spectra curves.

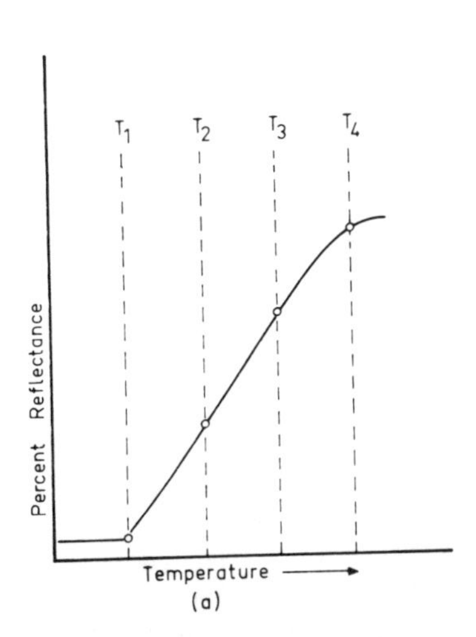

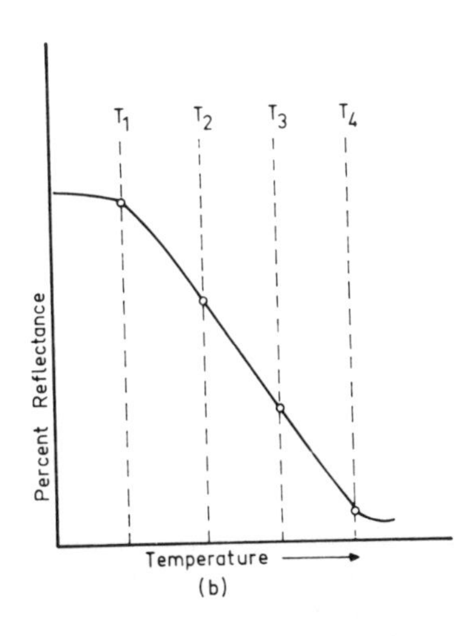

FIGURE 6.11a,b Dynamic reflectance spectra curves, at λ_1 and λ_2, respectively.

cedures can be similar to those described in this chapter in Section 1a. Applications of these techniques have been largely restricted to inorganic systems, even though this need not be the case and is coincidental with the particular research interests of the investigators. The spectral ranges to which these studies have been applied include the UV visible and near-infrared.

Two typical examples have been selected for a more detailed description of an actual investigation by HTRS and DRS techniques. One deals with the deaquation of copper sulfate ($CuSO_4 \cdot 5H_2O$)[67] and the other with solid-solid interaction phenomena between cobalt chloride and cobalt bromide adsorbed on potassium halides.[70]

The deaquation process is recorded in Figures 6.12 to 6.14. From Figure 6.12 it can be seen that the absorption maximum observed at room temperature at 680 nm is shifting to a higher wavelength upon heating, until it reaches a maximum of 715 nm at 135°C. At this temperature a quantitative deaquation of the $CuSO_4 \cdot 5H_2O$ to the monohydrate has taken place. The last water molecule would not be driven off until temperatures higher than 250°C have been reached. Upon inspection of the corresponding DRS curve (Figure 6.13), which was measured at a wavelength of 625 nm, one observes a major increase in reflectance, starting at 105°C and then leveling off to a slightly decreasing slope at about 125°C. The actual deaquation process could therefore be expected to occur between 105 and 120°C. Figure 6.14 also shows one of the rarer examples of a diffuse reflectance spectrum in the near-infrared region, which in studies of this nature can yield useful information, particularly when inorganic salts or coordination compounds are studied. In the case of $CuSO_4$, for example, the absorption maxima observed at room temperature are positioned at

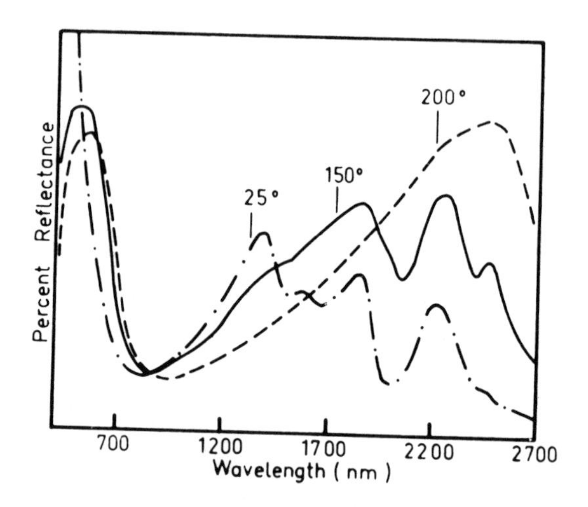

FIGURE 6.13 DRS curve of $CuSO_4 \cdot 5H_2O$ at 625 nm (heating rate 6.7°C/min).[67] (From Wendlandt, W. W., in *Modern Aspects of Reflectance Spectroscopy,* Wendlandt, W. W., Ed., Plenum Press, New York, 1968. With permission.)

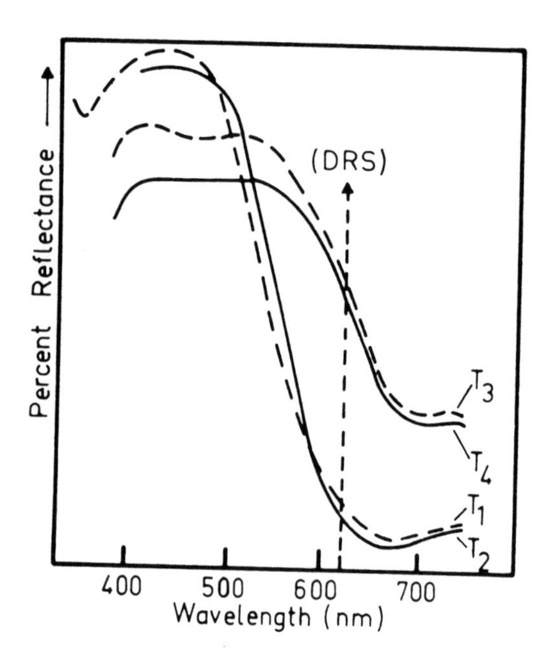

FIGURE 6.12 HTRS curves of $CuSO_4 \cdot 5H_2O$ in the visible wavelength region.[67] T_1, 25°; T_2, 100°; T_3, 135°; T_4, 160°C. (From Wendlandt, W. W., in *Modern Aspects of Reflectance Spectroscopy,* Wendlandt, W. W., Ed., Plenum Press, New York, 1968. With permission.)

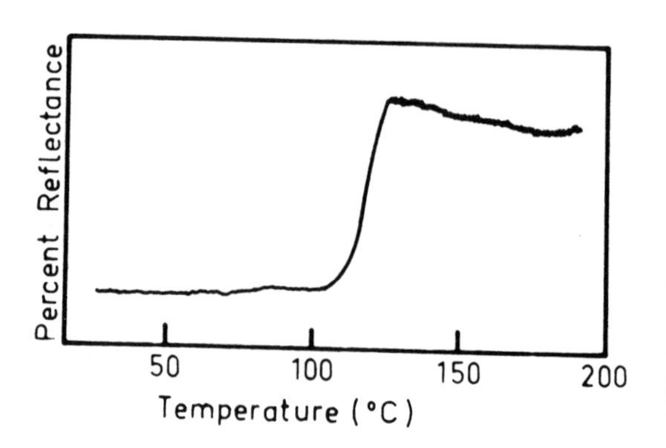

FIGURE 6.14 HTRS curves of $CuSO_4 \cdot 5H_2O$ in visible and near-infrared wavelength regions.[67] (From Wendlandt, W. W., in *Modern Aspects of Reflectance Spectroscopy,* Wendlandt, W. W., Ed., Plenum Press, New York, 1968. With permission.)

1510, 1675 and 2000 nm, respectively, whereas at 150°C a new band at 2400 nm is observed, and the first two bands have disappeared. At 200°C all the bands observed at ambient temperature have vanished and a single absorption maximum at about 2500 nm appears instead.

Other deaquation processes that have been studied in a similar manner include chromium(III) and cobalt(III) complexes of the type $[M(NH_3)_5 \cdot H_2O]X_3$ (X: Cl^-, Br^-, I^-, NO_3^-).[70,72] The deaquation processes observed with these systems were of the general reaction type as shown in Equation 6.06:

$$[M(NH_3)_5 \cdot H_2O]X_3 \rightarrow [M(NH_3)_5 \cdot X]X_2 + H_2O. \quad (6.06)$$

The solid-solid-type reaction between cobalt chloride and potassium chloride[70] can be interpreted from the data presented in Figures 6.15 and 6.16. Spectrum #1 in Figure 6.15 was obtained on rapidly increasing the temperature to 55°C. After standing for 30 min at 55°C, spectrum #2 was obtained; further heating to 110°C resulted in spectrum #3. The compound associated with the first spectrum was obviously not stable due to the loss of crystal water, and after 30 min $CoCl_2 \cdot 2H_2O$ was produced, as indicated by the appearance of spectrum #2, which has been found to be identical with the spectrum of pure $CoCl_2 \cdot 2H_2O$. The final solid-solid interaction between the cobalt salt and the KCl occurred on further heating to 110°C, at which temperature the third spectrum was produced. Spectrum #3 is identical with the spectrum of pure K_2CoCl_4. Again, additional information can be gathered on examination of the corresponding DRS curve produced at 675 nm (Figure 6.16). The changes in structure and composition for this system could be summarized as follows: $CoCl_2 \cdot 6H_2O$ (octahedral) → K_2CoCl_4 (tetrahedral) → $CoCl_2 \cdot H_2O$ → K_2CoCl_4. Evidence for this sequence is given by the dip observed in curve A at 70°C, which appears to be the result of an intermediate formation of the octahedral $CoCl_2 \cdot 2H_2O$ species. Thus, the DRS curve easily permits the interpretation of the instability observed at 55°C (Figure 6.16, curve 1). $CoCl_2 \cdot 6H_2O$ was later investigated without the presence of KCl,[56] and it was concluded that the spectrum observed at 55°C could be the result of a mixture of octahedral $CoCl_2 \cdot H_2O$ and tetrahedral $Co[CoCl_4]$. The final product obtained at 155°C was believed to be anhydrous octahedral $CoCl_2$.

Further investigations on octahedral and tetrahedral Co(II) complexes, carried out by Simmons and Wendlandt,[73] tend to support the conclusions

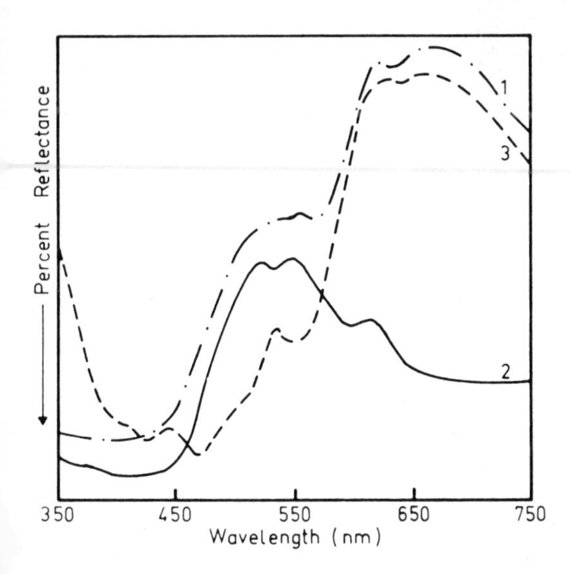

FIGURE 6.15 HTRS of the 90% $CoCl_2 \cdot 6H_2O$-10% KCl system.[70] 1, 55° 1 min; 2, 55° 30 min; 3, 110°C. (From Cathers, R. E. and Wendlandt, W. W., *Chemist-Analyst*, 53, 110 (1964). With permission of J. T. Baker Chemical Co.)

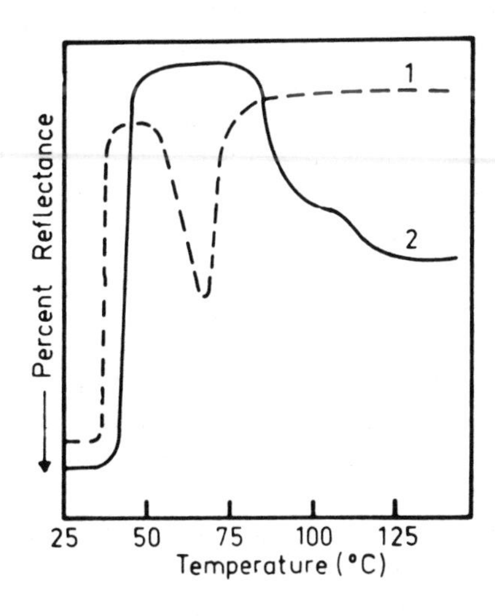

FIGURE 6.16 DRS curve of the 90% $CoCl_2 \cdot 6H_2O$-10% KCl mixture at 675 nm, heating rate 2°C/min. 1, in 90% KCl; 2, in 60% Al_2O_3.[70] (From Cathers, R. E. and Wendlandt, W. W., *Chemist-Analyst*, 53, 110 (1964). With permission of J. T. Baker Chemical Co.)

reached in the previously discussed work. The conversion of the a form of $Co(py)_2Cl_2$ to the β form, which upon heating involves a change from a polymeric to a monomeric structure, has also been investigated by Wendlandt.[74]

Another possible use of the DRS method is the investigation of thermochromic transitions. An example for an application has been reported by Wendlandt and George[75] and involves such compounds as $Cu_2[HgI_4]$, $Ag_2[HgI_4]$, and AgI. The first of these compounds, for example, exhibits a reversible color change from a red to a black modification upon heating. DRS studies showed a continuous decrease in reflectance when going from room temperature to about 80°C, with the most drastic color change taking place at about 70°C. Recently Wendlandt[76] discussed an extension of his studies of thermochromic systems. DRS was applied to the study of compounds such as $Tl_2[HgI_4]$, $Pb[HgI_4]$, HgO, HgS, and $Cu_2[Hg(SCN)_4]$. All compounds showed drastic reversible color changes when heated in the temperature range 50 to 200°C.

Wendlandt and Bradley have reported on the use of a dynamic reflectance spectroscopy (DRS) – gas evolution detection (GED) system in the study of the dehydration of $CuSO_4 \cdot 5H_2O$ and the deaquation and ananation of $[Co(NH_3)_5H_2O]Cl_3$.[77] The DRS sample cell described by Wendlandt and Dorsch[78] (see also Chapter 4) was connected to a Carle Model 100 Micro-Detector system, linked to one channel of a Leeds and Northrup four-channel recorder. The output from a Beckman DK-2A spectroreflectometer and also the output from a thermocouple in the DRS heater block were fed into other channels of the recorder. From 50 to 200 mg of sample was placed in the cell, and the system was flushed with helium until a stable recorder baseline was obtained. The flow rate was then set at from 40 to 60 ml/min and the cell was heated at 5 to 10°/min, while the temperature of the DRS cell, the DRS curve at the wavelength of measurement, and the GED curve were plotted simultaneously.

$CuSO_4 \cdot 5H_2O$ produced two GED peaks and two distinct increases in reflectance in the same temperature ranges as the GED peaks. The first GED peak was attributed to the reaction

$$CuSO_4 \cdot 5H_2O(c) \rightarrow CuSO_4 \cdot 3H_2O(c) + 2H_2O\ (1)$$

$$2H_2O\ (l) \rightarrow 2H_2O\ (g),$$

while the second peak was believed due to the reaction

$$CuSO_4 \cdot 3H_2O(c) \rightarrow CuSO_4 \cdot H_2O(c) + 2H_2O\ (g).$$

In the case of the cobalt complex, the peak was attributed to the reaction

$$Co(NH_3)_5H_2O\,Cl_3\,(c) \rightarrow Co(NH_3)_5Cl\,Cl_2\,(c) + H_2O\ (g).$$

In this reaction, water is removed from the coordination sphere and replaced by a chloride ion from the ionization sphere of the complex.

Using these techniques, Wendlandt and Bradley suggest that one should be able to correlate the DRS curves with the dissociation reaction producing volatile products and the structural change in the compound, a chemical reaction not involving volatile products or thermochromic behavior. DRS has the advantage over TGA and DTA in that it monitors only one reaction at a time, thus eliminating the effects of simultaneous side reaction s. Wendlandt has also studied the dehydration of $CuSO_4 \cdot 5H_2O$ using microreflectance techniques.[79]

Wendlandt and Bradley have also applied the techniques of HTRS and DRS to a study of the unstable aquopentammine complexes of cobalt(III) and chromium(III).[80] The cobalt chloride complex has a reflectance spectrum with an absorption band at about 485 nm at room temperature, but at 100° this is shifted to about 525 nm. The shift is caused by the deaquation and ananation reactions. The chromium chloride complex has a reflectance spectrum with an absorption maximum at about 475 nm at room temperature, but this shifts to 505 nm at 100°C. DRS curves measured at 600 nm showed that the deaquation and ananation reactions for this complex occur in the range 75 to 110°C. In addition to the chloride complexes, the bromide and nitrate complex salts of both chromium and cobalt were studied, as well as the iodide salt of chromium. All complexes were found to undergo deaquation and ananation reactions in the temperature range from 60 to 150°C, with the reaction being characterized by a shift in the reflectance minimum (absorption maximum) to longer wavelength. Thermal stabilities could not be determined as the temperature measurement data were not sufficiently sensitive.

Wendlandt and Bradley next investigated the

thermochromic behavior of complexes of the type $M_n[HgI_4]$, where M is Ag(I), Cu(I), Hg(II), Tl(I), or Pb(II).[81] These compounds display quite dramatic color changes during their thermochromic transitions and thus are ideal subjects for study by DRS and HTRS. As before, the measurements were made on a Beckman DK-2A spectroreflectometer using the specially designed heated sample holder.[78] The thermochromic transitions observed in complexes of the type $M_n[HgI_4]$ are attributed to an order-disorder change in the crystal lattice.

In the case of $Ag_2[HgI_4]$, the order-disorder transition involves three phases, α, β, and β'. The yellow low-temperature β-form has tetragonal symmetry with the mercury ion at the corners of a cubic unit cell and the silver ions at the mid-points of the vertical faces. As the temperature rises, the mercury and silver ions may exchange positions or may occupy the two extra lattice sites at the top and bottom face centers of the cube. The α-modification, which exists above 52°, has average face-centered cubic symmetry, with the mercury and silver ions completely disordered. The β'-form, which exists as an intermediate between the β- and α-modifications, has a tetragonal unit cell. HTRS curves show the transition from the yellow β-form to the red α-form. The transition temperature was not well-defined and the color change was found to be dependent on the heating rate.

The thermochromism of $Cu_2[HgI_4]$, while less thoroughly investigated than that of the analogous silver complex, probably involves the same type of transition mechanism. The red β-modification reflects strongly above 600 nm. During the transition to the brown-black α-form, the reflectance maximum shifted to longer wavelengths and decreased in magnitude. The transition was reversible, occurring in the range 50 to 75°C.

$Hg[HgI_4]$ is photosensitive and undergoes several transitions, becoming red above 160°C. The compound exhibits a transition to orange-red at 172.6°, deep red at 220.1°, and melts sharply at 224.4°C to form a red mass.[82] The HTRS curves recorded from 24 to 125°C reveal a small shift to longer wavelength on heating, with a shoulder peak at 500 nm also shifting to longer wavelength as the temperature rises. The transition from yellow-orange to red occurs gradually at lower temperatures but accelerates above 100°C. This transition is reversible.

$Pb[HgI_4]$ showed similar behavior to the mercury compound in its HTRS and DRS curves, for $Tl_2[HgI_4]$ revealed a gradual reversible transition from yellow to red in the range 23 to 150°C.

HTRS and DRS are uniquely suited for the study of the thermochromic transitions of these compounds. The ΔH of transition involved is too small for DTA or DSC studies to be used effectively and, as no weight losses are involved, TGA could not be used.

Gore and Wendlandt have studied the thermal dehydration of a number of 8-quinolinol chelate hydrates[83] and some corresponding 2-methyl-8-quinolinol chelate hydrates[84] by a number of thermochemical methods, including HTRS and DRS. While the reflectance spectra of the anhydrous chelates of both compounds were found to be quite similar, some significant differences were observed in the spectra of the corresponding chelate hydrates. The most striking of these occurred for the copper(II), nickel(II), and chromium(III) complexes. The 8-quinolinol complex of copper is bright green with a strong reflectance band at 520 nm. The 2-methyl-8-quinolinol complex, on the other hand, is dull bronze. The band at 520 nm does not appear, but there is a small hump in the curve at 620 nm. The 2-methyl cobalt complex also is quite different in appearance from that of the unsubstituted ligand. While the 8-quinolinol cobalt complex is a dull yellow in color, the 2-methyl complex is copper-bronze in color and the reflectance spectrum shows a sharp peak at 620 nm. The reflectance spectra of the nickel complexes also show a variation in the red region of the spectrum. A peak at 575 nm in the spectrum of the 2-methyl complex is not present in the spectrum of the unsubstituted ligand.

The changes in the spectra of the cobalt, chromium, and nickel complexes have been attributed to the presence in these ions of d-orbitals, which are available for bonding. Thus, changing the number of water molecules in the complex may result in a change in either the bonding or coordination or both. This could result in a change in the spectra. However, the transitions responsible for these spectra have not yet been definitely determined. Chang and Wendlandt have recently investigated the solid-solid and gas-solid reactions of the type $Co(NH_3)_4X_2Y_{3-n}$ (n = 1,2) complexes with NH_4X and HX.[85] Reflectance spectroscopy was used in the identification of the products.

Other investigators have also used diffuse reflectance spectroscopy at elevated temperatures to study thermochromic reactions in an isothermal mode of operation. Hatfield et al.[86] investigated thermochromism of bis(*N,N*-diethylethylenediamine)copper(II) perchlorate, which exhibits a change from red to blue after heating to 50°C. Kortüm[15] studied the polymorphic conversion of red HgI_2 at elevated temperatures. A conversion to a yellow form takes place at 140°C. The yellow form is metastable at room temperature as indicated earlier (see Section 1b). An isothermal study of the kinetics of interaction between Fe_2O_3 and CaO has been carried out by HTRS at temperatures between 500 and 900°C.[87] DRS should be a valuable tool for certain kinetic investigations, even though no information on such work is available as yet.

b. Low-temperature Reflectance Spectroscopy

Reflectance measurements at low temperatures have been carried out primarily in studies of inorganic systems.[88-92] The data obtained with this method have aided in the interpretation of complex spectra in terms of ligand field theory. Often sharpening of the structural features of reflectance spectra recorded at low temperatures (e.g., liquid-nitrogen temperatures) can be observed in comparison to similar spectra recorded at room temperature (see Figure 6.17). This effect can partly be attributed to the elimination of "hot bands" originating from vibrationally excited ground-state transitions. The technique of measurement of reflectance spectra at low temperatures does not pose any great problems. A typical cell for work at liquid-nitrogen temperatures has been described by Symons and Travalion[93] (see also Chapter 4, Section 3c).

3. Inorganic Systems

Various inorganic systems have been studied by reflectance spectroscopic techniques in connection with chromatographic work. They are discussed in Chapter 1, Section 3. The same technique has been applied to inorganic pigments, ceramic materials, inorganic building materials, and geological samples (see Chapter 5, Section 3). Many inorganic systems have been investigated in connection with studies of surface phenomena; they are discussed in Section 1b. The work discussed in the section on high- and low-temperature reflectance spectroscopy was solely concerned with inorganic materials, including metal chelates.

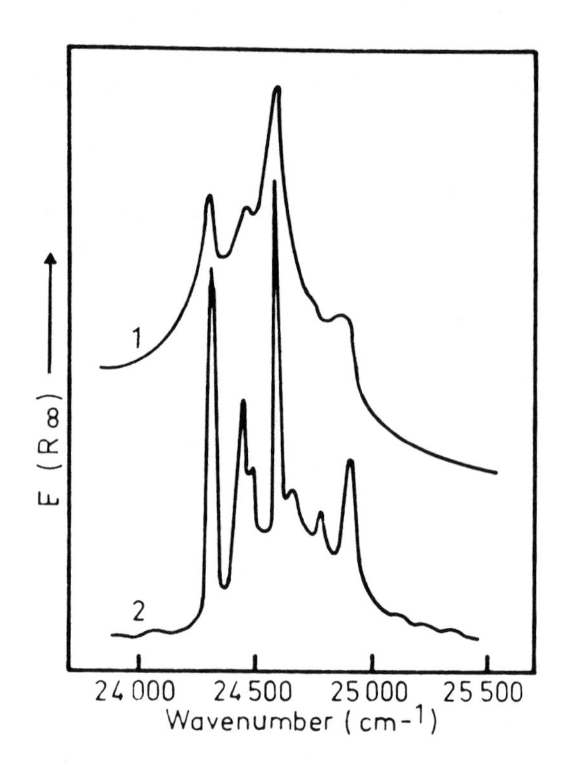

FIGURE 6.17 Reflectance spectra of $MnCl_2 \cdot 4H_2O$ at room temperature (1) and at 78°K (2).[5] (From Kortüm, G., *Reflexionsspektroskopie,* Springer-Verlag, Berlin, 1969. With permission.)

A comprehensive review (with 32 literature citations) on the application of diffuse reflectance spectroscopy to problems in inorganic chemistry has been presented by Clark.[94] He particularly emphasized the use of this technique to the structural analysis of metal complexes and other metal compounds whose transmission spectra cannot be measured easily in solution or where reflectance spectra could yield additional valuable information on the behavior of the system in the solid state. Interpretations of data in terms of crystal field and ligand field parameters have been attempted in many instances. In this field significant contributions have been made on the basis of reflectance spectroscopic data.[95-103]

Other applications of more analytical interest have been discussed in a number of publications. Rare earth-metal oxides and other rare earth compounds have been studied by this technique by White[104] and Ropp.[105] The latter work was concerned with Er, Ho, Tb, and Pr compounds studied in the UV and visible regions of the spectrum with a Cary Model 15 Spectrophotometer for high-resolution work, equipped with a dual-reflectance sphere attachment coated with barium sulfate. The kind of resolution one can obtain (1-2 Å) is quite astonishing (see Figure 6.18). Experimental conditions had to be controlled rigidly in this project. Inert gas flushing of the monochromator, reflectance spheres, and photomultiplier housing was required to improve the resolution. The barium sulfate standards were prepared carefully with a Bausch & Lomb powder press (Bausch & Lomb Co., Rochester, New York). The $BaSO_4$ was chosen as a standard due to its resistance to bleaching when exposed to UV light for prolonged periods of time. A high-intensity source (150-W Xe-lamp) was used. The author[105] does not mention anything about the preparation of the powder sample and particularly the particle size employed; from previous discussions this is known to be critical with respect to the resolution that can be obtained.

High-resolution reflectance spectra offer the advantage of identifying optically narrow and broad bands of the rare earths resulting from upper energy states and unperturbed 4f-transitions. Similar investigations have been reported by Loh,[106,107] who carried out reflectance measurements on single crystals.

The effect of pressure on the spectral reflectance of compacted powders has been studied by Schatz.[108,109] The systems investigated include rare earth and transition metal oxides, alkaline earth oxides, $BaSO_4$, Al_2O_3, SiO_2, and others (see also Chapter 3, Section 2). Schatz also studied the effect of particle size of similar materials at high pressures up to 35,300 psi on diffuse reflectance spectra (see Figure 6.19). He concluded that the spectral reflectance of powders under this high-pressure condition shows little dependence on the size of the powder particles, provided that the particles are large compared to the wavelength of the irradiated energy.

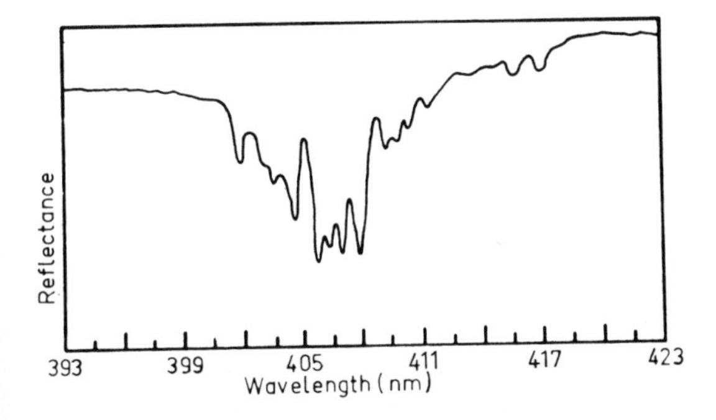

FIGURE 6.18 Relative reflectance of $ErBO_3$.[105] (From Ropp, R. C., *Appl. Spectrosc.*, 23, 235 (1969). With permission of the American Institute of Physics.)

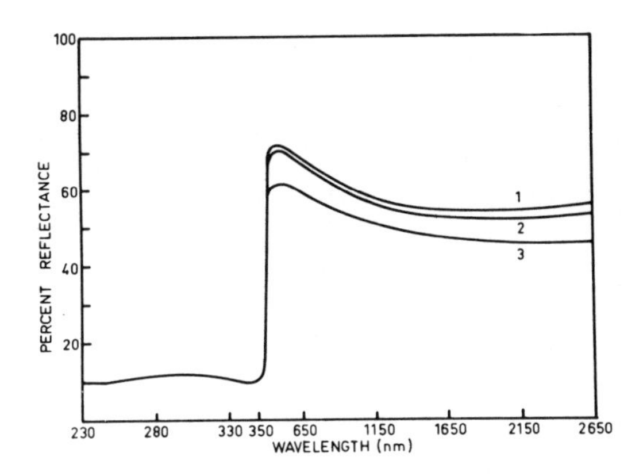

FIGURE 6.19 Spectral total reflectance (vs. MgO) of compacted TiO_2 powders as a function of particle size: particle size (mesh), 1, +150 to −325 +400 (37 to 105 μ); 2, −400 +500 (30 to 37 μ); 3, −500 (< 30 μ). Compacted at 35,300 psi.[109] (From Schatz, E. A., in *Modern Aspects of Reflectance Spectroscopy*, Wendlandt, W. W., Ed., Plenum Press, New York, 1968. With permission.)

Bulk uranium oxide has been investigated[110,111] over the UV and visible regions of the spectrum. A change in reflectance spectra was observed as a function of varying $UO_{2.66}$ to $UO_{2.00}$ content.

Griffiths et al.[29,30] measured reflectance spectra of a wide range of inorganic systems in an attempt to explore the analytical potential of diffuse reflectance spectroscopy in the UV region. LiF was chosen as a relatively neutral diluent for the compounds under investigation, which included HgI_2, K_2HgI_4, $K_2HgI_3 \cdot H_2O$, KI, CdI_2, H_5IO_6, KIO_3, KIO, $Na_2H_3IO_6$, $Na_2S_2O_4$, $Na_2S_2O_5$, $Na_2S_2O_6$, $BaCrO_4$, K_2CrO_4, NaO_3, and commercial sodium peroxide (see Figure 6.20). Despite the use of lithium fluoride as diluent for strongly absorbing samples, surface effects could not be excluded completely and were a serious interference for investigations of a purely inorganic nature. Dilution series of KIO_3 in LiF plotted as absorbance vs. concentration in moles/l. (see Figure 6.21) resulted in linear calibration curves. This is contrary to the present state of knowledge in the field, and the author feels that the data as they appear on the plot (Figure 6.21) do not warrant such a conclusion. The assumption that the plots would extend through zero is also quite poorly founded, since relative reflectance values were measured in this work, e.g., vs. filter

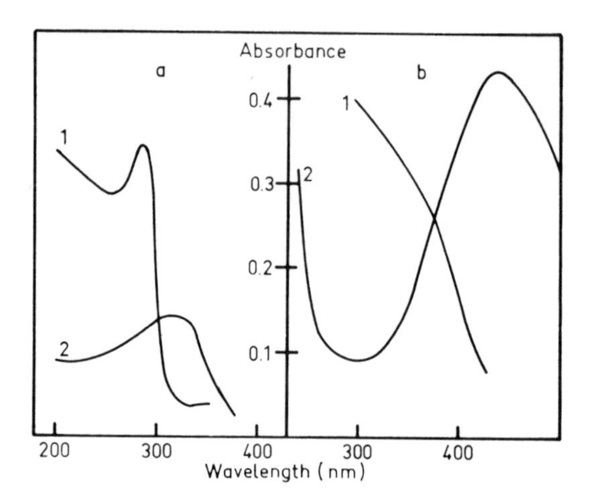

FIGURE 6.20 Diffuse reflectance spectra of salts homogeneously diluted with lithium fluoride. a: 1, potassium iodate (KIO_3); 2, sodium paraperiodate ($Na_2H_3IO_6$). b: 1, commercial sodium peroxide (Na_2O_2 + about 10% O_2); 2, sodium ozonide (NaO_3).[29] (From Griffiths, T. R., Lott, K. A. K., and Symons, M. C. R., *Anal. Chem.*, 31, 1338 (1959). With permission of the American Chemical Society.)

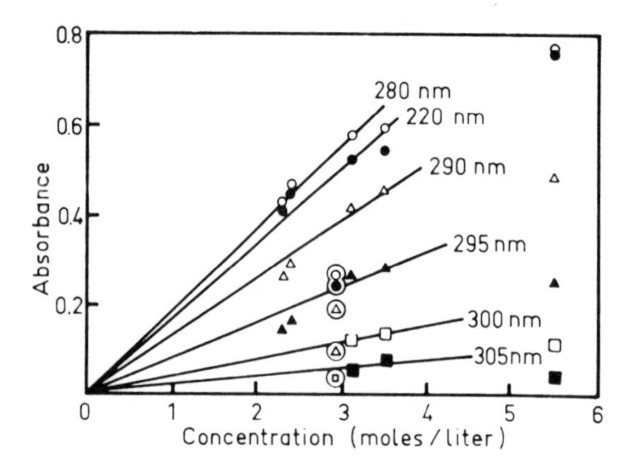

FIGURE 6.21 Plot of absorbance of KIO_3 against concentration in moles per liter at different wavelengths, with LiF as diluent.[30] (From Griffiths, T. R., *Anal. Chem.*, 35, 1077 (1963). With permission.)

paper as a standard, and no attempts were made to correct to absolute values. Deviations from zero would also be caused by the appreciable self-absorption of the diluent at the wavelength region used for the study (<300 nm).[112] Griffiths found the method quite useful for the investigation of unstable compounds such as sodium ozonide, superoxides, and peroxides, which are difficult to prepare as single crystals.

Spectral reflectance methods have also been used for a study of inorganic phosphors.[93,94] Increasing the amount of thallium about threefold in a KBr:Tl phosphor resulted in a pronounced drop (about 15%) in reflectance at 260 nm.[113] Other UV reflectance spectra were reported for KCl:Tl phosphors[113,114] and for NaCl:(Pb,Mn), and $CaSiO_3$:(Pb,Mn).[114] As expected, particle size of the powders was found to be critical.[113]

Near-infrared reflectance spectra have been reported by Goulden,[115] who observed strong absorption bands for $CaSO_4 \cdot H_2O$ at 1440 and 1930 nm, respectively. Lermond and Rogers[116] used the Beckman DU spectrophotometer equipped with the standard reflectance attachment to explore the means of direct analysis of difficultly soluble inorganic materials. The results of a study of the packing reproducibility of 0.10% ferric oxide samples diluted with barium sulfate and measured vs. magnesium carbonate as a reference standard are shown in Table 6.07. The reproducibilities obtained for repeated measurements of a single sample were between ±0.1 and ±0.2%, and for 5 samples measured against one another they were ±2% maximum on the 100-unit reflectance scale. Standardization of the sample preparation procedure, such as screening and packing, was found to be critical. Lermond and Rogers[116] also explored the use of differential reflectance techniques (see Chapter 2, Section 5), particularly in the case of strongly absorbing samples. Various calibration methods were investigated in connection with this work[96] and linear calibration curves were reported for concentration ranges of 0.05 to 14.3% Fe_2O_3 with plots $(R - R_m)$ vs. C^{-1}. R is the reflectance of the sample and R_m the reflectance of the sample with the highest concentration of colored component. Other relationships tested were the Kubelka-Munk function and $(R - R_m)$ vs. $C/(1 - C)$.

Several methods have been proposed for the simultaneous analysis of two or multi-component mixtures of inorganic compounds.[109,117-119] The principle of such a technique has been discussed in Chapter 2, Section 7. Kortüm and Herzog[117] have studied the simultaneous determination of the two titanium oxide modifications, rutile and anatase. There is considerable interest and demand for a method of this nature since rutile has a higher refractive index, and, if used as a pigment in the paint industry, it has a better covering power than anatase. The ratio of rutile vs. anatase is therefore a quality factor in commercial

TABLE 6.07

Reproducibility of Packing Powder Samples[116]

Wavelength (nm)	Trial 1	2	3
400	79.8	79.5	79.3
420	78.8	78.7	78.7
440	77.9	77.9	77.9
460	77.8	77.8	77.8
480	77.5	77.1	77.3
500	77.1	77.2	77.2
520	77.0	77.1	77.0
540	77.0	77.1	77.2
560	80.2	80.3	80.2
580	86.2	86.2	86.2
600	90.8	91.0	90.8
650	93.8	93.7	93.9
700	94.8	94.2	94.8

Samples of 0.10% Fe_2O_3 in barium sulfate, measured vs. magnesium carbonate standard

(From Lermond, C. A. and Rogers, L. B., *Anal. Chem.*, 27, 340 (1955). With permission of the American Chemical Society.)

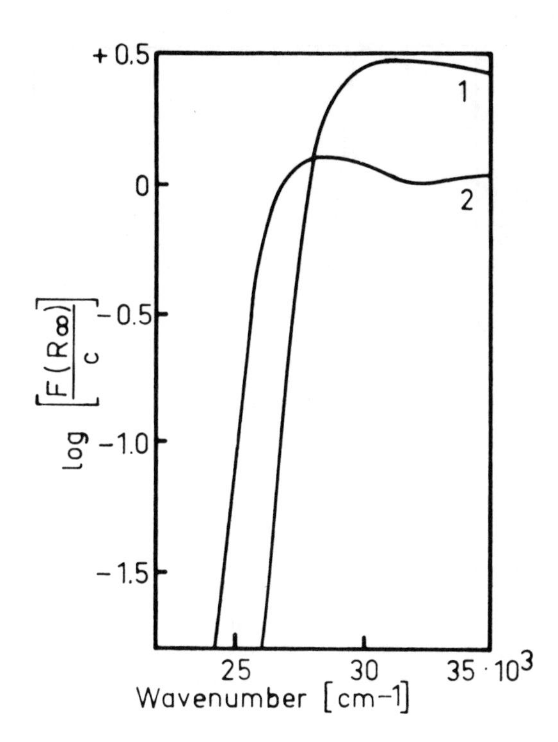

FIGURE 6.22 Reflectance spectra of anatase (1) and rutile (2) measured vs. NaF as a reference standard.[117] (From Kortüm, G. and Herzog, G., *Z. Anal. Khim.*, 190, 239 (1962). With permission of Springer-Verlag.)

products. The absorption maxima of the two components are sufficiently separated in the UV region of the spectrum (see Figure 6.22), so that a simultaneous analysis according to the principles discussed in Chapter 2, Section 7, is feasible. The Kubelka-Munk functions $F(R'_\infty)$ of various concentrations of the two titanium oxides diluted with MgO and measured vs. a suitable standard, such as magnesium oxide or barium sulfate, can be used for linear calibration curves. (Absolute reflectance values have been employed by Kortüm and Herzog.[117]) An accuracy of about ±2% was reported for this method, which is in close agreement with other findings for simultaneous analysis procedures.

Schatz[109] investigated powder mixtures of Al and Ti, SiO_2 and Al, ZrO_2 and Al_2O_3, and others. Doyle and Forbes[119] carried out similar studies with the Unicam SP500 spectrophotometer fitted with the standard diffuse reflectance attachment. The substances investigated were lead monoxide, silver iodide, and zinc oxide with absorption maxima at 515, 425, and 368 nm, respectively. Magnesium oxide and silica gel served as diluents.

Besides the Kubelka-Munk function, an empirical relationship $(A/R)^{1.383}$ was found to yield linear calibration curves and to give an additive function for the multi-component systems in analytically suitable concentration ranges (see Figure 6.23). Actual analysis results for mixtures of PbO and ZnO diluted with silica gel are presented in Table 6.08.

Powdered mixtures were also investigated by Fisher and Vratny.[120] A linear relationship 100-%R vs. logarithm weight % Fe_3O_3 in $MgCO_3$ was reported. Most of the work, however, dealt with inorganic and organic compounds adsorbed on strips of filter paper, such as copper ammine and iron thiocyanate on Whatman No. 2 paper. Again, fairly linear calibration curves 100-%R vs. logarithm of concentration were obtained over useful concentration ranges.

Occasionally, reflectance techniques are used to evaluate spot tests carried out on filter paper. In 1949 Winslow and Liebhafsky[121] recorded reflectance curves for the α-benzoin oxime spot test for copper, but they concluded that transmission measurements through the paper gave a more

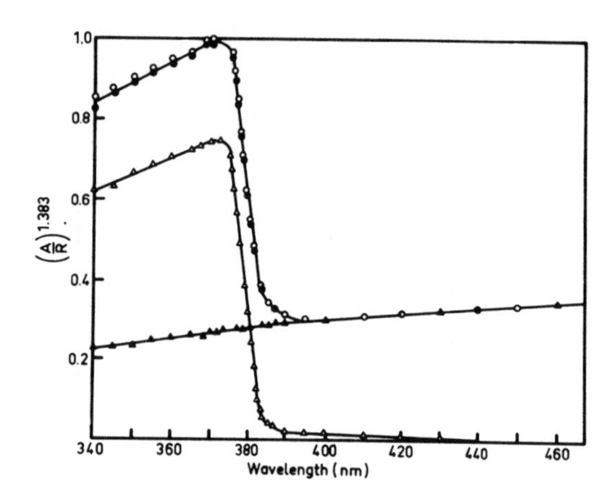

FIGURE 6.23 Additivity of absorption function for zinc oxide lead monoxide silicic acid mixture. ▲, spectrum of 2.02% PbO silicic acid; Δ, spectrum of 0.64% ZnO silicic acid; o, sum of spectra of 2.02% PbO silicic acid and 0.64% zinc oxide silicic acid; •, observed spectrum of 2.02% PbO 0.64% ZnO silicic acid.[119] (From Doyle, W. P. and Forbes, F., *Anal. Chim. Acta,* 33, 108 (1965). With permission of Elsevier Pub. Co.)

TABLE 6.08

Results of Analysis of Lead Monoxide-zinc oxide-silicic Acid Mixtures[119]

	% PbO		% ZnO	
Mixture	Actual	Found	Acutal	Found
1	6.98	7.15	3.05	3.26
2	5.93	5.69	2.60	2.62
3	4.95	5.22	2.16	2.29
4	3.94	4.08	1.72	1.87
5	2.90	2.82	1.27	1.28
6	1.91	1.99	0.83	0.86
7	7.65	8.24	1.54	1.71
8	3.86	4.10	1.48	1.58
9	3.20	2.98	1.51	1.49
10	2.02	2.11	1.49	1.62
11	0.98	1.01	1.49	1.55
12	0.54	0.56	1.45	1.53

(From Doyle, W. P. and Forbes, F., *Anal. Chim. Acta,* 33, 108 (1965). With permission of the Elsevier Pub. Co.)

suitable signal to concentration relationship for quantitative work. This may well have been the case at that time, since the instrumentation available for spectral reflectance measurements was not very advanced.

Other workers[122,123] investigated metal sulfides on filter paper, using the same approach. Linear plots were reported for peak height of reflectance scans plotted as a function of concentration of copper(II), nickel(II), and iron(III) sulfides in concentrations of 0.5-5 μg of metal.[122] However, this relationship is valid only if the color in the spot is uniformly distributed. In practice, such a homogeneous distribution of coloring material can hardly be achieved, due to the occurrence of irregular diffusion processes which are difficult to control. Reproducibility of the data was found to be strongly dependent on control of moisture co-adsorbed on the paper and on the use of very low concentrations.[123] Other reflectance spectroscopic methods for metal sulfides have been reported by Takagi et al.[124]

Some investigators[125] have used spectral reflectance to investigate arsenic in trace amounts in biological materials after isolation of the metal on paper strips. The cation was converted to mercuric arsenide directly on the paper, and the colored compound was measured. Mizuniva et al.[126] have published a method for the rapid determination of the total iron content in boiler water by means of diffuse reflectance spectroscopy. The water samples were digested with oxidizing acid and, after addition of ammonia, the colloid iron was filtered through a Millipore® membrane filter. After drying, the degree of coloration from the Fe_2O_3 was measured by reflectance at 320 nm. Between 5 and 3,000 ppb concentrations of iron were analyzed and a reproducibility equivalent to transmission spectroscopy was claimed.

Another, more recent method based on the combination of in situ reactions on filter paper and evaluation by reflectance methods for the determinations of mercury in air was described by Palalau.[127] After collection on the filter paper, the mercury is reacted with a copper(I)-iodide solution and dried at room temperature. The yellow-orange complex is evaluated quantitatively at 510 nm.

In some cases diffuse reflectance techniques have been applied after preconcentration steps by ion-exchange procedures.[128,129] Ermolenko et al.[128] concentrated nickel and manganese on cation-exchange paper consisting of oxidized cellulose. After reaction with suitable chelating reagents, the reflectance spectra were recorded. Detection limits of 0.01 μg metal-ion were reported for this method, and up to 100-fold excesses of other metals did not interfere.

The same evaluation technique serves conveniently for the recording of reflectance spectra of heavy metal chelates after a scavenging step with ion-exchange resins. As an example of such a procedure, Fujimoto and Kortüm[129] reported obtaining concentrating factors from 10^3 to 10^4 for complex ions such as $[Cu(H_2O)_4]^{2+}$ on Dowex 50 W-X8(H-form) or $[Co(II)(CNS)_4]^{2-}$ on Dowex 1-X8. The ion-exchange resin loaded with the complex is measured in air-dry condition by spectral reflectance (see Figure 6.24). Pure resin serves as a reference standard and the reflectance values are corrected to absolute values as discussed earlier (see Equation 6.03). The Kubelka-Munk theory was found to be valid up to the maximum exchange capacity of the resins (~ 30%) and an accuracy of ±2% was reported. These results certainly suggest the use of diffuse reflectance spectroscopy for a wide range of column chromatographic problems. An interesting application has been reported by Hoffman,[130] who used reflectance techniques in the near-infrared region of the spectrum to determine the humidity content of a variety of samples such as leather, paper, wool, gelatin, etc. The specific absorption maxima for H_2O at 1.93 and 1.7 μ were chosen for this analysis. Kubelka-Munk plots $F(R_\infty)$ vs. percent concentration of water were used for this purpose (see Figure 6.25).

The measurement of reflectance spectra in the short UV-region below 250 nm has been the subject of recent investigations. Fassler and Zimmerman reported that the presence of usable matrix substances as diluents is necessary for the quantitative evaluation of diffuse reflectance spectra in this region of the spectrum.[131] The absolute reflectivity of a number of compounds, including NaCl, LiF, CaF_2 and powdered quartz, was measured in the short UV-region as low as 180 nm. By using a standardized procedure for the preparation of the powders, a constant grain-size distribution with a comparable coverage of surface moisture may be achieved, so that the values obtained for R_∞^{abs} may generally be used as

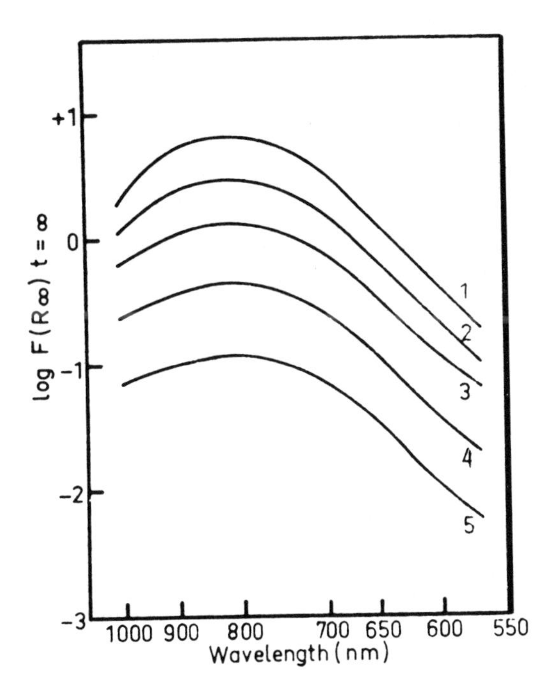

FIGURE 6.24 Typical reflectance spectra of $[Cu(H_2O)_4]^{2+}$ ions adsorbed on air-dry Dowex ion-exchange resin at various concentrations (g atom Cu^{2+}/g equiv. active group).[129] (1) c = 4.41 x 10^{-1}; (2) c = 2.22 x 10^{-1}; (3) c = 1.14 x 10^{-1}; (4) c = 3.68 x 10^{-2}; (5) c = 9.16 x 10^{-3}. (From Fujimoto, M. and Kortüm, G., *Ber. Bunsenges Phys. Chem.*, 68, 488 (1964). With permission.)

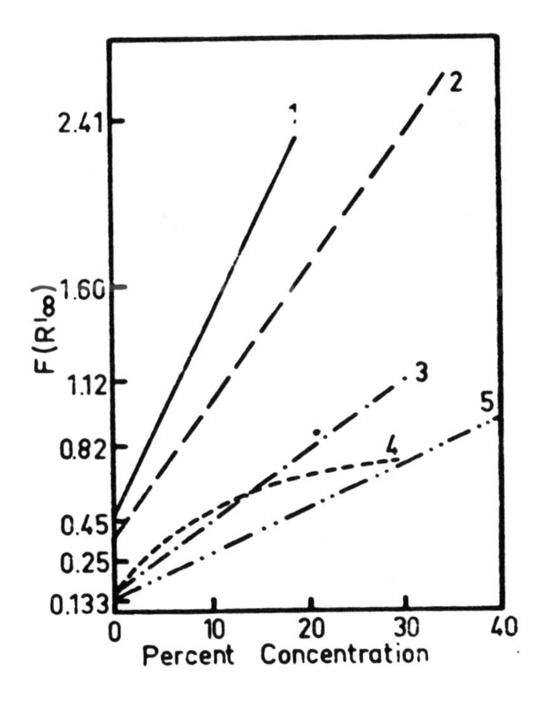

FIGURE 6.25 Kubelka-Munk plots of several materials measured at 1.93 μ as a function of the H_2O content in percent of the dry weight.[130] 1, gelatin; 2, wool; 3, starch; 4, flour; 5, paper. (From Hoffmann, K., *Chem.-Ing.-Tech.*, 35, 55 (1963). With permission of Verlag Chemie GMBH.)

standard values in the evaluation of the diffuse reflectance spectra.

Zimmerman has also studied the application of diffuse reflectance spectroscopy in the region below 250 nm.[132] Using Kubelka-Munk theory and Kortüm's dilution method as the basic method for measurement of powders usually resulted in band flattening in spectra recorded below 250 nm. Adsorbed water decreases the absolute reflectivity of the matrix substances in this region, and this, together with the scattering coefficients, produces the observed anomalies in the diffuse reflectance spectra. The standardized method for sample preparation was tested successfully for alkali iodide spectra down to 190 nm.

Fassler and Stodolski have studied the calculation of absorption coefficients from diffuse reflectance spectra.[133] Assuming the powder particles to be spherical and the incident light to be isotopic, the authors derived an equation for the calculation of absorption coefficients of densely packed powder monolayers from the reflectance measurements.

$$K_R = [(\exp 2kd/3) - 1] \,/\, [(\exp 2kd/3) - 1 + [2x_u - (1 - \bar{r}_2)]]$$

where K_R is the absorption coefficient of the monolayer; k is the absorption coefficient of a single particle of diameter d; x_u is the shading parameter; and $\bar{r}_2$ is the reflectance coefficient over all possible angles of incidence.

4. Organic Materials

Possible uses of diffuse reflectance spectroscopy for the investigation of organic compounds have been discussed in numerous papers. The major portion of such examples can be found in connection with chromatographic work, where dyes, amino acids, nucleotides, vitamins, alkaloids, pesticides, and many other compounds have been studied after separation on thin-layer and paper chromatograms[134,135] (see also Chapter 7).

Other applications involving organic systems have been discussed in Chapter 5, Section 3.

Much of the work discussed in the section on surface phenomena (Section 1b) involves organic compounds. Zeitlin and Niimoto,[3,24] for example, have reported UV spectra of 2,4-dinitrophenylhydrazones and other systems on filter paper. Zeitlin et al.[16-19,23] have investigated nitrophenols and nitroanilines as well as some organic dyes on a large number of adsorbents by spectral reflectance. Kortüm[5] has chosen compounds such as anthracene, 2-(2′,4′-dinitrobenzyl)-pyridine, spiropyran, *p*-dimethylaminoazobenzene, green *o*-carboxylic acid lactone, stilbene, and other benzene derivatives for investigations of surface phenomena by diffuse reflectance spectroscopy in the UV, visible, and near-infrared regions of the spectrum. The diffuse reflectance spectra of filter papers dyed with 1-arylazo-2-naphthols have also been reported.[136]

Schwab[27,28] has investigated many organic dyes of the benzene-derivative type on various adsorbents with catalytic properties. Guilmart[137,138] has made systematic studies of groups of compounds, e.g., acids and corresponding amides of the type $C_6H_5(CH_2)_nCOOH$ (n = 0 to 6). Reflectance measurements were made in the UV and visible regions of the spectrum. Pruckner et al.[139] have shown that reflectance spectra of organic substances concentrated on particulate adsorbents can be used for their identification. Schwuttke,[140] on the other hand, has shown that spectral reflectance can be used for the determination of concentrations of organic dyes scavenged from solutions by the batch-wise addition of starch.

Briegleb and Delle[141] have followed the reaction of picric acid with aromatic amines upon heating by means of reflectance techniques. Fisher and Vratny[120] developed a method for the determination of dyes, such as congo red or malachite green, adsorbed on various powder substances, and Frei and Zeitlin[18] investigated factors such as temperature, humidity, pH, etc., which affected the reflectance spectra of some dyes on alumina.

The reflectance spectra of adsorbed charge-transfer complexes have recently been shown to reveal multiple charge-transfer bands.[142] The perylene-chloranil complex was adsorbed on calcium fluoride; the components were then ground for several minutes, and the samples were allowed to stand for several days until the charge-transfer complex adsorption had reached a maximum value. The charge-transfer band can then be distinguished from the absorptions of the components that overlie it but that are constant. The formation rate constant for the charge-transfer complex calculated from the charge-transfer absorption was found to be dependent on both the components of the complex and the adsorbent.

Mehta et al. have studied the diffuse reflectance spectra of copper soaps to investigate their structures in the solid state.[143] As the soaps do not form single crystals or thin films, they were studied by diffuse reflectance in the powdered state. Measurements were carried out on a Uvispek H700 photoelectric spectrophotometer fitted with an annular ring-type diffuse reflectance attachment from 300 to 700 nm. Lithium fluoride was used as the reflectance standard, and precautions were taken to exclude moisture from the sample. The reflectance spectra of the soaps revealed two absorption bands; one, near 660 nm, is commonly observed for Cu(II) in the presence of a ligand, while the other, near 385 nm, is not commonly observed with copper salts, and so is of special interest in the study of the copper soaps. The presence of this band was taken to indicate that copper soaps exist as a binuclear molecule in the solid state having the general formula $Cu(RCOO)_2$, where R is the appropriate alkyl substituent.

The above examples by no means give an exhaustive treatment of applications in this area, and the possibilities of extending this method to other systems and problems are unlimited.

REFERENCES

1. **Lucchesi, P. J., Carter, J. L., and Yates, D. J. C.,** *J. Phys. Chem.,* 66, 1451 (1962).
2. **Basila, M. R.,** *J. Phys. Chem.,* 66, 2223 (1962).
3. **Zeitlin, H. and Niimoto, A.,** *Anal. Chem.,* 31, 1167 (1959).
4. **Kortüm, G. and Schreyer, G.,** *Angew. Chem. Int. Ed. Engl.,* 67, 694 (1955).
5. **Kortüm, G.,** *Reflexionsspektroskopie,* Springer-Verlag, Berlin, 1969, 264. (English translation: Springer-Verlag, New York, 1969.)
6. **Barnes, I. L., Goya, H., and Zeitlin, H.,** *Rev. Sci. Instrum.,* 34, 292 (1963).
7. **Weitz, E., Schmidt, F., and Singer, J.,** *Z. Elektrochem.,* 46, 47 (1941).
8. **Weitz, E., Schmidt, F., and Singer, J.,** *Z. Elektrochem.,* 47, 47 (1941).
9. **de Boer, J. H. and Houben, G. M. M.,** *Koninkl. Ned. Akad. Wetenschap., Ser. B., Proc.,* 54, 421 (1951).
10. **Zeitlin, H., Goya, H., and Waugh, L. T.,** *Nature,* 198, 178 (1963).
11. **McRae, E. H.,** *J. Phys. Chem.,* 61, 562 (1957).
12. **Brooker, L. S. and Sprague, R. H.,** *J. Am. Chem. Soc.,* 63, 3214 (1941).
13. **Zeitlin, H. and Goya, H.,** *Nature,* 183, 1041 (1959).
14. **Goya, H., Waugh, J. L. T., and Zeitlin, H.,** *J. Phys. Chem.,* 66, 1206 (1962).
15. **Kortüm, G.,** *Trans. Faraday Soc.,* 58, 1624 (1962).
16. **Zeitlin, H., Kondo, N., and Jordan, W.,** *Phys. Chem. Solids,* 25, 641 (1964).
17. **Zeitlin, H., Frei, R. W., and McCarter, M.,** *J. Catal.,* 4, 77 (1965).
18. **Frei, R. W. and Zeitlin, H.,** *Anal. Chim. Acta,* 32, 32 (1965).
19. **Zeitlin, H. and Lieu, V. T.,** *J. Catal.,* 4, 546 (1965).
20. **Kortüm, G., Vogel, J., and Braun, W.,** *Angew. Chem.,* 70, 651 (1958).
21. **Kortüm, G. and Vogel, J.,** *Chem. Ber.,* 93, 706 (1960).
22. **Kortüm, G., Braun, W., and Herzog, G.,** *Angew. Chem. Int. Ed. Engl.,* 2, 333 (1963).
23. **Frei, R. W., Zeitlin, H., and Fujie, G.,** *Can. J. Chem.,* 44, 3051 (1966).
24. **Zeitlin, H. and Niimoto, A.,** *Nature,* 181, 1616 (1958).
25. **Anthony, P. and Zeitlin, H.,** *Nature,* 187, 936 (1960).
26. **Zeitlin, H., Anthony, P., and Jordan, W.,** *Science,* 141, 423 (1963).
27. **Schwab, G. M. and Schneck, E.,** *Z. Phys. Chem.* (Frankfurt am Main), 18, 206 (1958).
28. **Schwab, G. M., Dadlhuber, B. C., and Wall, E.,** *Z. Phys. Chem.* (Frankfurt am Main), 37, 99 (1963).
29. **Griffiths, T. R., Lott, K. A. K., and Symons, M. C. R.,** *Anal. Chem.,* 31, 1338 (1959).
30. **Griffiths, T. R.,** *Anal. Chem.,* 35, 1077 (1963).
31. **Frei, R. W. and Zeitlin, H.,** *Can. J. Chem.,* 47, 3902 (1969).
32. **Frei, R. W., Liiva, R., and Ryan, D. E.,** *Can. J. Chem.,* 46, 167 (1968).
33. **Heit, M. L. and Ryan, D. E.,** *Anal. Chim. Acta,* 34, 407 (1966).
34. **Briegleb, G.,** *Elektronen-Donator-Acceptor-Komplexe,* Springer-Verlag, Berlin, 1961.
35. **Kortüm, G. and Vogele, H.,** *Ber. Bunsenges. Phys. Chem.,* 72, 401 (1968).
36. **Kortüm, G. and Grathwohl, M.,** *Ber. Bunsenges. Phys. Chem.,* 73, 500 (1968).
37. **Kortüm, G. and Koffer, H.,** *Ber. Bunsenges. Phys. Chem.,* 67, 67 (1963).
38. **Kortüm, G. and Grathwohl, M.,** *Ber. Bunsenges. Phys. Chem.,* 3, 1015 (1969).
39. **Kortüm, G. and Schlichenmaier, V.,** *Z. Phys. Chem.* (Frankfurt am Main), 48, 267 (1966).

40. Kortüm, G. and Friz, M., *Ber. Bunsenges. Phys. Chem.*, 73, 605 (1969).
41. Kortüm, G. and Delfs, H., *Spectrochim. Acta*, 20, 405 (1964).
42. Kortüm, G. and Bayer, G., *Z. Phys. Chem.* (Frankfurt am Main), 33, 254 (1962).
43. Kortüm, G., Kortüm-Seiler, M., and Bailey, S. D., *J. Phys. Chem.*, 66, 2439 (1962).
44. Kortüm, G. and Braun, W., *Liebigs Ann. Chem.*, 632, 104 (1960).
45. Kortüm, G., *Spectrochim. Acta Suppl.*, 534 (1957).
46. Kortüm, G., Theilacker, W., and Schreyer, G., *Z. Phys. Chem.* (Frankfurt am Main), 11, 182 (1957).
47. Kortüm, G. and Quabeck, H., *Ber. Bunsenges. Phys. Chem.*, 73, 1020 (1969).
48. Ishii, E., Mamiya, M., and Murakami, T., *Nippon Nogei Kagaku Kaishi*, 353 (1972).
49. Frei, R. W. and Frodyma, M. M., *Anal. Biochem.*, 9, 310 (1964).
50. Lach, J. L. and Bornstein, M., *J. Pharm. Sci.*, 54, 1730 (1964).
51. Lach, J. L. and Bornstein, M., *J. Pharm. Sci.*, 55, 1033 (1966).
52. Lach, J. L. and Bornstein, M., *J. Pharm. Sci.*, 55, 1040 (1966).
53. Bornstein, M., Walsh, J. P., Munden, B. J., and Lach, J. L., *J. Pharm. Sci.*, 56, 1410 (1967).
54. Bornstein, M., Lach, J. L., and Munden, B. J., *J. Pharm. Sci.*, 57, 1653 (1968).
55. Lach, J. L. and Bighley, L. D., *J. Pharm. Sci.*, 59, 1261 (1970).
56. McCallister, J. D., Chin, T.-F., and Lach, J. L., *J. Pharm. Sci.*, 59, 1286 (1970).
57. Hakusui, A., Matsunaga, Y., and Umehara, K., *Bull. Chem. Soc. Jap.*, 43, 799 (1970).
58. Kortüm, G. and Oelkrug, D., *Z. Phys. Chem.* (Frankfurt am Main), 34, 58 (1962).
59. Yatsimirsky, K. B., *Kinetic Methods of Analysis*, Pergamon Press, London, 1966.
60. Kortüm, G. and Braun, W., *Z. Phys. Chem.* (Frankfurt am Main), 18, 242 (1958).
61. Kortüm, G. and Braun, W., *Z. Phys. Chem.* (Frankfurt am Main), 28, 362 (1961).
62. Kortüm, G. and Braun, W., *Z. Phys. Chem.* (Frankfurt am Main), 48, 382 (1966).
63. Braun, W. and Kortüm, G., *Z. Phys. Chem.* (Frankfurt am Main), 61, 167 (1968).
64. Wendlandt, W. W., *Thermal Methods of Analysis*, John Wiley & Sons, New York, 1964, Chap. 10.
65. Wendlandt, W. W. and Hecht, H. G., *Reflectance Spectroscopy*, John Wiley & Sons, New York, 1966, Chap. 7.
66. Wendlandt, W. W., in *The Encyclopedia of Chemistry*, 2nd ed., Clark, G. L. and Hawley, G. G., Eds., Reinhold, New York, 1966, 357.
67. Wendlandt, W. W., in *Modern Aspects of Reflectance Spectroscopy*, Wendlandt, W. W., Ed., Plenum Press, New York, 1968, 53.
68. Wendlandt, W. W., Franke, P. H., Jr., and Smith, J. P., *Anal. Chem.*, 35, 105 (1963).
69. Wendlandt, W. W., *Science*, 140, 1085 (1963).
70. Cathers, R. E. and Wendlandt, W. W., *Chemist-Analyst*, 53, 110 (1964).
71. Wendlandt, W. W., *J. Inorg. Nucl. Chem.*, 25, 833 (1963).
72. Wendlandt, W. W., Robinson, W. R., and Yang, W. Y., *J. Inorg. Nucl. Chem.*, 25, 1495 (1963).
73. Simmons, E. L. and Wendlandt, W. W., *J. Inorg. Nucl. Chem.*, 28, 2187 (1966).
74. Wendlandt, W. W., *Chemist-Analyst*, 53, 71 (1964).
75. Wendlandt, W. W. and George, T. D., *Chemist-Analyst*, 53, 100 (1964).
76. Wendlandt, W. W., Paper presented at the 21st Pittsburgh Conf. Analytical Chemistry, Cleveland, Ohio, March 1970.
77. Wendlandt, W. W. and Bradley, W. S., *Thermochim. Acta*, 1, 143 (1970).
78. Wendlandt, W. W. and Dorsch, E. L., *Thermochim. Acta*, 1, 103 (1970).
79. Wendlandt, W. W. and Bradley, W. S., *Thermochim. Acta*, 1, 305 (1970).
80. Wendlandt, W. W., *Thermochim. Acta*, 1, 419 (1970).
81. Wendlandt, W. W. and Bradley, W. S., *Thermochim. Acta*, 1, 529 (1970).
82. Meyer, M., *J. Chem. Educ.*, 20, 145 (1943).
83. Gore, R. H. and Wendlandt, W. W., *Anal. Chim. Acta*, 52, 83 (1970).
84. Gore, R. H. and Wendlandt, W. W., *Thermochim. Acta*, 2, 93 (1971).
85. Chang, F. C. and Wendlandt, W. W., *Thermochim. Acta*, 3, 69 (1971).
86. Hatfield, W. E., Piper, T. S., and Klabunde, U., *Inorg. Chem.*, 2, 629 (1963).
87. Baistrocchi, R., *Ann. Chim.* (Rome), 49, 1824 (1959).
88. Kortüm, G. and Oelkrug, D., *Naturwissenschaften*, 53, 600 (1966).
89. Kortüm, G. and Schottler, H., *Z. Elektrochem.*, 57, 353 (1953).
90. Clark, R. J. H., *J. Chem. Soc.*, 417 (1964).
91. Hartmann, P. L., Nelson, J. R., and Siegfried, J. G., *Phys. Rev.*, 105, 123 (1957).
92. Oelkrug, D., *Ber. Bunsenges. Phys. Chem.*, 71, 697 (1967).
93. Symons, M. C. R. and Travalion, P. A., *Unicam Spectrovision*, 10, 8 (1961).
94. Clark, R. J. H., *J. Chem. Educ.*, 41, 488 (1964).
95. Asmussen, R. W. and Bostrup, O., *Acta Chem. Scand.*, 11, 745 (1957).
96. Asmussen, R. W. and Bostrup, O., *Acta Chem. Scand.*, 11, 1097 (1957).
97. Sintra, S. P., *Spectrochim. Acta*, 22, 57 (1966).
98. Anysas, J. A. and Companion, A. L., *J. Chem. Phys.*, 40, 1205 (1964).
99. Joergensen, C. K., Pappalado, P., and Rittershaus, E., *Z. Naturforsch.*, 19a, 424 (1964).

100. Balduin, M. E., *Spectrochim. Acta*, 19, 319, (1963).
101. Poole, C. P. and Itzel, J. F., *J. Chem. Phys.*, 39, 3445 (1963).
102. Jassie, L. B., *Spectrochim. Acta*, 20, 169 (1964).
103. Boudreaux, E. A. and Englert, J. P., in *Modern Aspects of Reflectance Spectroscopy*, Wendlandt, W. W., Ed., Plenum Press, New York, 1968, 47.
104. White, W. B., *Appl. Spectrosc.*, 21, 167 (1967).
105. Ropp, R. C., *Appl. Spectrosc.*, 23, 235 (1969).
106. Loh, E., *Phys. Rev.*, 154, 270 (1967).
107. Loh, E., *Phys. Rev.*, 158, 273 (1967).
108. Schatz, E. A., *J. Opt. Soc. Am.*, 56, 389 (1966).
109. Schatz, E. A., in *Modern Aspects of Reflectance Spectroscopy*, Wendlandt, W. W., Ed., Plenum Press, New York, 1968, 107.
110. Ackermann, R. J., Thorn, R. J., and Winslow, G. H., *J. Opt. Soc. Am.*, 49, 1107 (1959).
111. Companion, A. L. and Winslow, G. H., *J. Opt. Soc. Am.*, 50, 1043 (1960).
112. Frei, R. W., unpublished data.
113. Johnson, P. D., *J. Opt. Soc. Am.*, 42, 978 (1952).
114. Schulman, J. H. and Klick, C. C., *J. Opt. Soc. Am.*, 43, 516 (1953).
115. Goulden, J. D. S., *Chem. Ind.* (Lond.), 142 (1957).
116. Lermond, C. A. and Rogers, L. B., *Anal. Chem.*, 27, 340 (1955).
117. Kortüm, G. and Herzog, G., *Z. Anal. Khim.*, 190, 239 (1962).
118. Kortüm, G., in *Analytical Chemistry*, West, P. W., Macdonald, A. M. G., and West, T. S., Eds., Elsevier, New York, 1963, 307.
119. Doyle, W. P. and Forbes, F., *Anal. Chim. Acta*, 33, 108 (1965).
120. Fisher, R. B. and Vratny, F., *Anal. Chim. Acta*, 13, 588 (1955).
121. Winslow, E. H. and Liebhafsky, H. A., *Anal. Chem.*, 21, 1338 (1949).
122. Ayers, C. W., *Mikrochim. Acta*, 85 (1956).
123. Malissa, H., in *Analytical Chemistry*, West, P. W., Macdonald, A. M. G., and West, T. S., Eds., Elsevier, New York, 1963, 80.
124. Takagi, K., Nakano, E., and Lonemura, K., *J. Chem. Soc. Jap., Ind. Chem. Sect.*, 54, 706 (1951).
125. Kawashiro, I., Okado, S., and Kato, S., *Bull. Nat. Hyg. Lab.*, 15, 1957.
126. Mizuniva, F., Umino, T., and Sakai, K., *Jap. Anal.*, 16, 1373 (1967).
127. Palalau, L., *Rev. Chim.* (Bucharest), 19, 54 (1968).
128. Ermolenko, I. N., Longin, M. L., and Gavrilov, M. Z., *Zh. Anal. Khim.*, 17, 1035 (1962).
129. Fujimoto, M. and Kortüm, G., *Ber. Bunsenges. Phys. Chem.*, 68, 488 (1964).
130. Hoffmann, K., *Chem.-Ing.-Tech.*, 35, 55 (1963).
131. Fassler, D. and Zimmerman, G., *Z. Phys. Chem.* (Leipzig), 246, 33 (1971).
132. Zimmerman, G., *Z. Phys. Chem.* (Leipzig), 246, 181 (1971).
133. Fassler, D. and Stodolski, R., *Z. Chem.*, 11, 276 (1971).
134. Frei, R. W., in *Recent Progress in Thin-layer Chromatography and Related Methods*, Vol. 2, Niederwieser, A. and Pataki, G., Eds., Ann Arbor Science Publishers, Ann Arbor, Michigan, 1970, Chap. 1.
135. Frei, R. W., *CRC Crit. Rev. Anal. Chem.*, 2, 179 (1971).
136. Matsunaga, Y. and Miyajima, N., *Bull. Chem. Soc. Jap.*, 44, 361 (1971).
137. Guilmart, T., *Bull. Soc. Chim.*, 5, 1209 (1938).
138. Guilmart, T., *Compt. Rend.*, 207, 289 (1938).
139. Pruckner, F. and von der Schulenburg, M., *Naturwissenschaften*, 51, 45 (1951).
140. Schwuttke, G., *Z. Angew. Phys.*, 5, 303 (1953).
141. Briegleb, G. and Delle, H., *Z. Phys. Chem.* (Frankfurt am Main), 24, 359 (1960).
142. Junghaehnel, G., Gall, R., Goetz, H., and Proksch, G., *Z. Chem.*, 11, 271 (1971).
143. Mehta, V. P., Govil, R. C., and Nagar, T. N., *Z. Naturforsch.*, 25b, 310 (1970).

Chapter 7

APPLICATIONS IN CHROMATOGRAPHY

1. General Experimental Procedure

a. Chromatographic Separation

The use of reflectance spectroscopy for the evaluation of thin-layer or paper chromatograms does not limit the chromatographic separation procedures that can be employed. Many procedures for the resolution of mixtures on various types of adsorbents have been taken from the literature and used with little or no modification before reflectance spectroscopic analysis.

b. Detection of Spots

Techniques for detection of spots in reflectance work do not differ greatly. Unless the compounds are colored, spraying with a chromogenic reagent is necessary. If the resulting colored spots are to be analyzed by spectral reflectance, the instantaneous and quantitative formation of a stable and reproducible color is of prime importance, as can be seen from a study of color stabilities of ninhydrin complexes of amino acids on thin-layer chromatograms using the reflectance technique.[1] Occasionally it is possible to form a colored product by heating the thin-layer plate.[2] If the compound fluoresces, the plate is observed under UV light.[3] Compounds absorbing in the UV spectrum can be detected on a fluorescent background produced by incorporating a luminous pigment into the adsorbent,[4-6] or by just taking advantage of the natural fluorescence of these adsorbents.[7,8] They also lend themselves to direct scanning procedures. Compounds absorbing in the UV spectrum have been located by scanning the thin-layer plates with a Beckman Model DK-2 spectrophotometer set at the absorption maximum of the compound of interest.[4] The scanning is carried out by holding the chromatoplate, which is taped to a protective plastic shield, against the sample exit port of the reflectance attachment unit in such a way that the adsorbent along the path of chromatographic development is exposed to the impinging beam of light. As can be seen in Figure 7.01 the 0.3-cm-thick plastic plate, whose other dimensions match those of the chromatoplates, has a 3 x 18-cm window about which are spaced four 0.1 x 0.2-cm strips of plastic. A sudden decrease in reflectance occurs when the beam of light falls upon a spot containing the compound of interest. During the

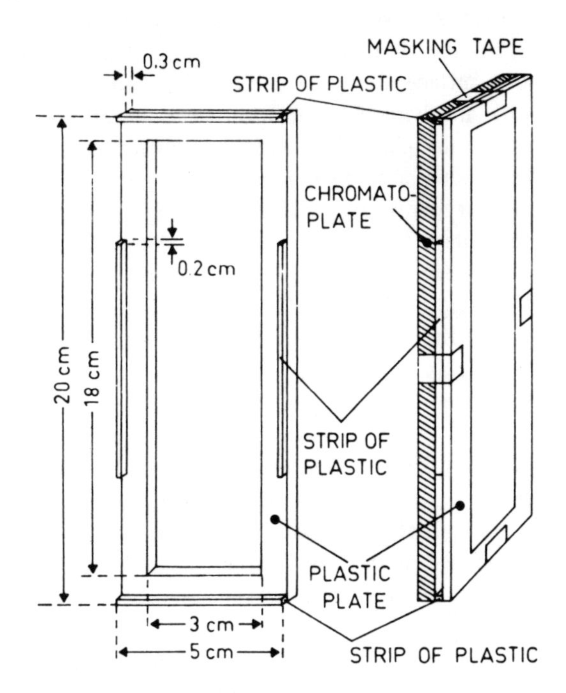

FIGURE 7.01 Assembly used for scanning chromatoplates.[4] (From Frodyma, M. M. and Lieu, V. T., *Anal. Chem.*, 39, 814 (1967). With permission of the Am. Chem. Soc.)

scanning process, the reflectance attachment is covered with a dark cloth to exclude outside light. When a particular compound has been located, its position is marked on the reverse side of the glass plate. The same approach has also been used successfully with a Spectronic 505 equipped with a reflectance attachment.

More-sophisticated devices for mechanical scanning, recording of reflectance, and detection of zones on chromatoplates have been discussed by Frei et al.,[7,8] Stahl and Jork,[9-16] Klaus,[17,18] and de Galan et al.[19] The new line of specialized recording instruments for chromatograms (Chapter 3, Section 4b) can be used conveniently for this purpose.

c. Recording of Reflectance Spectra

Once the resolved compounds have been located, their reflectance spectra are recorded in one of two ways. The direct recording of spectra is made possible by positioning the chromatoplates against the sample exit port of the reflectance attachment of a Beckman Model DK-2 or Spectronic 505 spectrophotometer in such a way that the light beam is centered on the spot of interest. A glass plate of identical size is taped on top of the thin-layer plates to protect the adsorbent surface during the recording process. For work involving UV radiation, it is necessary to replace the glass plate with the protective plastic plates illustrated in Figure 7.01 or with a paper mask of identical size, after punching holes where the spots are located. A sheet of thick nontransparent paper, resembling in color the plate coating, is inserted behind the plate to serve as a reflecting background. Spectra can be recorded also with the use of glass and, where necessary, quartz window reflectance cells.[20] When this procedure is followed, the spot of interest is excised and placed on top of 30 to 50 mg of adsorbent that has been removed from the same plate and introduced into an appropriate cell. After this sample has been compressed as described earlier, its reflectance is recorded in the usual manner. The reference standard in both procedures consists of adsorbent from the plate under investigation packed into the appropriate cell. For the procedure of recording spectra with the Zeiss Chromatogram spectrophotometer, see References 7, 10, 13, and 21. Automatic recording of spectra directly on the chromatogram can now be carried out in a matter of minutes with many of the double-beam instruments and with the Zeiss Chromatogram Scanner discussed in Chapter 2, Section 4b.

Newer instruments, such as the Farrand UV-VIS Chromatogram Analyzer, simplify the recording of spectra considerably. The original model of this instrument was not sold equipped with the monochromator drive motors, although provision was made for their addition. These motors are now available as options for this instrument. Using an instrument equipped with the motors, the recording of a reflectance spectrum is a relatively simple procedure. The proper auxiliary filters and slits are selected to give optimum response; the spot is centered in the slit; and the spectrum is measured.

The measurement of a reflectance spectrum using an instrument not equipped with monochromator drive motors is somewhat more time-consuming but is not difficult. The authors have generally found that in doing in situ reflectance measurements with the Farrand instrument, the analyzer monochromator may be removed from the system, and suitable auxiliary filters may be used in its place. This can save time when measuring spectra, as only one monochromator must then be adjusted. In making the measurements, the monochromator is set at a particular chosen wavelength; the recorder pen is adjusted to the baseline; and then the spot is scanned until a maximum recorder deflection is obtained. It is generally advisable to scan forwards and backwards across the spot and compare the maximum recorder deflections obtained in case of slight variations due to recorder lag or instrument noise. Once a uniform deflection is obtained and marked on the recorder chart, the recorder chart is advanced to correspond to the chosen wavelength scale; the monochromator is adjusted to the next setting chosen for measurement. After the baseline has been adjusted, the spot is scanned as before and the point of maximum deflection is again marked on the recorder chart. This procedure is repeated at successive wavelength settings until the spectrum has been recorded in the region of interest. An experienced operator can record a spectrum between, for example, 400 to 700 nm at 10 nm intervals in from 20 to 30 min.

d. Quantitative Measurement of Reflectance

Quantitative data have been obtained by the direct examination of chromatographic plates. By carefully adjusting the light beam of a Spectronic 505 or Beckman DK-2 spectrophotometer to the center of the spots, semiquantitative results with an accuracy of about 10% can be obtained. Tailing of the spots results in a drastic drop in accuracy.[22] Recently, spots no more than 0.75 cm in diam. chromatographed on Eastman chromatogram sheets have been cut out and placed in the light beam of a Beckman DU or Spectronic 20 reflectance attachment. The area of the impinging light beam is large enough to cover the entire spot, and it permits the measurement of reflectance in a single reading.[23] The accuracy and reproducibility of this rapid technique are superior to those of the method mentioned above, but light scattering and inhomogeneity of the spot-material distribution poses a problem.[24,25]

The quantitative measurement of reflectance by mechanical scanning and recording devices has been discussed in detail.[9-16] Jork[12] and Pataki[21] used this technique, and they have reported relative errors of between 3 and 4% for measurements in the UV regions.

Generally, the authors of this book have found that a greater degree of precision results in the evaluation of thin-layer chromatograms when the measurements are carried out on spots removed from the thin-layer plates.[22] The substance being analyzed is removed from the plate together with enough adsorbent to make up the analytical sample of predetermined weight (20 to 80 mg, depending on the adsorbent, provide an optimum thickness for reflectance measurements). The reflectance of this mixture is measured after it has been ground in a small agate mortar for a given period of time to ensure homogeneity and after it has been packed in an appropriate cell. The reference standard consists of adsorbent from the same plate that has been treated in the same way as the analytical sample.

The analysis can be speeded up by removing the analytical sample from the thin-layer plate with a circular aluminum planchet, which is manipulated by means of a cork stopper affixed to the planchet.[26] The size of the planchet employed is dictated by the thickness of the adsorbent layer and by the area to be excised. An assembly that has been used for this purpose is shown in Figure 7.02. Once the sample is cut from an adsorbent layer by exerting slight pressure on the inverted planchet, the most direct path between it and the nearest plate edge is cleared of adsorbent with a brush, and the planchet is moved along this path until the sample is deposited in the agate mortar. This way it is possible to remove a spot for analysis in less than a minute.

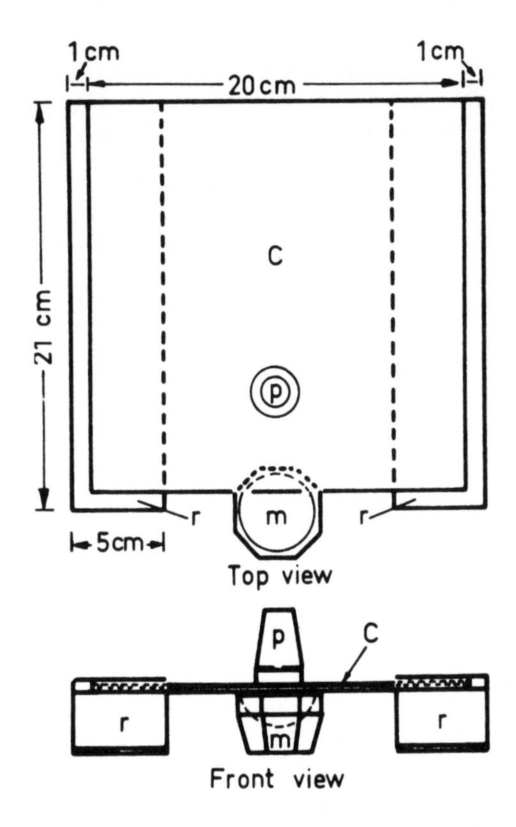

FIGURE 7.02 Assembly used to excise samples from chromatoplates. C, chromatoplate; m, agate mortar; p, planchet affixed to cork stopper; r, wooden rack.[28] (From Lieu, V. T., Frei, R. W., Frodyma, M. M., and Fukui, I. T., *Anal. Chim. Acta*, 33, 639 (1965). With permission of Elsevier Pub. Co.)

The suitability of this planchet technique as a means of speeding up the analyses of substances resolved on chromatoplates and absorbing in the visible or in the UV was tested by determining the deviation in the percent reflectance of samples of eosine B and salicylic acid prepared from center plates 2, 3, and 4. Plates 1 and 5, which in the coating procedure were the starting and the end plates, were eliminated because of poorer quality. Table 7.01 gives the precision attained with different thicknesses of these plates for 4- and 5-membered sets of samples. For individual plates of all thicknesses, average relative standard deviations of 0.3 and 0.4 reflectance units were obtained for eosine B and salicylic acid, respectively. A value of 0.5 reflectance unit was obtained for both compounds when the precision was computed for all plates of all thicknesses. The first values represent the level of precision obtained in the application of the reflectance technique to TLC using the weighing procedure, whereas the last value exceeds this level only slightly. It can therefore be concluded that the surfaces laid down by a commercial applicator are uniform enough, within certain limitations, to permit this procedure to be used in routine analyses without sacrificing the precision inherent in the reflectance technique.

If direct scanning techniques are used, it is important to standardize coating procedures in order to obtain uniform and reproducible plate coatings. Reflectance spectroscopy is, however, somewhat less sensitive to variations in layer thickness than are transmission techniques.[11,12] The effects of fluctuations in layer thickness, layer quality, and other experimental conditions on the reproducibility of reflectance measurements were investigated by Klaus[27] and Huber.[28]

In general, the disadvantage of the somewhat longer time of analysis involved in using the spot-removal technique is offset by its improved precision. The major advantages, however, partic-

TABLE 7.01

Relative Standard Deviation of Percent Reflectance of Samples of Eosine B and Salicylic Acid Adsorbed on Silica Gel and Prepared by means of the Planchet Technique[26]

Gate setting (MM)	Mean reflectance (% R)			Rel. S.D.* (% R)			Mean reflectance (% R)	Rel. S.D. (% R)
	Plate 2	Plate 3	Plate 4	Plate 2	Plate 3	Plate 4	All Plates	All plates
Eosine B								
0.75	84.5	83.2	83.8	0.3	0.5	0.6	83.5	0.6
0.50	83.1	82.6	82.5	0.4	0.2	0.2	82.7	0.5
0.25	84.6	84.4	84.1	0.2	0.2	0.3	84.3	0.4
					Av. rel. S.D. for individual plates 0.3			Av. rel. S.D. for all plates 0.5
Salicylic acid								
0.75	78.2	78.8	78.0	0.4	0.3	0.5	78.3	0.5
0.50	78.8	78.2	78.0	0.3	0.5	0.3	78.3	0.5
0.25	77.3	77.5	76.8	0.3	0.3	0.4	77.5	0.5
					Av. rel. S.D. for individual plates 0.4			Av. rel. S.D. for all plates 0.5

*S.D. (standard deviation) = $\frac{\sqrt{\Sigma(x - \bar{x})^2}}{n - 1}$

(From Lieu, V. T., Frei, R. W., Frodyma, M. M., and Fukui, I. T., *Anal. Chim. Acta,* 33, 639 (1965). With permission of Elsevier Pub. Co.)

ularly in comparison with mechanical scanning techniques, are its simplicity and the ability to better control experimental conditions, such as humidity and the homogeneity of the sample.

Although the authors of this book have used only Beckman and Bausch & Lomb instruments in their studies, the cells and procedures described earlier can doubtless be adapted easily to any other diffuse reflectance attachment available commercially.

The major disadvantage of the spot-removal technique is the fact that it cannot be automated or applied to PC and TLC sheets, and, with the advent of more and more commercial chromatogram scanners, which are available also with electronic readout equipment, it is felt that for handling a large number of samples the latter technique will become more popular. The spot-removal technique will retain its advantages, however, for research purposes where the investigation of adsorption properties of these systems is important for the actual design of an analytical method.

e. Reflectance vs. Transmittance

As mentioned previously, there has been considerable discussion about the relative merits of measurements in the reflectance and transmittance modes in situ on thin-layer chromatograms. While several groups have advanced arguments favoring transmittance, many authors have favored reflectance measurements. Some of the factors that are involved and the arguments that have been presented are treated here.

Reflectance has been shown to be less sensitive to layer variations than transmittance in several papers by Jork.[11,12] Other workers, while admitting the validity of the claim that reflectance is less susceptible to surface irregularities than is transmittance, prefer transmittance, as they believe it to give more accurate results for material within the layer.[29]

The case for transmittance measurements has been put forward by Touchstone et al.[30] and Goldman and Goodall.[31-33] Touchstone and coworkers studied the operation of the Schoeffel SD3000 densitometer in both the reflectance and transmittance modes of operation and found that the double-beam mode of operation provided an improvement in background and a more stable baseline. This will be discussed in more detail later. A comparison of the peak areas obtained upon scanning the azo-derivatives of estrone, estradiol-17, and estradiol by both transmittance and reflectance in the dual beam mode at 410 nm revealed that the peaks obtained for transmittance were approximately 2.5 times greater than those obtained for reflectance. In the reflectance mode, measurements of a number of steroids by fluorescence quenching and by means of their absorbance showed the quenching measurements to give larger peak areas, but a comparison of transmittance and reflectance for fluorescence quenching again revealed the transmittance mode to be superior. There are several points that should be noted about this comparison, however. The instrument Touchstone's group was using was basically designed as a densitometer, so it should not be particularly surprising that better response was obtained in the transmittance mode. Furthermore, the work was done with naturally colored compounds, so the interferences caused by spray background were avoided. An argument in favor of reflectance is that, being less sensitive to minor variations in the layer, it should be more precise. No conclusions about the precision obtained with reflectance and transmittance can be drawn from the work reported in this paper. Finally, the authors did not address themselves to the problems associated with the measurement by transmission of compounds that absorb in the UV-region of the spectrum.

Goldman and Goodall derived a theoretical basis for the measurement of compounds separated on thin-layers by reflectance and transmission.[31] From their considerations based on the Kubelka-Munk theory, they concluded that measurement by transmission offered advantages over reflectance and derived an equation for transmission that was a simplified form of the Kubelka-Munk equation. Their equation includes two basic effects – the background absorbance and the curvature of response. The expression was shown by experimental measurements with Sudan III on silica-gel thin-layers to be valid. The instrument used for the measurements was a modified Joyce, Loebl Chromoscan.

Goldman and Goodall next further modified the Chromoscan to operate as a flying-spot instrument.[32] The chromatographic slide was moved in a sawtooth motion while the light spot was stationary, so that the angular distribution of the incident light and the collection angle of the photomultiplier could be kept constant. As the methodology produced a large number of calculations (about 900 per analysis), a computer was used to handle the data. These workers next had an instrument constructed for measurements by transmission in the UV-region.[33] The design included a powerful UV-source and a very sensitive detector. A flying-spot design using a square wave motion was incorporated in the scanning principle, and data processing was again carried out by computer. The instrument was used to make measurements down to 239 nm, but limits of precision are not available based on the study.

Goldman and Goodall[33] have concluded that while in situ absorptiometry is not a particularly simple analytical technique, it has the advantage that it makes possible some analyses that were not previously possible and can replace other slower methods, which have been used in the past for certain analyses. The same may, however, be said for reflectance measurements. In absorptiometry, the adsorbent layer must be very carefully prepared to prevent such layer defects as small holes produced by bubbles or an inhomogeneous crystallization of the binder. The adsorbent's absorption to scattering ratio must be accurately determined, and developing solvents that absorb in the UV-region must be completely removed from the plate prior to measurements in the UV. The data processing used by these workers is also a source of possible error. It should be noted that the problems in layer preparation just described are of considerably less consequence in working with reflectance. Novacek has reviewed the work of Goldman and Goodall and supports their findings.[34]

The use of both reflectance and transmittance modes for in situ determinations with a Zeiss Chromatogram spectrophotometer has recently been reported.[35] This instrument is designed to operate in both modes and uses the same light-collecting device for both modes of operation. In the experiment, cobalt was chromatographed on Eastman Chromagram precoated cellulose sheets (6064), and the developed chromatograms were sprayed with 4-(2-thiazolylazo)-resorcinol (TAR), 0.1% in 95% ethanol. The cobalt((II)-TAR complexes appear as purple spots on a yellow background. A chromatogram containing eight spots of complexed cobalt (0.02 μg/spot) was scanned at 580 nm in both reflectance and transmittance modes. In both cases, the same slits and scanning speeds were used. Relative standard deviation in the transmittance mode was 4%, while that calculated for the reflectance measurements was about 2%. Figure 7.03 shows calibration plots of peak area vs. concentration for both modes of operation.

The relationship

$$A^2 = kc \tag{7.01}$$

is a simplified form of the Kubelka-Munk relationship

$$F(R) = (1-R)^2/2R = k/s \tag{7.02}$$

where k is the absorption coefficient and s is the scattering coefficient (constant under the experimental conditions). Applying function (Equation 7.01) to the transmission results yields a plot that approaches linearity over approximately the same concentration range (0.02 to 0.8 μg/spot) as the reflectance measurements. The transmittance plots, however, do not extend through the origin, nor do Beer's Law plots of absorbance vs. concentration improve the transmission results. Thus, despite the findings of Touchstone et al.,[30] Goldman and Goodall,[31-33] and Novacek,[34] these experiments display the superiority of reflectance measurements, even though the comparative measurements were made at 580 nm where the transparency of the chromatogram sheets is relatively good.

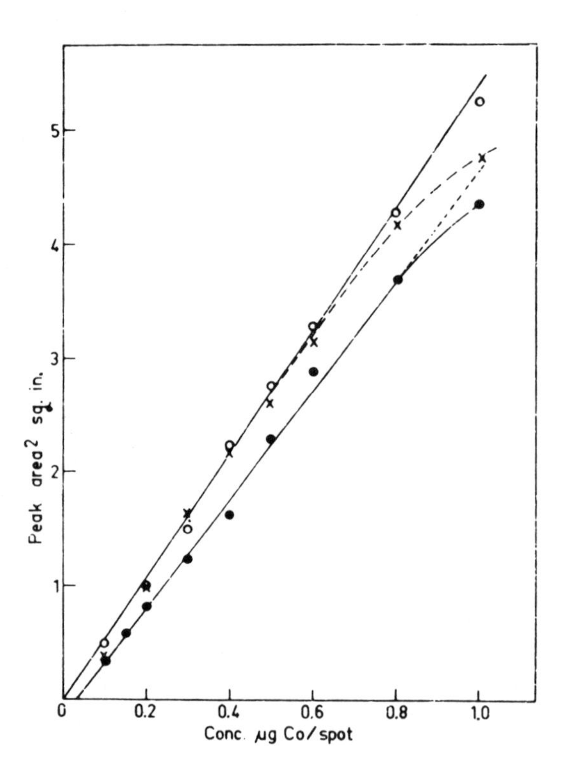

FIGURE 7.03 Calibration curves peak area squared vs. concentration, obtained by reflectance spectroscopy without backing sheets (x-x), for infinite layer thickness (2 backing sheets) (O-O), and by transmission spectroscopy.[35] (From Frei, R. W., *J. Chromatogr.*, 64, 285 (1972). With permission of Elsevier Pub. Co.)

As previously mentioned, a major argument favoring diffuse reflectance over transmittance has been that reflectance is less susceptible to variations in the thin layer and therefore should offer better reproducibility than transmittance. These experiments with the Zeiss instrument, which uses the same collection device for diffusely reflected, scattered, and transmitted light, makes a comparison of results obtained in the various modes more meaningful. In the arrangements used, all parameters can be kept constant for both modes of operation, but the slanting of the collection tube to eliminate specular reflectance actually produces a slight bias in favor of the transmittance measurement. Thus, one must consider the possibility that other experiments,[30-34] which have shown transmission to be more sensitive than reflectance, simply reveal an inefficient collecting device for reflectance in the instruments used in the experiments.

The actual transparencies of thin-layer adsorbents was investigated by Jork.[11] At 500 nm, only about 2.2% of the incident light is transmitted through a silica-gel layer 300 μm thick, while at a layer thickness of 160 μm, which is nearer normal layer thicknesses used in chromatography, 5 to 10% of the incident light is transmitted. From these considerations, it becomes immediately evident that reflectance should offer greater sensitivity if an efficient collecting device is used, although admittedly transmitted light poses fewer problems in the design of an efficient collecting system. Most adsorbents exhibit an appreciable UV-absorbance, so that much less light is transmitted through layers in this region of the spectrum than at longer wavelengths. Thus, while it has been demonstrated that transmittance may be used in the UV-region,[33] one cannot help but observe that the same measurements may be made by reflectance using much less elaborate equipment.

Other researchers have also concluded that reflectance offers advantages over transmittance.[13,36-38] The relative merits of the two techniques have also been discussed in a recent review.[39] Boulton and Pollack[39] observed that reflectance should be favored when the adsorbent has strong scattering and a high transmittance loss. Seiler and Möller surveyed methods for the direct in situ determination of compounds that absorb in the UV-visible range, including spot area measurement, densitometry, and reflectance.[37] While simple spot area measurements sometimes can provide useful information with a minimum of expense, reflectance measurements were found to give the most precise results, with relative standard deviations from 3 to 7% when samples were compared with standards of the same approximate concentrations.

Treiber et al. have found that considerable improvement in the direct spectrophotometric evaluation of chromatograms can be achieved by measuring the reflectance and the transmittance of the chromatogram simultaneously.[40] Studies were made using a Vitatron Model VFD 500 densitometer, a Schoeffel Model Sd 3000 densitometer, and a Zeiss Chromatogram spectrophotometer. The comparative studies with the Zeiss instrument in the transmission, reflection, and transmission-reflection modes are quite revealing. The results of scanning a chromatogram containing 50 ng of testosterone as testosterone-2,4-DNPH and a non-polar reaction product are shown in Figures 7.04 to 7.06.

Figure 7.04 shows the result in the transmission mode. Peak A is the testosterone-2,4-DNPH and peak B is the reaction product. The upper line is the scan of the area of the chromatogram containing the spots, while the lower plot is the spray

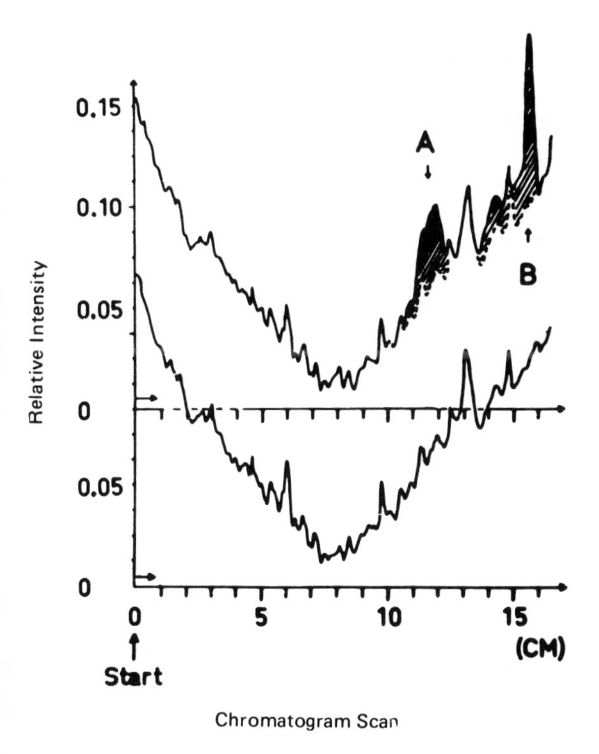

FIGURE 7.04 Scan of chromatogram containing 50 ng testosterone-2, 4-DNPH (A) and a nonpolar reaction product (B) using the Zeiss Chromatogram spectrophotometer in the transmission mode. Bottom curve is baseline. (From Treiber, L. R., Nordberg, R., Lindstedt, S., and Stollenberger, P., *J. Chromatogr.,* 63, 211 (1971). With permission of Elsevier Pub. Co.)

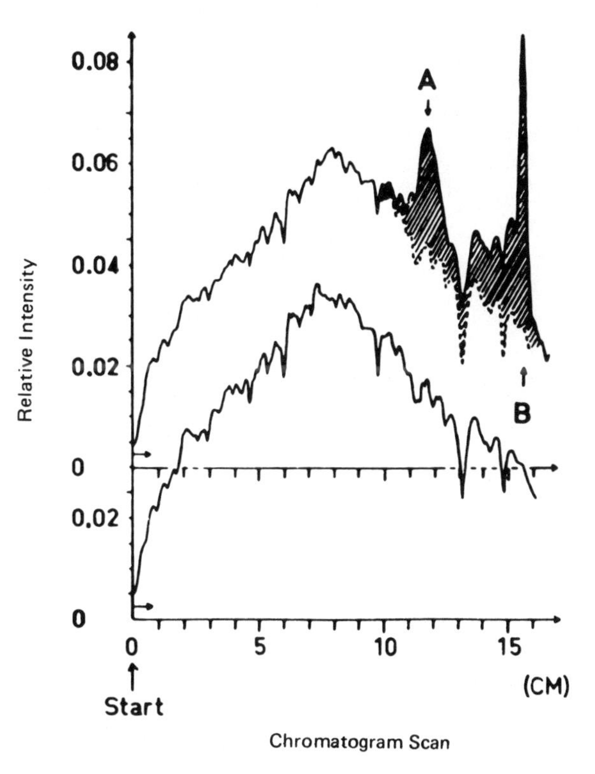

FIGURE 7.05 Scan of chromatogram containing 50 ng testosterone-2, 4-DNPH (A) and a nonpolar reaction product (B) using the Zeiss Chromatogram spectrophotometer in the reflectance mode. Bottom curve is baseline.[40] (From Treiber, L. R., Nordberg, R., Lindstedt, S., and Stollenberger, P., *J. Chromatogr.,* 63, 211 (1971). With permission of Elsevier Pub. Co.)

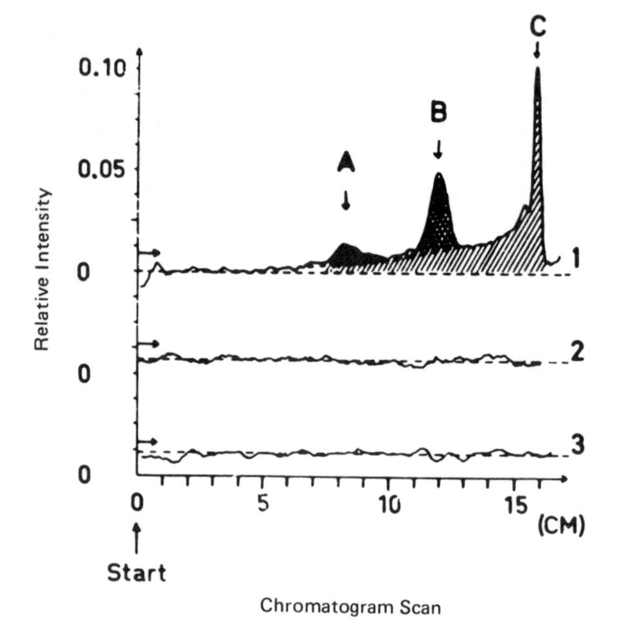

FIGURE 7.06 1. Simultaneous measurement of transmittance and reflectance using Zeiss Chromatogram spectrophotometer; B = 50 ng testosterone-2, 4-DNPH, C = a nonpolar reaction product, A = unreacted 2,4-dinitrophenylhydrazine. 2. Baseline for simultaneous measurement of reflectance and transmittance. 3. Instrument stability.[40] (From Treiber, L. R., Nordberg, R., Lindstedt, S., and Stollenberger, P., *J. Chromatogr.*, 63, 211 (1971). With permission of Elsevier Pub. Co.)

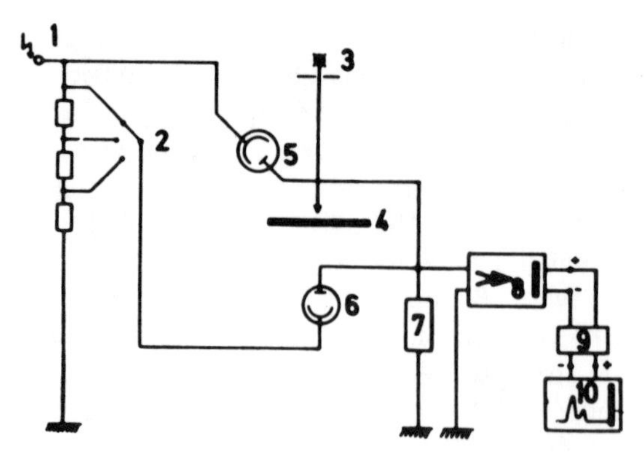

FIGURE 7.07. Block diagram of the instrument setup for simultaneous measurement of reflectance and transmittance. 1, high voltage source; 2, circuit for adjustment of optimum transmittance-reflectance ratio; 3, light source; 4, sample; 5, photomultiplier for reflectance; 6, photomultiplier for transmittance; 7, common work resistance; 8, amplifier; 9, scale expander; 10, recorder.[40] (From Treiber, L. R., Nordberg, R., Lindstedt, S., and Stollenberger, P., *J. Chromatogr.*, 63, 211 (1971). With permission of Elsevier Pub. Co.)

background. Figure 7.05 shows the same scan in the reflectance mode. Again, the spray background is quite significant. However, simultaneous measurement of the chromatogram, as shown in Figure 7.06, results in a very stable baseline response and enables the measurement of an additional spot, which was not previously distinguishable. The schematic block diagram of the instrument setup used in these experiments for the simultaneous measurement of transmittance and reflectance is shown in Figure 7.07.

This method of operation offers several advantages over measurements in a single mode. There is an increase in sensitivity and reproducibility, a constant baseline, and, experimentally, a linear relationship between the recorded peak area and the concentration of material. Recently, Treiber has reported a new function for use in simultaneous measurement of reflectance and transmittance in situ on thin-layer chromatograms.[41]

f. Problems in Preparation of Samples

The optimization of conditions for measurement of a sample is a primary concern to an analyst. In working with in situ determinations on thin-layer chromatograms, there are two sources of possible error that must be considered – the chromatographic process and the measurement process. Errors in the measurement process are generally inherent in the operation of the instrument, and the analyst's task is to determine the optimum conditions for a particular measurement, thus minimizing the error. Possibilities for introduction of error in the chromatographic process are much greater and much more difficult to bring under control.

(i) Error Analysis

A complete error analysis with a simulated, infinite layer thickness and using scanning with an optical slit has recently been reported.[35] The study was carried out on a Zeiss Chromatogram spectrophotometer using the cobalt-TAR system for the in situ reflectance study. The optimum concentration of sample for best accuracy for systems that adhere to the Kubelka-Munk theory is deducible from the relative error term dc/c. From the Kubelka-Munk equation, the error in c may be written as

$$dc = \frac{K'(R_\infty - 1)dR_\infty}{2R_\infty{}^2}, \qquad (7.03)$$

so that the relative error in c may be represented as

$$\frac{dc}{c} = \frac{(R_\infty + 1)dR_\infty}{(R_\infty - 1)R_\infty}. \quad (7.04)$$

If we assume a reading error of 1%R, then dR = 0.01, so that

$$\frac{dc}{c} \times 100 = \frac{(R_\infty + 1)}{(R_\infty - 1)R_\infty} = \% \text{ error in c.} \quad (7.05)$$

Plotting dc/c X 100 (from Equation 7.05) as a function of %R yields an error analysis curve such as that depicted in Figure 7.08 (●-●). The slope of a calibration curve may also be used as a basis for such a plot. Plotting standard deviation (%) vs. concentration of cobalt per spot produced a curve such as that shown in Figure 7.08 (●-●). In this case where the measurements have been made in situ, results are obtained as peak areas rather than as reflectance values; so meaningful reflectance values could not be obtained. Performing such an error analysis reveals the optimum working range for the analyst to use. In the example given, the optimum range for the analysis of cobalt as the TAR complex is 0.2 to 0.8 μg/spot.

(ii) Optimizing Instrumental Parameters

Little can be said about the measures an analyst should take to get the best results from a particular instrument, as each of the various instruments on the market has its own particular requirements. However, a few observations that should be generally applicable can be made.

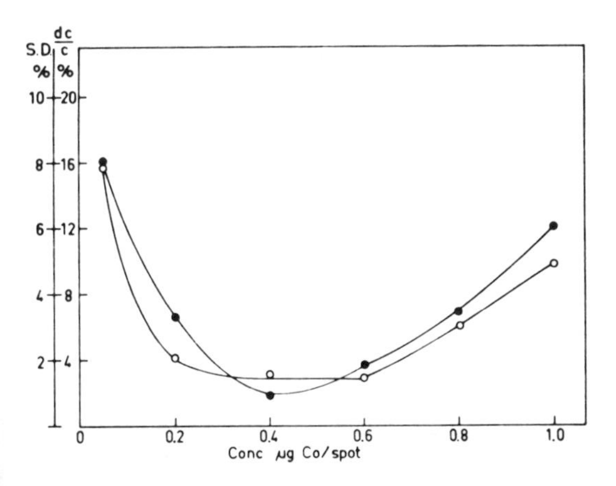

FIGURE 7.08 Error analysis curves. Variation of % S.D. (O-O) and % dc/c (●-●) as a function of cobalt concentration.[35] (From Frei, R. W., *J. Chromatogr.*, 64, 285 (1972). With permission of Elsevier Pub. Co.)

Optimum conditions, of course, require that the measurement be made at the maximum absorbance of the compound. This means, depending on the instrument being used, selection of the correct monochromator setting or filter for the measurement. The proper slit arrangement must be chosen to give the largest recorder deflection, and the best amplification factor should be chosen so as to obtain measurable on-scale peaks from all samples in the calibration range. Other factors that should be considered also include the selection of proper lamps and detectors for the wavelength region in which the measurement is being made. A most important consideration in reflectance work is, that the chromatogram being measured be part of the optical system, so care must be taken to see that the chromatogram plate is properly aligned. Different instruments contain various provisions to meet this requirement. In the Zeiss, for example, the detector is lowered manually to the surface of the plate before measurement. This allows the measurement of plates of varying thicknesses, but the operator must be careful to lower the detecting head to the same position for each determination. In the Farrand instrument, on the other hand, the chromatogram sits on a spring-loaded plate that rises to meet a set of stops. Thus, due to the stops, no matter what thickness of plate is used, the surface will always be in the same place for the measurement.

Double-beam measurements have been shown to be quite superior to single-beam measurements, particularly when the plate being measured has a high spray background.[35] Figure 7.09 depicts various conditions of measurement. Plot (a) shows the background for reagent obtained on the Zeiss in the transmission mode at 580 nm with the TAR spray reagent, while plots (b) to (d) show reflectance measurements on the Zeiss with (b) a black chromatogram stage, (d) two clean sheets backing the chromatogram to simulate an infinite layer, and (c) a black chromatogram stage with white lines. The importance of the backing sheets is especially obvious for measurements on chromatogram stages containing grid markings. Plot (e) shows the results obtained for work with a Farrand chromatogram analyzer with the two backing sheets in the double-beam mode. The improvement in background stability is quite dramatic. A considerable improvement in reproducibility has also been noted for a comparison of measurements in the single- and double-beam

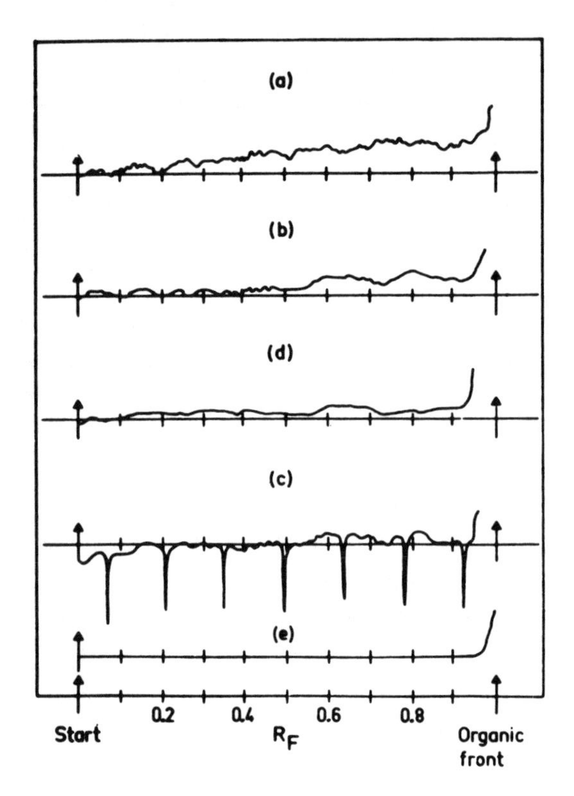

FIGURE 7.09 Scans of reagent background (580 nm) for cobalt-TAR complex at different conditions with the Zeiss Chromatogram Spectrophotometer (a-d): (a) by transmission measurement; (b) on black chromatogram stage; (c) on black stage with white lines; (d) with infinite layer thickness (2 empty sheets); (e) double-beam recording with the Farrand UV-VIS Chromatogram Analyzer.[35] (From Frei, R. W., *J. Chromatogr.*, 64, 285 (1972). With permission of Elsevier Pub. Co.)

modes with the Farrand instrument.[42] Measurements in the single-beam mode produced a relative standard deviation of 3.5%, while the same spots measured in the double-beam mode produced a relative standard deviation of 1.5%. For plates that do not require spraying and especially for measurements in the UV-region, the improvement noted for the double-beam mode is less obvious.[35]

(iii) Optical and Electrical Noise

Boulton and Pollack have approached the problem of optical noise in photoelectric densitometers from a theoretical point of view.[43] Although the publication deals with instruments operating in the transmission mode, many of the factors influencing sensitivity, stability, and reproducibility should be similar to those encountered in using reflectance instruments. Data obtained from measurements are generally in the form of an electrical signal in spectrophotometric work. The information contained in the signal is determined by the ratio of the amplitude of the useful component of the signal to that of the sum of all unwanted components contained in the signal. These undesirable components arise from two sources – random signals (noise) and a deterministic signal arising from nonlinearities in the transfer function of the system. Practically speaking, it is the so-called "signal-to-noise" ratio that determines accuracy and instrumental working limits.

Two main types of noise are encountered in work with photodensitometers – electrical noise arising mainly in the photodetector elements and optical noise due to fluctuations in the optical transfer parameters of the medium.[37] Outside signals, which may cause interferences, can be largely excluded by proper design. The two types of noise being considered are of a random nature and are not correlated. Their combined effect may be represented by the square root of the sum of their respective squared amplitudes

$$E_{n_t} = \sqrt{E_{el}^2 + E_{opt}^2} \quad . \qquad (7.06)$$

When the two components are of unequal amplitude, the stronger one becomes the more important in determining the signal noise.

Most electrical noise arises in the photodetector, which is a photomultiplier tube in most instruments, and may be represented as the Noise Equivalent Power (NEP) of the detector.[39] The intensity of a light beam that, upon striking the detector, produces a signal equal to the amplitude of the noise signal generated within the detector is determined by the NEP. Most tubes have an NEP value of 10^{-16}W, although tubes with an NEP of 10^{-20}W are available. The signal sensed by the detector is proportional to the incident light, and making this value sufficiently large reduces the importance of the electrical noise, which is constant.

The optical noise is a function of the medium and acts through its effect on the incident light beam. As the medium exerts much less effect in reflectance measurements than it does in transmission measurements, it should be of considerably less importance. Optical and electrical noise are decreased by decreasing the electrical bandwidth of the detector system, which also results in a slower response. The most efficient means of reducing optical noise, however, is to use a double-beam system. The signal from the wavelength at which the chromogen is being scanned is referenced to the signal obtained through scanning the same area of the chromatogram at a wavelength at which the chromogen is practically nonabsorbing. Thus, the first beam carries both the desired signal plus noise, while the second beam carries only noise. A differential amplifier may then be used to separate the two signals.[39] An alternative is a ratio-forming system, which is regarded as superior to difference-forming by Boulton and Pollack.[39] Other publications by Boulton and Pollack deal in detail with optical noise in double-beam difference systems[44] and with the use of semiconductor photodetectors.[45]

Pollack and Boulton have also considered other sources of error in quantitative thin-layer chromatographic determinations.[46,47] They have discussed special conditions for the simulation of the optical characteristics of plane parallel isotopic media where the general solutions either degenerate or may be considered simplified.[46] While the first case considered deals with a scattering medium with vanishing absorbance, the second assumes high scattering and medium absorbance and applies in the cases of both transmittance and reflectance. The determination of the basic constants of the solutions by optical measurements on the blank adsorbent and the character of the noise in both transmittance and reflectance modes are also considered. Results show that reflectance is less susceptible to optical noise than transmittance, and thus reflectance measurements are recommended when the adsorbent has high scattering power and low to medium absorption at high optical density, conditions usually encountered in work on thin-layers. The influence of finite spectral or spatial width of the scanning window in introducing errors in photodensitometers and reflectometers has also been dealt with.[47] The results show that relatively wide spectral windows may be used without producing excessive errors.

(iv) Chromatographic Factors

Many of the sources of error in quantitative work on paper and thin-layer chromatograms have been discussed in a recent book.[48] Sources of error outside the instrument have been divided into three types by Getz.[38] The first of these is the influence of irregularities in the adsorbent layer. While reflectance measurements are not usually adversely affected by variations in the layer thickness of the plate and other surface irregularities, plates that have cracked or flaked are generally unsuitable for measurement.

A second source of error is in the spotting technique. Errors arising from various means of delivering the sample to the chromatogram have been discussed previously,[48] but several observations may be made about proper procedure. It is generally advisable that the sample be delivered in as small a volume as possible and that all samples be spotted in the same volume so as to standardize the error in delivery volume.[38] Spot diameters at the origin should also be kept as small as possible to prevent the spots from becoming too diffuse as the chromatogram is developed. Finally, care must be taken not to mar the surface of the adsorbent at the origin during spotting, as this can have very adverse effects on the chromatography. Automatic spotting devices and plates containing a blotter pad at the origin, which begins all the spots from the same point and protects the adsorbent from being marred, offer promise of greatly reducing errors associated with spotting, while channeled plates should help prevent lateral diffusion of the spots during chromatography.

A third problem in the chromatographic process arises from the delivery of the chromogenic agent to the surface of the chromatogram. This is most commonly done by spraying the plate, although the chromogenic agent is sometimes included in the developing solvents for the chromatogram. Spraying techniques require both skill and patience on the part of the analyst. In the author's experience, there are two main problems associated with spraying the chromatogram. The first of these is to obtain a uniform coating of spray reagent over the surface of the plate and thereby prevent large fluctuations in background. The second is to deliver an optimum amount of reagent to the plate. Too little reagent results in low spot intensities and poor reproducibilities, while excessive spraying can cause severe background problems. The results of uneven spraying are usually revealed by a variation in intensity between spots in the center of the plate and those at the edges. Getz has suggested dipping the plates to apply the chromogenic reagent.[38] This may prove useful in some instances, but the authors have had some unfortunate experiences in attempting this procedure. Apart from the obvious problem of controlling the amount of reagent applied in a dipping procedure, some commercial plates, as well as plates prepared in the laboratory, tend to crack and flake after dipping.

Jork has studied the influence of chromatographic factors on reflectance measurements on thin-layer chromatograms.[49] He has shown that a stationary phase consisting of thin layers of finely granulated adsorbents having small pore diameters can increase sensitivity by a factor of 10. He also found that no relationship existed between the direction in which the layer was applied, the direction of chromatography, and the direction of scanning. The applications of samples and optimum conditions for measurements were also discussed.

Waksmundzki and Rozylo have also studied the effect of the adsorbent layer in quantitative TLC.[50,51] The area of the spot and the elimination of tailing are important considerations in such measurements. These workers have studied the relationship of the spot areas of separated substances to the specific surface area of the adsorbent and have found that the kind of adsorbents and solvents used affect the spot area of the separated substances.[50] The relationship between the peak area of the elution curve and the specific surface area of the adsorbent and the kind of solvent used was also investigated.[51]

(v) Homogeneity of Sample

If a system is to adhere to the Kubelka-Munk theory, the adsorbing material must be uniformly distributed over the illuminated area.[35] While this may best be accomplished by removing the spot from the plate and mixing it prior to packing, thus achieving an infinite layer thickness and sample homogeneity, the process is time-consuming and seems somewhat redundant if the laboratory is equipped with a chromatogram-scanning spectrophotometer. One way of avoiding this problem is to illuminate only a small area of the spot at one time, as in the flying-spot system proposed by Goldman and Goodall,[32] in which a spot of light oscillates over the chromatographic spot. In the small area covered by the spot at any given time, it may be assumed that the sample is homogeneous. Such systems can, however, have the disadvantages of slow scanning and considerable data processing.

Studies have been carried out on the Zeiss instrument in which the reflectance of a spot was measured in a single measurement using an adjustable light beam slightly larger in cross section than the largest spot in the dilution series.[35] Previous studies by Braun and Kortüm[24,52,53] investigated the problem in conjunction with paper chromatography. The results indicated that the measured reflectance depends on the irregularity of the spot and also on the amount of empty space around the spot included in the scan. In these experiments, infinite layer thickness was simulated by backing the spot with several clean sheets of chromatogram paper.

The Kubelka-Munk equation (7.07)

$$F(R_\infty) = \frac{(1 - R_\infty)^2}{2R_\infty} = \frac{\epsilon c}{s} \qquad (7.07)$$

was modified for the single-beam approach used by these workers to give equation 7.08

$$F'(R_\infty) = \frac{b^2}{2n(1 - b)} - \frac{F(R'_\infty)}{n}, \qquad (7.08)$$

which is proportional to the absorbing material in the spot. Using the mean observed diffuse reflectance $\overline{R}_\infty$ of the complete area of the spot and surrounding material covered by the light beam and the diffuse reflectance of the paper $\overline{R}'_\infty$, the term b may be computed from the equation (7.09)

$$b = 1 - n\overline{R}_\infty + (n - 1)R'_\infty \qquad (7.09)$$

where n is the ratio of the area scanned to the actual spot area.

Using the cobalt-TAR system, this relationship was tested on silica-gel thin-layer sheets backed with two clean sheets to simulate infinite layer thickness. The light beam was adjusted to 11 mm diam, and a monochromator slit of 0.3 mm was used. The results are shown in Figure 7.10.

The plot of percent reflectance vs. concentration shows a loss of linearity at higher concentration (⊙-⊙), while plotting the Kubelka-Munk function against concentration (O-O) results in deviations from linearity at low concentrations. This may be attributed to the nonhomogeneity of the system.[51] When the modified function was plotted (Equation 7.08), the reflectance measurements were made against a background adjusted to 100% reflectance so that the last term in equation 7.08 could be neglected, giving equation 7.10,

$$F'(R_\infty) = \frac{b^2}{2n(1 - b)} = \text{prop. c.} \qquad (7.10)$$

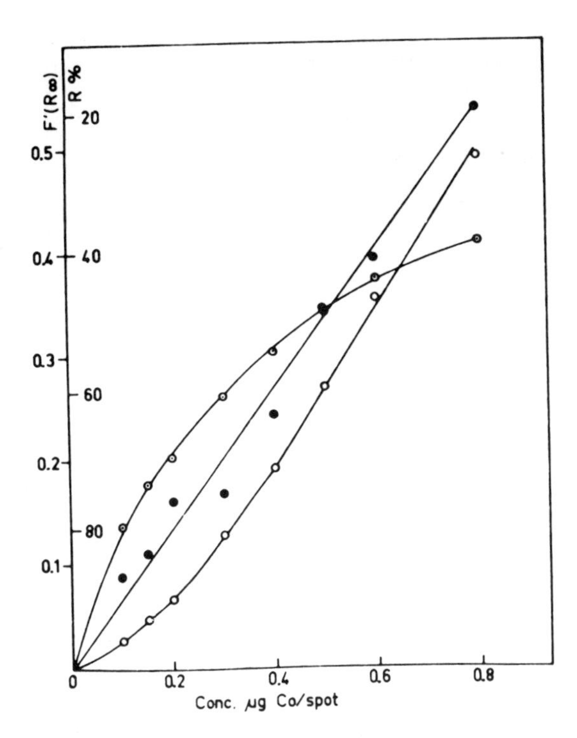

FIGURE 7.10 Calibration curves for cobalt-TAR complex obtained by reflectance spectroscopy with adjustable beam. Kubelka-Munk values (O-O); modified Kubelka-Munk values (•-•); % reflectance (⊙-⊙) as a function of concentration of cobalt. Measurements performed at 580 nm with adjustable beam d = 11 mm.[35] (From Frei, R. W., *J. Chromatogr.*, 64, 285 (1972). With permission of Elsevier Pub. Co.)

Plotting the modified function vs. concentration resulted in linear plots up to the maximum concentration studied (•-•), 0.8 μg Co/spot.

g. Conclusions

From the preceding considerations, it may be concluded that diffuse reflectance spectroscopy is preferable to in situ transmission measurements for the evaluation of chromatograms. In the UV-region, transmission measurements become particularly unattractive. A slit-scanning reflectance spectrometer thus seems to be the instrument that offers the greatest versatility for chromatographic work. Based on theoretical principles alone, the flying spot and spot-removal techniques may seem advantageous but offer no real advantages in practice. The use of a double-beam reflectance scanner with infinite layer simulation may clearly be used for quantitative work. Indeed, so may the other methods discussed, though with a loss in reproducibility.

Indeed, the instruments now available for chromatogram scanning are much more accurate than the chromatography, so it is not really with the instrumentation that problems in reproducibility generally arise. Better prepared plated, automatic spotting procedures and chromatography under carefully controlled, reproducible conditions should all combine to produce improvements in the results of in situ determinations.

2. Analysis of Pesticides

Thin-layer chromatography is a widely recognized technique for the separation of pesticides, and many chromatographic systems and colorimetric spray reagents for these compounds have been reported in the literature.[54] Pesticides may be detected under ultraviolet light either through their own UV-absorption or against an indicator that fluoresces under UV-light on thin-layer chromatograms. Pesticides also may be detected by a color reaction involving either reaction in solution prior to chromatography or by a reaction on the chromatogram with a spray reagent. There are three major areas of investigation in which in situ reflectance measurements on thin-layer chromatograms have been used for the analysis of pesticides that will be discussed in this chapter: residue analysis, formulation analysis, and degradation studies. These analytical applications are discussed in the following sections.

a. Residue Analysis

Taking advantage of the nondestructive detection of the compounds by fluorescence quenching,[5] scientists investigated the use of UV reflectance techniques for the evaluation of triazines from thin-layer plates.[6] Triazine stock solutions (2 mg/ml) were prepared in spectroscopic grade methanol. The 20 x 5 x 0.35-cm plates were coated with a layer of Camag silica gel DF-5 with organic fluorescence indicator 0.25 mm thick by means of a Desaga Model Sii applicator. In some cases the Camag silica gel was replaced by a 2% zinc sulfide mixture in Merck silica gel H. The triazines were applied on the chromatoplates by means of a 10-μl Hamilton microsyringe. After separation by the ascending technique with a 2% methanol-in-chloroform mixture, the plates were dried at 60°C for 10 min, and the spots were observed under an UV lamp.

The reflectance spectra were recorded with a Spectronic 505 spectrophotometer equipped with

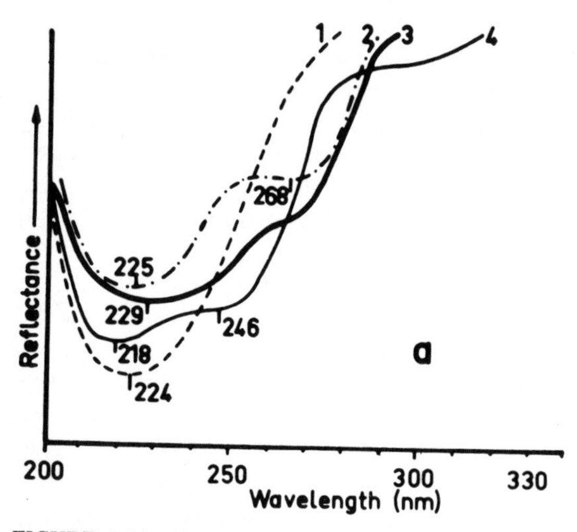

FIGURE 7.11 Reflectance spectra of 1, prometon; 2, propazine; 3, prometryne; 4, simazine adsorbed on silica gel.[6] (From Frei, R. W. and Nomura, N. S., *Mikrochim. Acta,* 565 (1968). With permission of Springer-Verlag.)

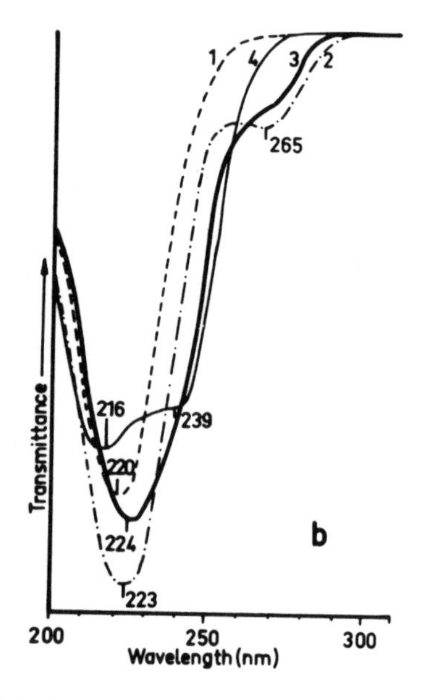

FIGURE 7.12 Transmission spectra of 1, prometon; 2, propazine; 3, prometryne; 4, simazine in methanol.[6] (From Frei, R. W. and Nomura, N. S., *Mikrochim. Acta,* 565 (1968). With permission of Springer-Verlag.)

a standard reflectance attachment for the UV and visible regions of the spectrum. The preparation of the chromatoplates for reflectance measurements and the packing of removed spots in cells suitable for quantitative work in the UV region have been described under General Experimental Procedure, Chapter 7 Section 1d. It was found that the spectra of the triazines could be used for identification purposes. Reflectance spectra of triazines

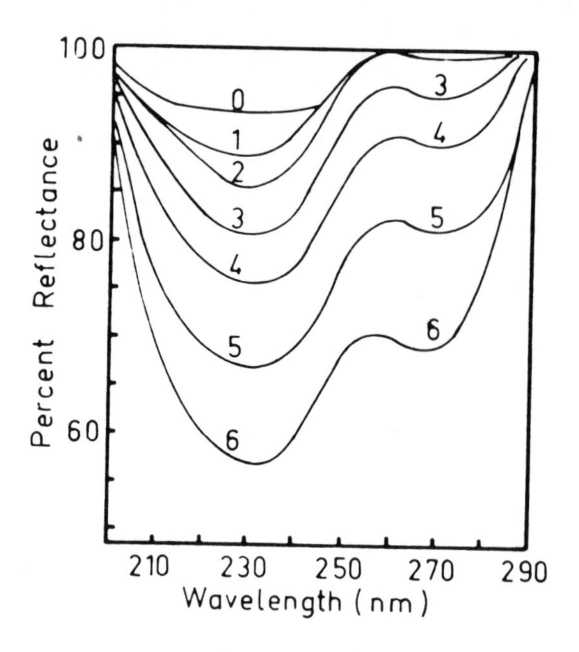

FIGURE 7.13 Reflectance spectra of various concentrations of atrazine adsorbed on silica gel. 0, base line; 1, 1, μg per sample; 2, 2 μg; 3, 5 μg; 4, 10 μg; 5, 20 μg; 6, 40 μg.[6] (From Frei, R. W. and Nomura, N. S., *Mikrochim. Acta,* 565 (1968). With permission of Springer-Verlag.)

commonly used for large-scale weed control are recorded in Figure 7.11.

A comparison of the reflectance spectra with the corresponding transmission spectra of the triazines dissolved in methanol shows some resemblance (see Figure 7.12). In general, the reflectance spectra have broader maxima because of increased light scattering and exhibit a more or less pronounced bathochromic shift.

The method was tested for its suitability for quantitative analysis using atrazine. Figure 7.13 shows the spectra of various concentrations of atrazine packed in suitable cells. The spectra reveal two maxima suitable for analysis at 230 and 270 nm. The use of the second maximum offers the advantage of higher radiation energy output, but it shows a lack of sensitivity. A small bathochromic shift is usually observed with increasing concentration. However, a single wavelength could still be used for the construction of a calibration curve that covers the concentration range of interest. Linear plots were obtained for up to 50 μg of herbicide per sample by plotting the Kubelka-Munk function for diffuse reflectance $[(1 - R)^2/2R]$ against concentration. Empirical functions, such as (2 - log %R), plotted against the square root of concentration were also used successfully.

The precision of this technique was determined by analyzing six 10-μg samples of atrazine simultaneously. The standard deviation of the reflectance readings was found to be 0.50 when measured at 270 nm and 0.84 at 230 nm. The decrease in precision can be attributed to the decrease in radiation energy output at a lower wavelength and to the increased self-absorption of the silica gel: hence more amplification is needed, and as a result, increased fluctuation of the meter needle is encountered. The corresponding accuracy of the method was found to be about 3% for a 10-μg sample. Samples stored in a desiccator for several days showed no significant changes in measured percent reflectance. The sensitivity of the technique ranges between 1 and 3 μg (at a 50% accuracy level) for all triazines investigated.

In a later publication dealing with in situ analysis of triazines,[55] the UV reflectance technique was compared to other methods. Relative percent errors for four methods are shown in Table 7.02.[5,6,55,56] From these data we can see the superiority of diffuse reflectance spectroscopy. For the fluorescence quenching mode, the loss in accuracy can be attributable to background fluctuations. If direct scanning techniques are used for these two methods (I and II), the sensitivity of fluorescence quenching is at best equal, but usually somewhat poorer than for reflectance techniques (see Figure 7.39). Methods III and IV are strictly semiquantitative. Method IV may be more timesaving than methods I to III, and, since no electronic equipment is needed, it may have some merit as a field method.

An *s*-triazine (prometryne) was used as a test substance to investigate parameters for the determination of substances from thin-layer plates by UV reflectance spectroscopy.[28]

The fiber-optics thin-layer scanner described by Beroza et al.[57] (see Chapter 4, Section 4a) was used in the determination of some chlorinated and thiophosphate pesticides. Chromatography of chlorinated pesticides (aldrin, dieldrin, endrin, DDT, DDD, DDE, lindane, heptachlor, heptachlor epoxide, methoxychlor, and toxaphene) was carried out on prepared alumina plates. (The plates were coated with a slurry containing 30 g of alumina, in 45 ml of ethanol containing 0.125 g of silver nitrate.) The pesticide solutions were applied to the plates with a 1 μl micropipette, spotting as many times as necessary to obtain the desired concentration of pesticide, and the plates were then developed 10 cm in solvent system containing 2% acetone in hexane. After drying, the plates were irradiated under a UV-lamp for 10 min to produce colored spots (light brown or grey to black) on a white background. Exposure to steam, followed by irradiation for another 30 sec produced spots of optimum intensity. The plates were then scanned by reflectance; peak areas were

TABLE 7.02

Quantitative Analysis of Atrazine by Four Different Methods[55]

Amount of atrazine added (μg): x_1	Amount of atrazine found (μg): x_2 I	II	III	IV
10.0	9.7	9.2	11.2	12.1
6.0	5.9	5.8	6.7	5.6
4.0	3.8	3.8	4.6	4.5
Relative % error $\frac{x_1 - x_2}{x_1}$ 100%	3.3	5.6	12.9	13.4

I. Ultraviolet reflectance spectroscopy[6]
II. Fluorescence quenching: fluorometric[55]
III. Fluorescence quenching: spot-area measurement[56]
IV. Fluorescence quenching: visual comparison of spots[5]

(From Frei, R. W. and Freeman, C. D., *Residue Rev.*, 1214 (1968). With permission of Springer-Verlag.)

recorded; and the results were evaluated planimetrically.

The thirteen chlorinated pesticides chosen for study were chromatographed singly and in mixtures at concentrations ranging from 0.1 to 32 μg/spot. Plots of peak area vs. concentration were generally found to be linear over a concentration range of approximately 0.2 to 10 μg. It was found necessary to run standards on all plates due to variations in intensity of the spots obtained from one plate to another. Above 10 μg, a change in slope was observed that was attributed to a saturation of the reflectivity due to too great a concentration of pesticide in a limited area of the plate. This could be overcome by streaking the material to obtain a more diffuse spot. Six replicate samples of methoxychlor were determined at a concentration of 1 μg/spot with a relative standard deviation of 11.34%, while six scans of a single spot of methoxychlor on the same plate showed a relative standard deviation of 1.09%. Similar experiments were conducted for 2 μg samples of heptachlor epoxide. Reproducibility for six spots was 16.2% (rel. S.D.), while the relative standard deviation for 6 scans of a single spot was 2.1%. The results indicate that the instrument in this case is much more accurate than the chromatography and color development procedures used.

The thiophosphate pesticides Abate®, Diazinon®, fenthion, malathion, parathion, Prefar®, and Prefaroxon® were separated on Florisil® layers containing a calcium sulfate binder, using as solvent 30% ethyl ether in benzene. The colors were developed by spraying the plates with *N*,2,6-trichloro-*p*-benzoquinoneimine (TCQ) and, after partial drying, with a solution of 1 part 37% HCl in 4 parts distilled water. Drying in an oven at 110° for 10 min produced the colors. Maximum absorption for these spots was at about 540 nm and determinations were carried out in the same concentration range as for the chlorinated pesticides. The method was tested for lettuce extracts containing 0.05 ppm Prefar and 0.10 ppm Prefaroxon. The analysts found 0.04 ppm Prefar and 0.09 ppm Prefaroxon, values which are in reasonable agreement with the known amounts of pesticide added to the extracts. The importance of this work, in the authors' opinion, is that the feasibility of using in situ reflectance methods for determinations on thin-layer chromatograms was demonstrated even though the work was performed under highly unfavorable circumstances. It must be noted that the instrument used for the work was an experimental prototype, so that somewhat better results might be expected with more recent chromatogram scanners. More important, however, is that the variation found in the analysis of Prefar and Prefaroxon from the known values corresponds quite closely to the irreproducibility demonstrated for the chromatographic procedures. Thus, much better accuracy might be achieved with more reproducible color reagents.

Getz also has used the fiber-optics scanner described by Beroza et al.[57] to determine organochlorine and organophosphorus pesticides. Standard solutions of *p,p*′-DDT, *p,p*′-DDD, and *p,p*′-DDE were spotted on commercial silica-gel sheets, developed with 2,2,4-trimethylpentane and, after drying in air, dipped in a solution of 1% silver nitrate in methanol containing 2% phenoxyethanol. The spots became visible upon exposure to UV-light. The results were similar to those described by previous workers.[57]

Studies were also carried out on broccoli extracts that had been subjected to prior cleanup. Methyl Trithion®, methyl parathion, and carbophenothion were spiked in the extracts to give concentrations of 0.5, 1.0, 2.5, and 5.0 μg/ml. Commercially prepared silica-gel plates again were used in the study. Following chromatography, the spots were made visible by dipping the chromatogram in a 2% *p*-nitrobenzylpyridine solution in chloroform. After the plates were heated at 110° for 10 min, they were dipped in a 10% tetraethylpentamine solution as described by Watts.[58] As an alternate, the plates may be dipped in the silver nitrate-bromphenol blue reagent to produce color.[59] Nearly linear calibration plots were obtained with both reagents in the range 1 to 5 μg/spot. Under optimum conditions, partial resolution of a mixture of nine organophosphorus pesticides in a mixture was achieved. While all compounds could be resolved qualitatively in the scan, some of the peaks overlapped too closely for quantitative measurements of some of the compounds separated to be made under the experimental conditions. However, it is unlikely that such a mixture would be encountered in a natural sample.

Experiments were also carried out with a number of carbamate pesticides using the ninhydrin chromogenic reagent.[38] A reasonably linear response was obtained for the analysis of carrot extracts fortified with Landrin® in the range 0.5 to 4.0 μg/spot. As a result of these experiments, Getz suggests that quantitative TLC methods may now be used as a supplement to present GLC methods. Thin-layer methods are advantageous in the analysis of the more polar pesticides and, in any event, most analysts should feel more secure if a second simple, reliable quantitative procedure is available.

The analysis of various isomers of the insecticide 1,2,3,4,5,6-hexachlorocyclohexane (HCH) has been reported by Petrowitz and Wagner.[60] Determinations were made with a Scheeffel SD3000 spectrodensitometer in both the reflectance and transmittance modes of operation. The samples were spotted on silica-gel plates and developed in hexane. The separated compounds were visualized by spraying the plate with *o*-toluidine, followed by exposure to UV-light (254 nm) for 10 min. Reflectance measurements were carried out at 510 nm while transmittance measurements were made at 380 nm. A larger linear working range (5 to 40 μg/spot) was found in the reflectance mode than in the transmittance mode, where the slope changes at above 20 μg/spot. Determinations of γ-HCH were also carried out in the range 3 to 10 μg/spot by reflectance with a linear calibration plot being obtained.

More recently, reagents suitable for use in the quantitative determination of pesticides on thin-layer plates have been investigated by Getz and Hill.[61] The studies included an evaluation of the selectivity, sensitivity, and reproducibility of these compounds.

b. Formulation Analysis

Kynast has reported a number of studies of the use of a diffuse reflectance-TLC method for the analysis of pesticidal formulations.[62-65] In the initial part of the study, reflectance measurements were used in the analysis of biscarbamates in herbicidal formulations.[62]

Measurements were made on a Zeiss Chromatogram spectrophotometer using the UV-absorbance of the compounds or after color formation. The herbicide phenmedipham was analyzed in the formulation Betenal®. Chromatography was carried out on Kieselgel F_{254} thin-layer plates using as solvent chloroform-isopropyl ether (9:1). Reflectance measurements were made at 240 nm. Technical mixtures of phenmedipham were also studied. In addition to the carbarmate phenmedipham, this mixture contained a number of other carbamate and aniline compounds. In this case, the material was diazotized and coupled with β-naphthol. The colored spots were measured at 495 nm. The quantitative determinations at 240 nm produced relative standard deviations of ±3% for the mean value of eight-spot tests. Kynast concluded that the method should be useful in process control and offered sufficient specificity for stability studies and the determination of impurities.

Further studies were carried out on the extension of the methodology to the analysis of impurities in pesticidal formulations.[63] Results indicated that the procedure could be successfully applied to the direct, specific, and quantitative analysis of all impurities in the pesticides tested that could be separated by thin-layer chromatography. It was found that impurities present in amounts as small as 0.1% could be detected quantitatively. Reported relative standard deviations were about 5%. It was determined that the method could be useful in the determination of impurities in pesticidal formulations that might cause unwanted effects in the properties of the formulation.

To optimize the method, computerized data processing was investigated.[64,65] The program devised should be universally applicable in such determinations and this one was tested in a monitoring program for the quality control of pesticidal formulations. The results showed some improvement in accuracy over manual evaluation, as well as a considerable saving in time. The investigations indicated that time and expense could be saved in routine analysis of the active ingredients and impurities in formulations, while at the same time requiring less highly trained personnel.

Kossman has applied the methods reported by Kynast to the analysis of chlorphenamidine and formetanate in the acaricidal product Fundal®.[66] As both of these compounds have low volatilities and undefined degradation patterns in gas chromatography, GLC methods are unsuitable for their determination in mixed formulations. Several liquid-liquid partitioning steps were used to remove interfering plant extractives, so that an aliquot equivalent to 20 g of crop material could be spotted on the chromatogram. Chromatography was carried out on silica-gel thin-layers using a benzene-diethylamine (95:5) solvent system. After the chromatogram had been dried, the plate was scanned at 270 nm on a Zeiss Chromatogram spectrophotometer, and the peak areas of the extracts were compared with standards. Sensitivities of less than 0.5 μg/spot were observed for both compounds and, with proper cleanup, this corresponds to a level of 0.02 ppm of the pesticides in the crop material. Further specificity may be obtained by diazotization and coupling of the aromatic amines with β-naphthol, but the additional treatment results in a loss in accuracy. A variety of crop materials were tested with mean recoveries in excess of 80%. Loss of material was attributed mainly to the cleanup steps. Relative standard deviations for both compounds present in the range 0.2 to 1.3 ppm were from 4 to 11%. Thus, the method was considered sufficiently accurate for the purposes of the analysis. Again, it was noted that a linear relationship between peak area and concentration existed in a relatively narrow range, thus making a pretest necessary to establish the amount of material that should be spotted to obtain optimum response.

From these examples it may be seen that the determination of the components of pesticidal formulations by in situ reflectance spectroscopy of the materials separated on thin-layer chromatograms has some excellent possibilities for analytical use, particularly with the introduction of automated techniques.

c. Degradation Studies

A comprehensive study of nondestructive spray reagents for the detection of pesticides and their artifacts on thin-layer chromatograms has recently been undertaken.[67-73] It was found that pesticides (and their degradation products) that contained an aromatic ring-system could be detected through the formation of electron-donor-acceptor complexes with a suitable spray reagent. The use of both donor[67,72] and acceptor[67-73] reagents was investigated. Detection limits in the microgram range were reported.

Initial studies[67] revealed that best results were obtained with the electron-acceptors 2,4,7-trinitro-9-fluorenone (TNF), 2,4,5,7-tetranitro-9-fluorenone (TetNF), 9-dicyanomethylene-2,4,7-trinitrofluorene (CNTNF), and 9-dicyanomethylene-2,4,5,7-tetranitrofluorene (CNTetNF). As the formation of pi-complexes often results in the appearance of new spectral bands in the visible region of the spectrum,[74] the use of reflectance spectroscopy in the analysis of the complexed pesticides was investigated.[67-71,73] In a qualitative study, the reflectance spectra of the CNTNF-complexes of some selected carbamate and related pesticides were measured in situ on cellulose and silica-gel thin-layers.[68] On the more active silica-gel adsorbent layer, a hypsochromic shift of about 10 nm was usually found in the absorption maximum from those measured on cellulose thin-layers. Reflectance spectra of the CNTNF-complexes of the carbamate insecticides carbaryl and Mobam®, the thiourea fungicide thiophanate-methyl, and α-naphthol (a major degradation product of carbaryl) are shown in Figure 7.14. Reflectance absorption maxima for some common carbamate and urea pesticides are given in Table 7.03.

A quantitative study was carried out using the Mobam-CNTNF complex on both cellulose and silica-gel thin-layers.[69] The complex is bright red in color, with an absorption maximum at about 490 nm on silica gel and 500 nm on cellulose (see Spectrum, Figure 7.15; see also Table 7.03).

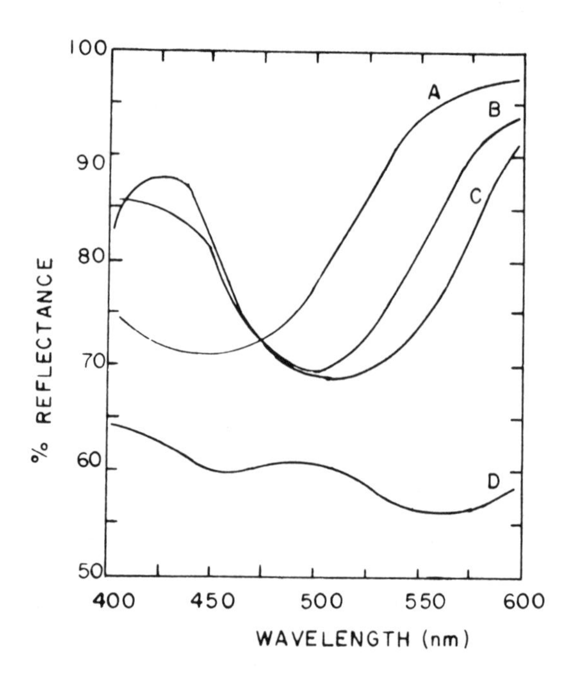

FIGURE 7.14 Reflectance spectra of pesticide complexes measured in situ on cellulose thin-layer chromatograms; A, thiophanate-methyl-CNTNF; B, carbaryl-CNTNF; C, Mobam-CNTNF; D, α-naphthol-CNTNF.[68] (From MacNeil, J. D., Frei, R. W., and Hutzinger, O., *Int. J. Environ. Anal. Chem.*, 1, 205 (1972). With permission of Gordon & Breach Inc.)

TABLE 7.03

Reflectance Absorption Maxima for Some Common Carbamate and Urea Pesticides Measured in situ on Chromatographic Thin-layers as the Pesticide-CNTNF pi-Complexes[68]

Pesticide	λ_{max}^{a}(cellulose) nm	λ_{max}^{a}(silica gel) nm
Carbaryl	500	490
Mobam	500	490
Diuron	500	490
Linuron	490	480
Maloran	490	470
Barban	490	470
Thiophanate	470	450

a λ_{max} is reproducible to within ±5 nm.

(From MacNeil, J. D., Frei, R. W., and Hutzinger, O., *Int. J. Environ. Anal. Chem.*, 1, 205 (1972). With permission of Gordon and Breach, Science Publishers, Inc.)

TABLE 7.04

Reproducibility of the Determination of the Insecticide Mobam as the Mobam-CNTNF Complex by in situ Reflectance Measurements on Chromatographic Thin-layers[69]

Adsorbent	Concentration Mobam (μg)	Av. Rel. % S.D.*
Cellulose	10	8.1
	5	7.1
	1	12.1
Silica gel	10	10.7
	5	7.6
	3	21.2

*Average relative standard deviation (%) for four plates, nine spots of pesticide per plate (at each concentration).

(From Frei, R. W., MacNeil, J. D., and Hutzinger, O., *Int. J. Environ. Anal. Chem.*, 2, 1 (1972). With permission of Gordon and Breach, Science Publishers, Inc.)

Detection limits on both adsorbents were about 1 μg/spot, but experiments revealed that while quantitative results could be obtained on cellulose thin-layers at this concentration, results on silica gel were at best semiquantitative at concentrations of less than 3 μg/spot. Reproducibility studies were carried out on both adsorbents, and the results are shown in Table 7.04. In these studies four plates, each containing nine spots of pesticide at the same concentration, were developed, sprayed, and measured at each of the concentrations reported. The relative standard deviation for each plate was then calculated, and this value was averaged for the four plates studied. This is the value given in the table.

Somewhat better reproducibilities were found on cellulose than on silica gel, but in both cases the optimum response seemed to be at a concentration of about 5 μg/spot. A linear response (peak area vs. concentration) was obtained over the range studied. Experiments were also carried out over a larger concentration range.[71] At concentrations above 25 μg/spot of the pesticide, some curvature was noted in the plot of reflectance vs. concentration. (The initial quantitative studies were performed on a Farrand UV-VIS Chromatogram Analyzer, while the studies over the larger concentration range were carried out on a Spectronic 505 equipped with diffuse reflectance attachment. The values for peak area obtained for the Farrand instrument are equivalent to the reflectance values for the total spot obtained on the Spectronic 505.) It was found that plotting reflectance vs. the square root of concentration (an approximation of the Kubelka-Munk function) resulted in a linear response for the complexes studied from 5 to 50 μg/spot.

The authors feel that this is an important consideration, which should receive some attention from other workers in this field. It has been fairly common practice to plot peak area vs. concentration and to make no attempt to extend the linear working range once this function becomes nonlinear. This function, which is in fact an approximation of the Beer-Lambert Law for transmittance measurements, is a hangover from earlier work with densitometers and, while it often is useful for measurements within a narrow concentration range, other functions should be examined for broader application. A comparison of the calibration curves reported by other workers, which have been discussed in previous sections of this chapter, leads the authors to believe that a recalculation of their data using peak area vs. the square root of concentration might extend the range of linear response for their measurements.

The stability of the pi-complexes was also studied on both silica-gel and cellulose thin-layers using the Mobam-CNTNF complex.[69] The results are shown in Figure 7.15. No change in the intensity of the absorption of the complex on the

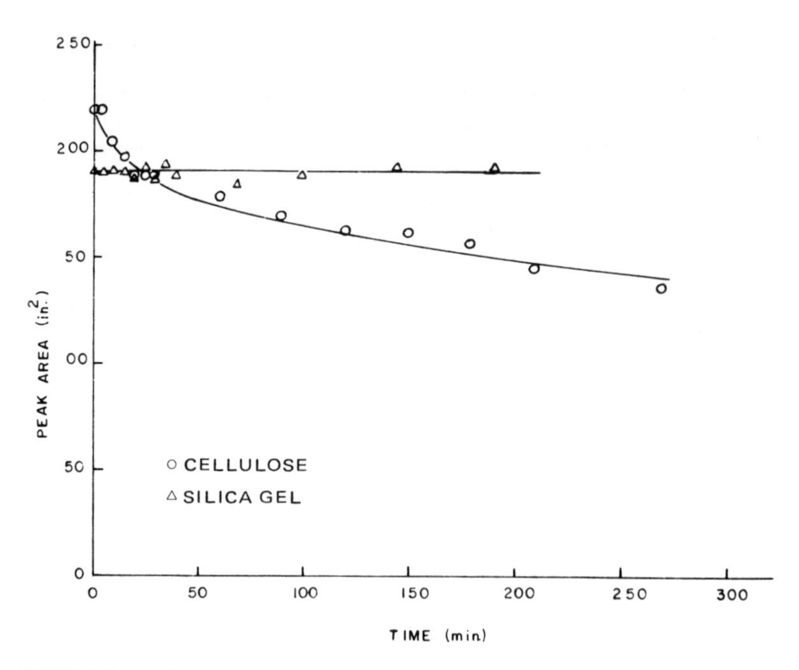

FIGURE 7.15 Stability study of Mobam-CNTNF complex on cellulose (o) and silica-gel (Δ) thin-layers.[69] (From Frei, R. W., MacNeil, J. D., and Hutzinger, O., *Int. J. Environ. Anal. Chem.*, 2, 1 (1972). With permission of Gordon & Breach Inc.)

cellulose layer was noted during the study. In fact, the author has kept chromatograms (cellulose), which have been sprayed with these reagents in the laboratory, for weeks with little visible change. On silica gel there is a gradual loss in intensity, so the optimum time for measurement (see Figure 7.15) is about an hour after spraying, when little change in response is noted.

The major source of error was in the chromatography and spraying. A single spot of Mobam-CNTNF complex (10 μg) was scanned six times with a relative standard deviation of 1.6%, so, as in studies by other workers, the instrumental error is relatively insignificant.

Several studies of the application of the methodology were undertaken to determine the usefulness of the approach in degradation studies.[70,73] In the first study, the system chosen was the photolysis of the insecticide methoxychlor.

Samples of methoxychlor (1 mg/ml, 400 ml) were irradiated at 310 nm. Aliquots were removed for analysis at regular intervals up to 12 hr. Mass spectral analysis revealed that the methoxychlor decomposed into DME (1,1-dichloro-2,2-bis(*p*-methoxyphenyl)ethane) and methoxychlor olefin (1,1-dichloro-2,2-bis(*p*-methoxyphenyl)ethylene), as well as a number of polar products. The quantitative experiments involved only the decomposition of methoxychlor and the formation of DME and methoxychlor olefin.

Four reactor samples were spotted on a silica-gel thin-layer (Eastman Chromagram plastic-backed silica-gel sheets 6061), together with five standards to give a calibration curve. Up to 1 hr, 25 μl aliquots of reactor samples were spotted, together with standards from 5 to 25 μg. For the remaining samples, 50 μl aliquots were spotted, together with standards from 2 to 10 μg. After development in hexane-benzene (2:1), the plates were sprayed with the electron-acceptor reagent 2,4,7-trinitro-9-fluorenone to visualize the spots and then measured on a Farrand UV-VIS Chromatogram Analyzer. Methoxychlor and DME were compared to methoxychlor standards for analysis, while methoxychlor olefin was determined with reference to a standard of that compound.

There is little difference in the donor abilities of methoxychlor and DME. Both complexes of these compounds with TNF show maximum absorption at 430 nm, while the absorption maximum for the methoxychlor olefin-TNF complex is at 440 nm due to the improved donor ability of the olefin. These wavelengths were used for the quantitative determinations.

After irradiation for two hr, less than 10% of the starting concentration of methoxychlor remained in the samples. After irradiation for one hour, DME appeared at a concentration about 15% of the initial methoxychlor concentration and increased gradually over the period of the experiment to about 20% after 12 hr. Methoxychlor olefin was not found in concentrations greater than 10% of the initial concentration of methoxychlor in any of the samples and was decreasing in concentration in the later samples. Duplicate plates were run for all the samples, with an agreement between replicates of about 3%. From these studies it appears that the most persistent compound resulting from the photolysis is DME. The results of the experiment are shown in Figure 7.16.

The experiment also shows that a relatively simple reflectance-TLC approach may be used successfully in the analysis of a pesticide and in its decomposition products. A further example was performed in a study of the hydrolysis of some common carbamate pesticides. The purpose of this study was to determine the effect of pH on the decomposition rates of these compounds in water. Many of the carbamate pesticides are quite water soluble, so a knowledge of their persistence in water is important.

Studies were carried out on the insecticides carbaryl and Mobam and the herbicides IPC and CIPC at the following pH's: 2,7,8,9. In each case 25 mg of the pesticide were dissolved in 500 ml of

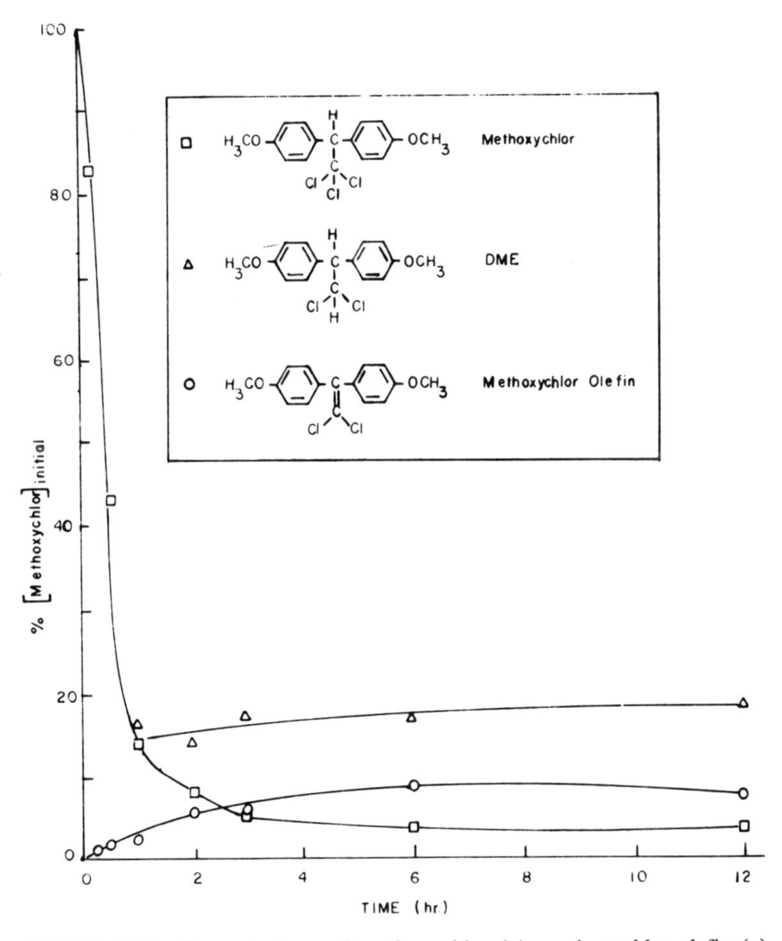

FIGURE 7.16 Concentrations of methoxychlor (□), methoxychlor olefin (○) and DME (Δ) in reactor samples relative to initial methoxychlor concentration (1 mg/ml) following irradiation at 310 nm.[70] (From MacNeil, J. D., Frei, R. W., Safe, S., and Hutzinger, O., *J. Assoc. Anal. Chem.*, 55, 1270 (1972). With permission of the Assoc. Off. Anal. Chem.)

distilled water buffered to the appropriate pH value. Aliquots of 50 ml were removed at regular intervals over a period of 5 weeks, extracted into ether, reduced, dried, and taken up in 2 ml of acetone, which was then evaporated to 1 ml. The samples were chromatographed with standards, and the decomposition of the pesticides, together with the appearance of major decomposition products, was determined.

Results showed that the decomposition rates of the insecticides (*N*-methyl carbamates) were accelerated in water as the basicity increased and that the compounds were stable at the acid pH studied. The herbicides (*N*-phenyl carbamates) showed no signs of decomposition at any of the pH's tested.

As the electron-donor-acceptor complex formation is reversible, the pesticide and complexing agent may be separated by slowly heating the complex in the probe of a mass spectrometer. Mass spectra of the pesticide carbaryl and also of the complexing agent 9-dicyanomethylene-2,4,5,7-tetranitrofluorene obtained from a carbaryl-CNTetNF complex are shown in Figure 7.17.[68] This enables the analyst to identify many of the products formed in these degradation studies from microgram quantities of material separated on chromatographic thin-layers and proved particularly useful in the study of the photochemical degradation of methoxychlor.[70]

This experiment, while quite simple in approach, was designed to show the need for simple model systems as an operational approach in the study of problems involving the persistence and decomposition of pesticides in the environment. The authors believe that such experiments may prove extremely useful as a preliminary measure in planning and implementing long-term research projects by pointing to the approach most likely to prove productive.

d. Future Outlook

While conversations with other scientists involved in pesticide research have indicated that there exists a considerable feeling against in situ photometric determinations of materials separated on chromatographic thin-layers, the authors believe that the coming years will see a substantial change in attitudes. Admittedly, gas chromatography is invaluable for residue determinations. However, new pesticides are more polar in structure and thus create problems in gas chromatographic analyses. Gas chromatography will probably be the primary method of residue analysis for many years, but other methods will also have to be used to solve some problems. The examples quoted in this chapter have shown the feasibility of quantitative reflectance-TLC methods for residue analysis, formulation analysis, and degradation studies. As more laboratories become equipped with TLC-scanners, it is felt that much more work will be done to demonstrate the usefulness of such techniques.

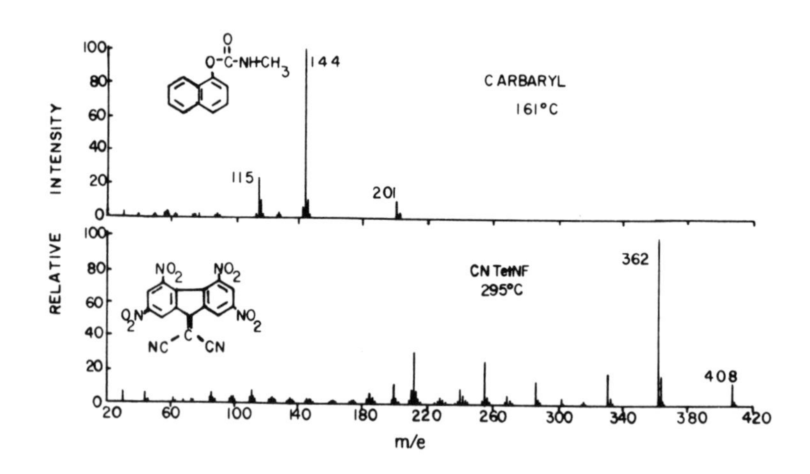

FIGURE 7.17 Mass spectra of carbaryl and CNTetNF obtained from carbaryl-CNTetNF complex on cellulose.[68] (From MacNeil, J. D., Frei, R. W., and Hutzinger, O., *Int. J. Environ. Anal. Chem.*, 1, 205 (1972). With permission of Gordon & Breach Inc.)

3. Visible Reflectance Spectroscopy of Other Organic Systems

The numerous applications of transmission densitometry[75-77] for the quantitative evaluation of paper and thin-layer chromatograms should be easily adaptable to diffuse reflectance spectroscopy. This subject is not discussed in this book.

Spectral reflectance has been used for the in situ identification and determination of many substances after resolution on thin-layer chromatograms[77-79] and occasionally on paper[80] and on electrophoresis strips.[81,82] For the sake of convenience, applications based on this technique are discussed in the following sections in their approximate chronological order of development, starting with visible and UV reflectance spectroscopy of organic systems and concluding with reflectance spectroscopy of inorganic systems, which, at least with regard to TLC, is the latest area to be investigated.

a. Dyes and Pigments

Yamaguchi et al.[83,84] reported the reflectance spectroscopic study of some food dyes on filter and chromatography paper. Linear relationships were observed for lower concentrations of dye if 2-log R was plotted vs. the square root of concentration. Kortüm et al.[24,85] investigated the dye malachite green on chromatographic paper in order to assess the use of spectral reflectance in the field of paper chromatography. They found that the major difficulty was the production of spots with a fairly homogeneous distribution of the absorbing component. However, with careful preparation of the dye spots they were able to obtain reproducible calibration curves for the Kubelka-Munk function of R plotted vs. concentration of the dye. Ten layers of the paper were needed to achieve infinite layer thickness.[85] For practical paper chromatography, the investigation of the spots in one single measurement would entail the same kind of errors found by Frodyma et al.[22] for dyes separated on thin-layer plates. However, in a later study, Braun and Kortüm[24,25] showed that, with appropriate correction for spot size and self-absorption of the paper, a suitable form of the Kubelka-Munk equation could be found, which resulted in linear calibration curves in a reasonable concentration range of the dye. The resulting spectrum and the calibration curve, which was obtained by measuring the reflectance at the absorption maxima at 21,000 cm,$^{-1}$ are depicted in Figures 7.18 and 7.19. As can be seen in Figure 7.19, a linear relationship up to about 3 μg/spot was obtained. The arrows point to measurements of equal concentration of dye but of different spot size, and again with the use of the modified function a reasonably good agreement of the two values is observed. The equations derived by these workers and their application to scanning thin-layer chromatograms have been discussed in Section 1f (v) of this chapter. Butler et al.[80] discussed in detail the use of the Joyce, Loebl

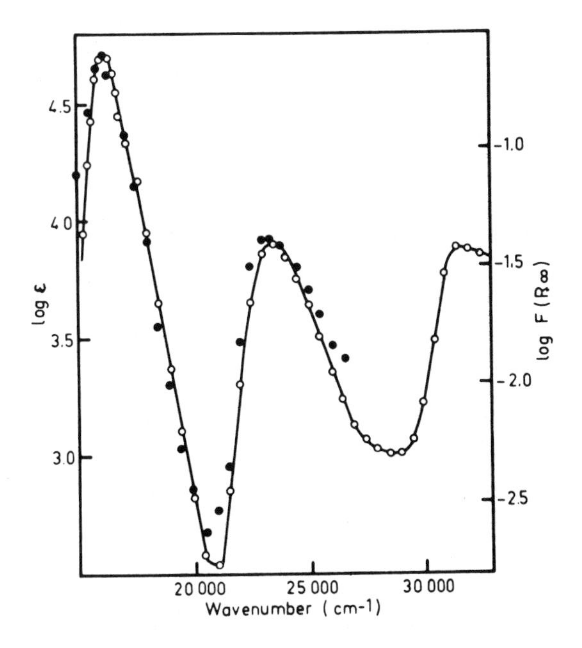

FIGURE 7.18 Spectra of malachite green.[24] o--o--o transmittance spectrum in methanol, •--•--• reflectance spectrum on paper. (From Braun, W. and Kortüm, G., *Zeiss Inform.* 16, 27 (1968). With permission.)

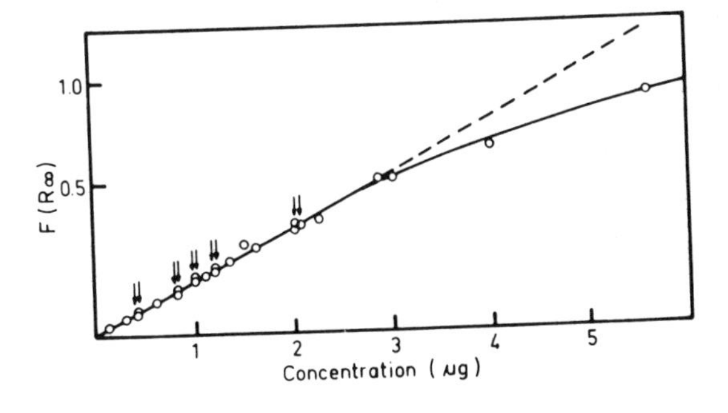

FIGURE 7.19 Calibration plot $F(R_\infty)$ (as determined by Equation 7.08) vs. concentration.[24] (From Braun, W. and Kortüm, G., *Zeiss Inform.*, 16, 27 (1968). With permission.)

Chromoscan for the evaluation of dye spots on paper chromatograms.

It is not surprising that the first application of reflectance spectroscopy to TLC was in the field of dyes.[22] This system, which is stable and does not require detection that involves spraying or scanning procedures, readily lent itself to the technique.

Stock solutions containing 50 mg of the dyes studied (aniline blue, eosine B, basic fuchsin, malachite green, naphthol yellow S, and rhodamine B) per 100 ml of solvent were applied as spots by means of a 10-μl Hamilton microsyringe. Except for the aqueous eosine B, the solvent used was 95% ethanol. The 10 x 7 x 0.15-cm plates were cut from ordinary window glass and were coated with adsorbent by distributing the adsorbent-water mixture with a glass rod that rested on one thickness of masking tape affixed to the ends of the plates. This technique gave a uniform coating, 0.2 to 0.3 mm thick. The plates were dried at 180°C for two hr and were stored in a desiccator. Merck aluminum oxide G and silica gel G were used as adsorbents.

The dyes were chromatographed in *n*-butanol:ethanol:water (80:20:10 by volume) by the ascending technique according to Mottier,[86] and the plates were dried at 100°C for 15 min. Direct spectral examination of these plates was carried out with a Beckman DK-2 spectrophotometer according to the experimental procedure discussed previously. Figure 7.20 shows the reflectance spectra of these dyes adsorbed on alumina. With proper precautions it is possible to obtain spectra suitable for identification purposes. The general shape of reflectance and transmittance spectra is similar. Some peak broadening is observed for reflectance spectra and a shift of the absorption maxima (see Figure 7.21) generally occurs, depending on the adsorbent. The operation can provide semiquantitative results.

A precision of about 3% can be attained by carrying out the reflectance measurements on spots removed from the chromatoplate and packed in appropriate cells (see General Experimental Procedure, Section 1d).

When this procedure was used as the basis for a student experiment[87] in an undergraduate course in quantitative analysis, the data obtained indicated that this technique can be used successfully even by people who have no specific training or prior experience in it. The data reported in Table

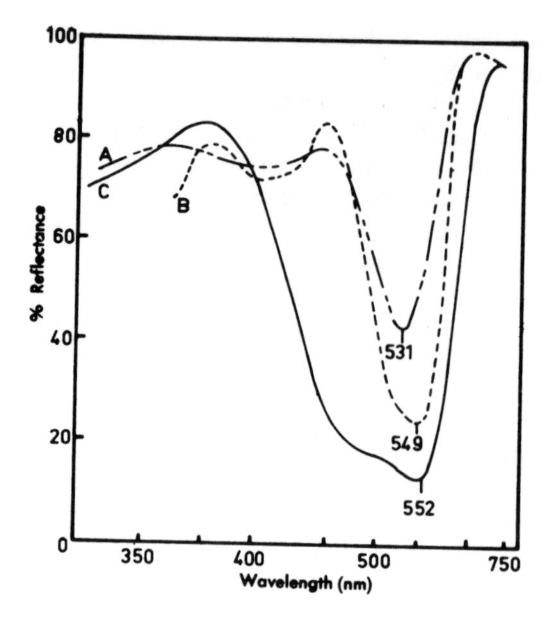

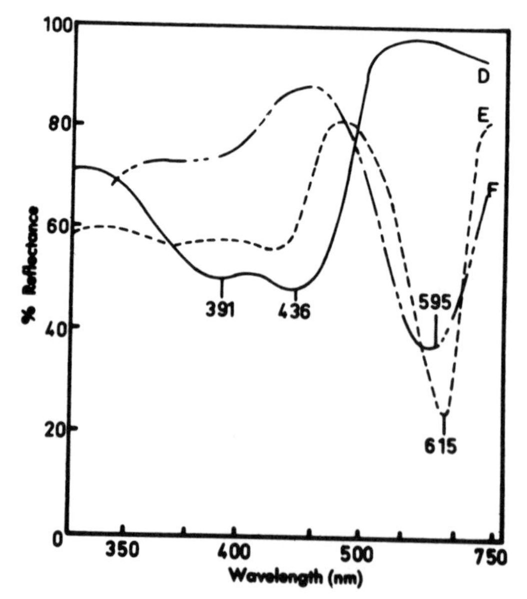

FIGURE 7.20 a. Reflectance spectra of dyes adsorbed on alumina. A, eosine B; B, rhodamine B; C, fuchsin. b. Reflectance spectra of dyes adsorbed on alumina. D, naphthol yellow S; E, malachite green; F, aniline blue.[22] (From Frodyma, M. M., Frei, R. W., and Williams, D. J., *J. Chromatogr.*, 13, 61 (1964). With permission of Elsevier Pub. Co.)

7.05 were obtained either by using a calibration curve or by an algebraic method. Since the concentrations of the unknowns all fell on the linear portion of the calibration curve, it was possible to determine their concentration by means of the equation

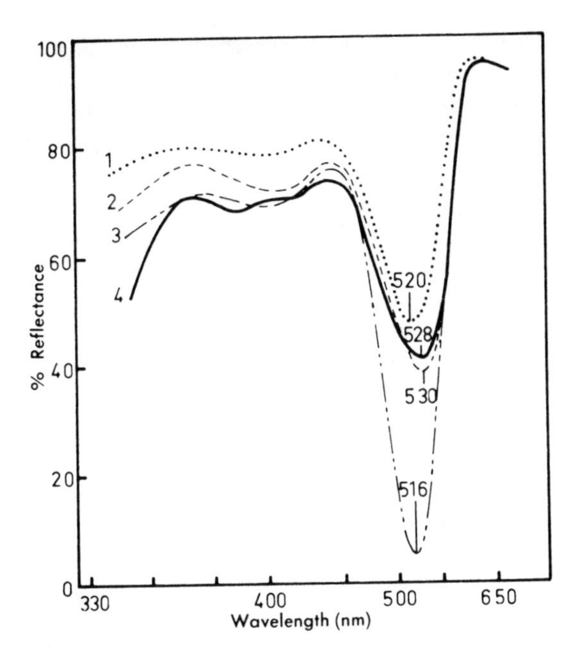

FIGURE 7.21 Reflectance spectra of eosine B adsorbed on filter paper, alumina, and silica gel, compared with the transmittance spectrum of an aqueous solution of the dye. 1, silica gel (Merck TLC grade); 2, filter paper (Whatman No. 42); 3, transmittance spectrum; 4, alumina (Merck TLC grade).[22] (From Frodyma, M. M., Frei, R. W., and Williams, D. J., *J. Chromatogr.*, 13, 61 (1964). With permission of Elsevier Pub. Co.)

$$C_u = \frac{C_s F(R)_u}{F(R)_s}, \tag{7.11}$$

where C_u and C_s represent the concentrations of the unknown and standard solutions, and $F(R)_u$ and $F(R)_s$ stand for the Kubelka-Munk functions of the unknown and standard. Ninety percent of the dyes were correctly identified by the students on the basis of reflectance spectra without resorting to R_f values.

The usefulness of reflectance spectroscopy for the examination of various effects (such as humidity, regeneration temperatures, and pH) on the behavior of dye systems adsorbed on chromatographic adsorbents has been demonstrated.[88]

The effect of regeneration temperatures on the adsorption of eosine B on alumina is shown in Figure 7.22. A shift to higher wavelength accompanied by an increase in color intensity is observed at higher regeneration temperatures. This is attributed to a higher adsorbent-adsorbate interaction and is consistent with earlier observations.[89] The importance of standardizing activation procedures before chromatography is further supported by Figure 7.23, which demonstrates the variations of Kubelka-Munk plots for a dilution

TABLE 7.05

Accuracy and Precision of Student Determinations, Using Spectral Reflectance of Dyes on Thin-layer Plates[87]

	Method	Crystal Violet	Fuchsin	Rhodamine B
Graphic method	Average accuracy in % deviation from the true value	5.7	5.7	3.8
	Range % deviation for a set of eight results	3.5–9.5	0.3–9.6	1.3–8.0
Algebraic method	Average accuracy in % deviation from the true value	7.8	7.1	5.8
	Range % deviation for a set of eight results	0.0–15.0	0.6–15.0	2.0–10.8

(From Frodyma, M. M. and Frei, R. W., *J. Chem. Educ.*, 46, 522 (1969). With permission of the American Chemical Society, Division of Chemical Education.)

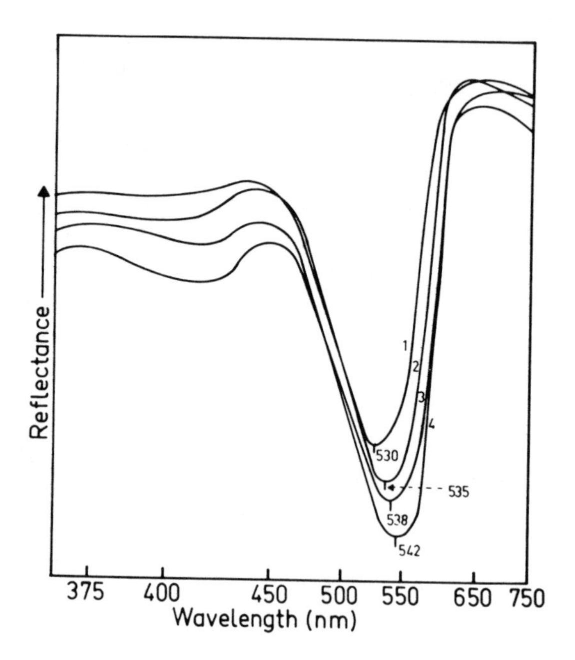

FIGURE 7.22 Reflectance spectra of eosine B adsorbed on 1, alumina (air-dried); 2, alumina (preheated to 200°C); 3, γ-alumina (preheated to 800°C); 4, α-alumina (preheated to 1100°C).[88] (From Frei, R. W. and Zeitlin, H., *Anal. Chim. Acta,* 32, 32 (1965). With permission of Elsevier Pub. Co.)

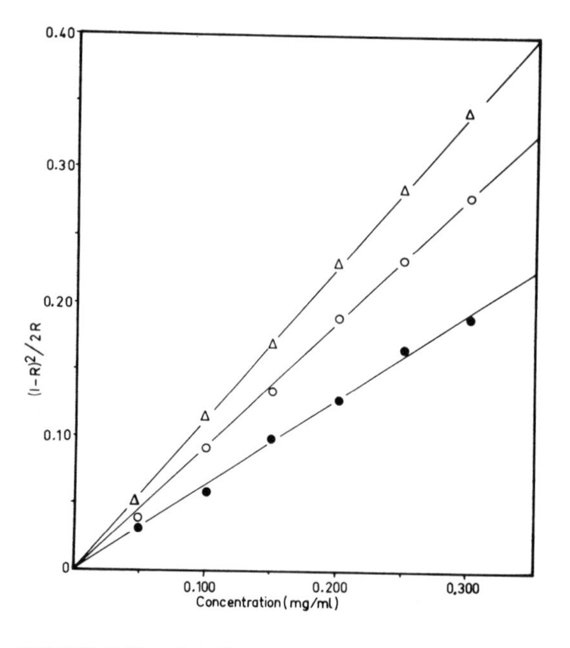

FIGURE 7.23 Kubelka-Munk plots for a dilution series of rhodamine B adsorbed on • alumina (air-dry state); O alumina (preheated to 200°C); Δ alumina (preheated to 1100°C).[88] (From Frei, R. W. and Zeitlin, H., *Anal. Chim. Acta,* 32, 32 (1965). With permission of Elsevier Pub. Co.)

series of rhodamine B adsorbed on alumina of different activity.

The advantage of using dye systems for the experimental verification of some fundamental studies in this field has been realized in several other investigations.[17,19,31,32,90-92]

Yamaguchi et al.[93] discussed the use of diffuse reflectance spectroscopy for the quantization of plant pigments. They studied, for example, the relationship between reflectance and the amount of spotted chlorophyll in paper-partition chromatography. The relationship 2-log R vs. $\sqrt{C}$ was found to be most suitable.

Garside and Riley[94] reported a combined thin-layer chromatographic reflectance spectroscopic study of pigments of the chlorophyll and carotenoid variety in particulate matters in sea water. A comparison of various filtration methods showed that Whatman GF/C glass-fiber filters covered with a layer of magnesium carbonate gave the best recovery. After extraction with acetone and methanol, the extracts were evaporated in vacuum and chromatographed on silica gel G according to a method by Riley and Wilson.[95] Chlorophyll *c* remains at the starting point and is separated from the spot of origin by subsequent development with a mixture of light petroleum, ethyl acetate, and dimethylformamide in the ratio 1:1:2 by volume. In situ measurement of the pigment spots by reflectance spectroscopy was chosen as a means of evaluation of the chromatograms. The major advantages to conventional elution and transmission spectroscopic evaluation of the spots are high speed and better sensitivity. A Joyce, Loebl Chromoscan was used in connection with the thin-layer scanner attachment in reflectance mode for the in situ work. An actual scan of the chromatogram after separation of six components is shown in Figure 7.24. An Ilford No. 601 filter (maximum transmission at 430 nm) was chosen for best monochromaticity condition of the measuring light beam; the peak areas were integrated with a mechanical disk and ball integrator. The identification of the pigment spots was made from the record (Figure 7.24) by measuring the distance between peak and point of origin and dividing this into the distance moved by the chlorophyll *a* peak. The corresponding ratios (R_p values) with averages and standard deviations are recorded in Table 7.06. Six replicate runs were used to test the reproducibility of the chromoscan integrator readings, and a spread of less than ±2%

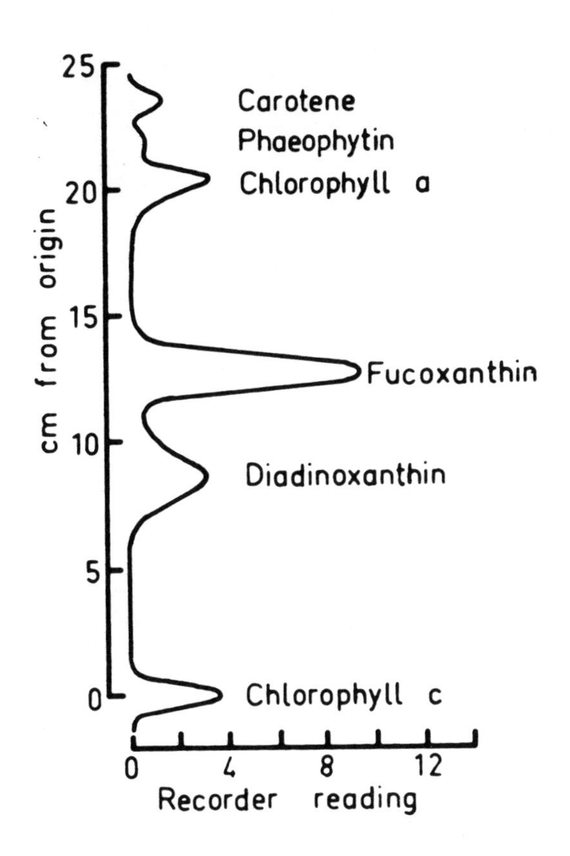

FIGURE 7.24 Chromoscan trace for thin-layer chromatogram of pigments from senescent *Phaeodactylum tricornutum.*[94] (From Garside, C. and Riley, J. P., *Anal. Chim. Acta,* 46, 179 (1969). With permission of Elsevier Pub. Co.)

TABLE 7.06

Averages and Standard Deviations for R_p Values and Calibration Values for Pigments[94]

Pigment	R_p	ng of pigment/unit integrator reading*
Chlorophyll *a*	1	47.4 ± 0.4
Chlorophyll *b*	0.82 ± 0.02	14.5 ± 0.7
Chlorophyll *c*	0.00	22.4 ± 0.5
Phaeophytin *a*	1.07 ± 0.04	42.8 ± 0.4
β-Carotene	1.17 ± 0.04	11.8 ± 0.4
Fucoxanthin	0.64 ± 0.03	29.1 ± 1.0
Lutein	0.55 ± 0.02	33.6 ± 0.9
Violaxanthin	0.41 ± 0.01	23.8 ± 1.2
Diadinoxanthin + dinoxanthin	0.44 ± 0.02	19.2 ± 0.9
Neoxanthin	0.21 ± 0.02	26.7 ± 0.7

*On the most sensitive setting of the instrument.

(From Garside, C., and Riley, J. P., *Anal. Chim. Acta,* 46, 179 (1969). With permission of Elsevier Pub. Co.)

was found. The reproducibility from one plate to another was estimated from six replicate chromatograms; the percentage of the total integrated area corresponding to one particular pigment never varied more than ±1%. The average calibration values for several pigments are shown in Table 7.06.

The following procedure was suggested by Garside and Riley[94] in order to calibrate the chromoscan. Silica-gel plates were spotted with known μl volumes of extracts of a range of phytoplankton species. After development, the plates were scanned with the instrument at its highest sensitivity. The amount of pigment in each spot was determined by extracting it with 90% acetone, diluting to a suitable volume, and determining the absorbance at the appropriate wavelength. The following $E_1{}^{1\%}{}_{cm}$ values were employed in calculating the weights of pigments: chlorophyll *a,* 911 at 663 nm; *b,* 525 at 645 nm; *c,* 1400 at 446 nm; phaeophytin *a,* 1310 at 409 nm; β-carotene, 2505 at 451 nm; fucoxanthin 1040 at 449 nm. Because reliable data were not available for the other xanthophylls the value of 1000 at the absorption maximum was assumed for these pigments. It was found that for each of the pigments the integrator reading was a linear function of the amount of pigment, up to a recorder reading of about 80% of full-scale deflection. The sensitivity of the method was reported to be from 0.03 μg for β-carotene to about 0.14 μg for chlorophyll *a,* based on the assumption that an integrator reading of 3 is significant. For reliable results no more than 30 μg should be loaded on the plates.

In order to assess the precision of the method, six replicate analyses were carried out on 100-ml portions of a composite phytoplankton culture. The mean weights and standard deviations found for the various pigments were chlorophyll *a,* 7.87 ±0.33 μg; chlorophyll *b,* 1.68 ±0.05 μg; chlorophyll *c,* 0.76 ±0.04 μg; phaeophytin *a,* 1.41 ±0.13 μg; carotene 1.04 ±0.04 μg; lutein 4.67 ±0.17 μg; violaxanthin 0.67 ±0.05 μg; neoxanthin 1.23 ±0.11 μg. Thus, for the chlorophylls and major xanthophylls the coefficient of variation of the method did not exceed 5%.

Three cultures of marine phytoplankton (including a senescent one of *Phaeodactylum tricornutum*) were examined by the proposed method. The pigments were also determined spectrophotometrically in 90% acetone extracts of

TABLE 7.07

Comparison between Pigment Analyses Carried Out by the Method Discussed and Those Obtained by Spectrophotometry of the Eluted Spots[94]

	Reflectance method		Elution and photometry	
Pigment	μg/100 ml	Percentage of total carotenoids	μg/100 ml	Percentage of total carotenoids
*Phaeodactylum tricornutum** (Plymouth No. 100)				
Chlorophyll *a*	2.13		2.07	
Chlorophyll *b*	0.00		0.00	
Chlorophyll *c*	0.99		0.72	
Phaeophytin *a*	0.17		0.16	
Carotene	0.18	3.4	0.17	3.2
Fucoxanthin	3.82	72.8	3.77	73.4
Diadinoxanthin	1.25	23.8	1.20	23.4
Ratio of chlorophyll *a*:*c*	1:0.46		1:0.34	
Dunaliella primolecta (Plymouth No. 81)				
Chlorophyll *a*	0.94		0.92	
Chlorophyll *b*	0.26		0.25	
Chlorophyll *c*	0.00		0.00	
Carotene	0.11	1.9	0.12	2.2
Lutein	4.06	72.8	3.99	74.4
Violaxanthin	1.03	18.6	0.94	17.6
Neoxanthin	0.31	5.7	0.31	5.8
Ratio of chlorophyll *a*:*b*	1:0.28		1:0.27	
Olisthodiscus sp. (Plymouth No. 239)				
Chlorophyll *a*	2.32		2.11	
Chlorophyll *b*	0.00		0.00	
Chlorophyll *c*	1.23		1.13	
Carotene	0.59	18.1	0.57	16.1
Fucoxanthin	2.07	63.6	2.41	68.2
Diatoxanthin	0.60	18.3	0.55	15.7
Ratio of chlorophyll *a*:*c*	1:0.53		1:0.53	

*A senescent culture was used.

(From Garside, C., and Riley, J. P., *Anal. Chim. Acta,* 46, 179 (1969). With permission of Elsevier Pub. Co.)

the individual spots eluted from the chromatograms.[95] The results of these two methods are presented in Table 7.07; they show good agreement. In conclusion, it can be said that the in situ TLC method, with its advantages mentioned earlier, is suitable for such problems and will also be adaptable to shipboard use. Quantitative data cannot be produced only for the photosynthetic pigments but will be available also for the individual xanthophylls. A complete analysis can be performed in about one hour.

b. Amino Acids

Reduced tailing, increased sensitivity, and greater speed and resolution have made TLC an important tool in amino acid analysis.[96] Improved in situ evaluation techniques of such chromatograms were desirable, and reflectance spectroscopy showed much promise. The various studies of reflectance spectroscopy applied to amino acids resolved on thin-layer plates have been reviewed.[21,97]

With the use of a standard ninhydrin spray reagent, no significant differentiation was observed for the reflectance spectra of a group of amino acids. The use of a modified spray reagent recommended by Moffat and Little,[98] however, permitted the author to identify complex mixtures of amino acids rapidly by using a combination of reflectance spectra, visual appearance, and R_f values of the spots.[20]

The solvent system used by the author to separate mixtures of 18 amino acids on silica-gel layers was suggested by Brenner and Niederwieser[99] and consisted of *n*-butanol:acetic acid:water (60:20:20). The spray reagent was a combination of two solutions:

I. 50 ml of a 0.2% ninhydrin solution in absolute EtOH, 10 ml AcOH, 2 ml 2,4,6-collidine.
II. 1% $Cu(NO_3)_2 \cdot 3H_2O$ solution in absolute EtOH.

The solutions are mixed in the ratio 50:3 shortly before spraying. The sensitivity is increased by substituting 0.4% for 0.2% ninhydrin solution.

The spots are removed for reflectance measurement with a Beckman DK-2 spectrophotometer. The preparation of the samples has been described above. Observations are reported in Tables 7.08 and 7.09.

A reflectance spectroscopic study of color stabilities of ninhydrin complexes of adsorbed amino acids[1] revealed that a ninhydrin reagent recommended by Bull et al.[100] gave the most stable color formation. The results of a time study carried out at room temperature over a three day period are represented in Figure 7.25. The spray reagent consisted of *n*-butanol (90 g), phenol (10 g), and ninhydrin (0.4 g) and was chosen for a quantitative investigation of amino acids on thin-layer plates by spectral reflectance.[101] The color intensity produced by this spray was superior to that produced by other reagents, as can be seen from calibration plots in Figure 7.26. The absorption maxima of the reflectance spectra were around 510 to 520 nm over a large concentration range for the 14 amino acids investigated. Very little peak shift was observed with varying concentrations. The same calibration curve could be used

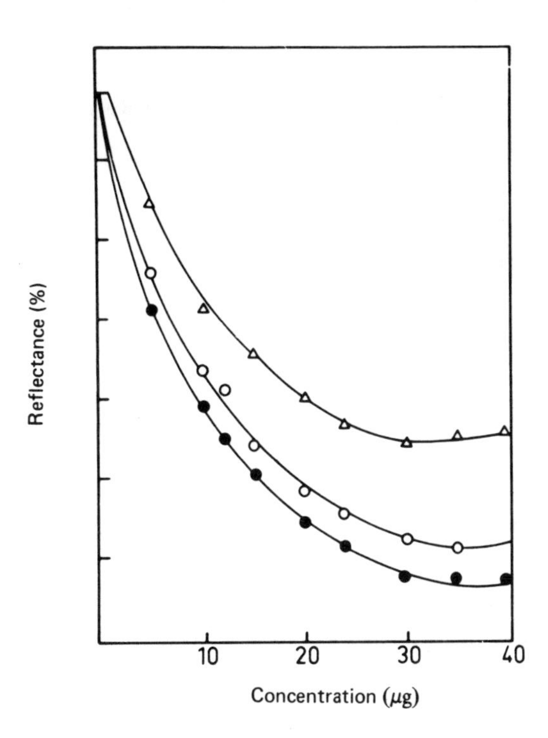

FIGURE 7.25 Percent reflectance at 520 nm of ninhydrin complex of leucine adsorbed on silica gel as a function of concentration and time. •–• after 2 hr; O–O after 24 hr; Δ–Δ after 3 days at room temperature (28°C).[1] (From Frei, R. W. and Frodyma, M. M., *Anal. Biochem.*, 9, 310 (1964). With permission of Academic Press Inc.)

for amino acids of similar structure, such as valine, norvaline, leucine, isoleucine, and norleucine, without affecting the precision significantly. The limiting factors for the precision attained with this procedure were the incomplete reaction of the amino acids with the ninhydrin reagent and the leaching out of the acids during the spraying operation.

Since the degree of precision achieved was less than that attained when the same procedure without the chromogenic step was applied to a stable system,[22] it was felt that the method would improve considerably if the color were developed without sprays. Therefore, it was decided to investigate the possibility of adapting the nonspray method suggested by El-Khadem et al.[102] for the identification of amino acids and sugars separated on paper chromatograms. By adding the detecting reagents to the solvent mixtures, the author[103] succeeded in eliminating not only the spraying operation but also the drying step preceding it, at

TABLE 7.08

Color, Absorption Maxima, and R_f Values for Some Amino Acids[20]

Amino acid	Concentration (μg/spot)	Color	Absorption maxima (nm)	DK-2	Visual	R_f value
Asparagic acid	5	Light blue with yellow ring, changes to blue-violet after 5 min				0.21
	4	As above			0.2	
	3	Pale violet				
Glycine	5	Intensive orange with violet center	484			0.22
	4	Orange	475			
	3	Orange	475			
	2	Orange	475	1	0.1	
Isoleucine	5	Dark violet	539/478/413			0.46
	3	Violet	534/477/413			
	2	Violet, pink after 24 hr	532/478	1	0.3	
Leucine	5	Dark violet	543/480			0.47
	3	Violet, pink after 7 days	542/480			
	2	Violet-pink	540/474			
	1	Pink		2	0.3	
Lysine HCl	5	Intensive yellow-brown, brown after 7 days	536/480/397			0.05
	3	As above		3	1	
	1	Light brown				
Phenylalanine	5	Brown with yellow ring, yellow-pink after 24 hr				0.49
	2	Brown-yellow			2	
Proline	5	Intensive brown-yellow, brown after 7 days			1	0.19
Threonine	5	Violet with yellow ring, pink after 5 days	537/478/417			0.25
	3	As above	529/475/412	3	1	
Tryptophane	5	Brown with light-blue ring, blue fades after 5 min	422			0.56
	3	As above				
	2	Pale blue, brown after 5 min				
	1	Brown			1	
Valine	5	Dark violet	537/480/409			0.35
	3	Blue-violet, pink-violet after 7 days		3	0.2	

(From Frei, R. W., Fukui, I. T., Lieu, V. T., and Frodyma, M. M., *Chimia*, 20, 23 (1966). With permission.)

the same time notably increasing the precision of the method.

The amino acids used for this study (DL-alanine, L-arginine, L-glutamic acid, glycine, L-leucine, L-lysine, DL-methionine, DL-phenylalanine, DL-serine and DL-valine) were of Calbiochem A Grade purity. Stock solutions of the acids containing 500 mg in enough distilled water to make 50 ml of solution were applied as spots by means of a Hamilton microsyringe, in 5-μl increments. The 20 x 5 x 0.35-cm plates used for one-dimensional analysis as well as the 20 x 20 x

TABLE 7.09

Color, Absorption Maxima, and R_f Values for Some Amino Acids[20]

Amino acid	Concentration (μg/spot)	Color	Absorption maxima (nm)	DK-2	Visual	R_f value
Alanine	5	Blue-violet, fades to red-violet after 24 hr	540/480/412.			0.26
	2	As above	532/486/416	2		
	1	As above	528/473		0.2	
Arginine HCl	10	Light violet, fades to pink	547/483/419			0.08
	5	As above	535/484	5	3	
Cystine	8	Pink-gray	447			0.16
	6	Pink-gray	455			
	5	Pink-gray		5	3	0.06
Glutamic acid	5	Dark violet with pale orange ring, brown after 24 hr	532/482/409			0.27
	3	As above	536/481/416	3		
	2	As above			1	
Histidine HCl	5	Gray with yellow ring	455			
	3	Yellow, brown-gray after 24 hr	455			
	0.5	As above	455	0.5	0.5	
Methionine	10	Violet with yellow ring, pink after 24 hr	535/478/421			0.40
	8	As above	538/483/410			
	5	As above	537/480	5		
	2	As above			1	
Serine	5	Dark violet with yellow ring, fades to pink	540/478/415			0.47
	3	As above	540/478/415			
	2	As above	537/479/429	2		
	1	As above			0.3	
Tyrosine	5	Orange with yellow ring, light yellow after 24 hr			2	0.47

(From Frei, R. W., Fukui, I. T., Lieu, V. T., and Frodyma, M. M., *Chimia,* 20, 23 (1966). With permission.)

0.35-cm plates used for the two-dimensional resolutions were coated with Merck silica gel G. After resolutions were achieved, the plates were heated in a mechanical convection oven at 60°C for 30 min to dry them and to develop the colors.

Both one- and two-dimensional chromatograms were used for investigating the applicability of four solvent mixtures: (1) *n*-propyl alcohol: water: acetic acid (64:36:20); (2) *n*-butyl alcohol:water: acetic acid (60:20:20); (3) phenol:water (75:25); and (4) *n*-propyl alcohol:34% ammonia (67:33). The first three solvent systems were employed in conjunction with one-dimensional analyses carried out by the ascending technique. Systems (3) and (2), (4) and (1) were paired off during the two-dimensional analyses, with the first of each pair being used for the initial development. Chromatograms were dried at 60°C for 30 min before development in the second dimension. For two-dimensional separation, 0.2% ninhydrin was added to the second solvent system. The ninhydrin-containing solvents had to be absolutely free of ammonia. If an ammonia-containing solvent was used for the first dimension, all ammonia had to be removed by drying sufficiently before the

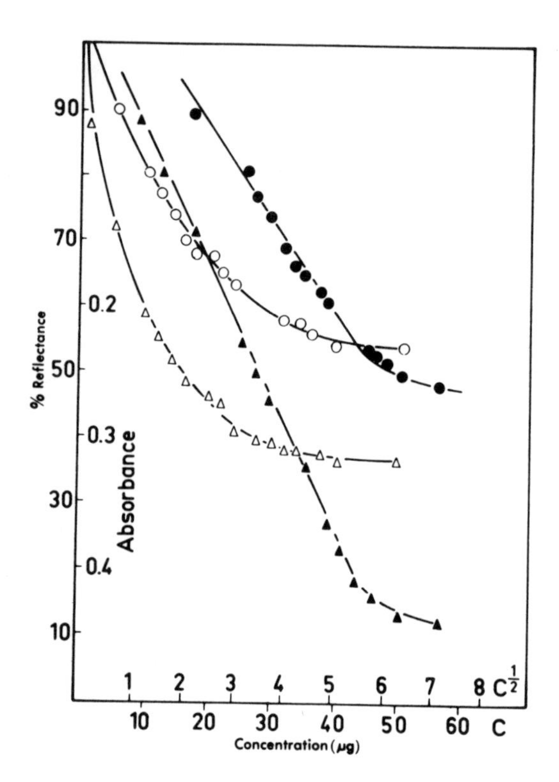

FIGURE 7.26 Percent reflectance and absorbance at 520 nm of ninhydrin complex of leucine adsorbed on silica gel as a function of concentration. Modified Brenner-Niederwieser spray;[99] O--O % reflectance vs. C; ●--● absorbance vs. $C^{1/2}$. Bull et al. spray;[100] Δ--Δ % reflectance vs. C; ▲--▲ absorbance vs. $C^{1/2}$.[101]

second development. Otherwise, a brownish or purplish tinge was imparted to the entire plate.

Successful resolutions of mixtures of the 10 amino acids were achieved in 10 hr or less, during which the solvent fronts were permitted to move 18 cm in each dimension by the ascending technique.

The amino acids were identified by using R_f values or, in ambiguous situations, by simultaneously running standards for comparison purposes. A Beckman Model DU spectrophotometer fitted with a standard attachment for the measurement of diffuse reflectance was used for the quantitative evaluation of the spots, which were scraped from the chromatographic plates and prepared as 40-mg analytical samples. The cells used to hold the samples and reference material, as well as the procedure followed in preparing material for examination, have been described earlier (see General Experimental Procedure, Chapter 7 Section 1d).

The reproducibility that can be expected with the method was determined by chromatographing four 5-μg replicates of each acid over a distance of 15 cm in one dimension by the ascending technique and preparing them for analysis according to the procedure outlined in General Experimental Procedure, Section 1d. When solvent mixture (1) (with 0.2% ninhydrin) was used to develop the plates, an average standard deviation of 0.45% R was obtained for the 10 sets. Table 7.10 summarizes the results of this experiment: the largest standard deviation found for any one set was the 0.84% R value observed with lysine. Similar data were obtained using solvent mixture (2), also 0.2% with respect to ninhydrin.

These results represent a considerable increase in precision over that attained by spray methods. An average standard deviation of 1.45% R and a maximum standard deviation for a single set of 2.32% R were found in a previous study conducted with three 30-μg replicates of the same 10 acids. The two studies differ principally in the solvent systems used and in the fact that in one the ninhydrin was applied as a spray. Since the results of this research indicate that precision changes of the magnitude discussed are not observed when the solvent systems are varied, one must ascribe the increase in reproducibility to the elimination of the spraying operation. By increasing the ninhydrin concentration to 0.4% in solvent

TABLE 7.10

Reproducibility of Reflectance Readings Obtained at 515 nm for Different Spots of the Same Concentration of Amino Acids Chromatographed in One Dimension, Using Solvent Mixture No. 1 (0.2% Ninhydrin)[103]

Amino acid	Range (% R)	Mean (% R)	S.D.* (% R)
Alanine	72.8–73.6	73.0	0.39
Arginine HCl	80.3–81.6	80.8	0.59
Glutamic acid	80.5–81.7	81.2	0.51
Glycine	77.3–77.9	77.5	0.27
Leucine	73.8–74.5	74.2	0.30
Lysine HCl	79.0–80.9	79.8	0.84
Methionine	78.3–79.4	78.8	0.58
Phenylalanine	81.3–82.6	82.3	0.71
Serine	75.5–76.4	76.1	0.41
Valine	74.2–74.7	74.4	0.27
			Av. S.D. 0.49

*See footnote to Table 7.01.

(3), the average standard deviation for the ten sets was 0.53% R and the sensitivity improved from 1.0-0.5 μg/spot.

As expected, there was some decrease in reproducibility when the ten amino acids were chromatographed in two dimensions, although the precision was still considerably better than that achieved in one dimension with the use of sprays. The results obtained when four 5-μg replicates were chromatographed in the first dimension with solvent mixture (3) and in the second dimension with solvent mixture (2) to which 0.2% ninhydrin had been added gave a 0.77% R average standard deviation; the standard deviation of none of the sets was in excess of 1.18% R.

The probable relative error in the measurement of the concentrations of alanine, leucine, serine, and valine was determined by using the precision data obtained with four 5-μg replicates of the amino acids (listed in Table 7.10) and of the calibration curves for these same acids. A similar investigation was carried out to ascertain the relationship between the probable relative error and the concentration of glycine. In this instance, solvent (1), which was 0.2% with respect to ninhydrin, was used in conjunction with 4 replicates of acid at each concentration investigated. Data relative to these two studies are presented in Tables 7.11 and 7.12, respectively. The change in measured concentration equivalent to the deviations observed for the various acids was obtained from the appropriate calibration curves and is expressed as a probable percent relative error in concentration. For the 5 acids at the 5-μg concentration, this figure ranged from a low value of 2.8% for valine to a high of 5.0% for serine. In the case of glycine, minimal values were obtained in the intermediate concentration range. The relatively large 9.0% value observed at the 2-μg concentration may be attributed to the fact that this concentration approaches the 1-μg sensitivity limit for glycine, as well as to the increased contribution of volumetric and gravimetric errors associated with such operations as the preparation of the standard solutions. At the opposite end of the scale, the 7.0% figure found for 20-μg concentrations can be ascribed to the flattening of the calibration curve that occurs at high concentrations. Of the amino acids investigated, this effect is particularly noticeable in the case of glycine.[101] Quantitative analysis by reflectance spectroscopy of amino acids resolved on thin-layer plates has also been investigated by Pataki,[21] using the Zeiss Chromatogram spectrophotometer.

The example of glycine sprayed with a ninhydrin spray similar to the one used by Frodyma and Frei[101] is shown in Figure 7.27. The glycine was chromatographed on silica gel, using propanol:water (7:3). The analysis was made at about 510 nm. The calibration curve was obtained by plotting peak areas vs. $\sqrt{C}$ (C, concentration in μg/spot). The reproducibility that can be expected with this approach is seen in Table 7.13.

Heathcote and Haworth[104] have reported on in situ amino acid analysis (ninhydrin complexes) on

TABLE 7.11

Probable Relative Error in the Measurement of the Concentrations of Some Amino Acids[103]

	Alanine	Leucine	Serine	Valine
Range (% R)	72.8–73.6	73.8–74.5	75.5–76.4	74.2–74.7
Mean (% R)	73.0	74.2	76.1	74.4
S.D.* (% R)	0.39	0.30	0.41	0.27
Equivalent change in measured concentration of acid (μg)	0.18	0.15	0.25	0.14
Probable % relative error	3.6	3.8	5.0	2.8

*See footnote to Table 7.01.

(From Frodyma, M. M. and Frei, R. W., *J. Chromatogr.*, 17, 131 (1965). With permission of Elsevier Pub. Co.)

TABLE 7.12

Probable Relative Error in the Measurement of the Concentration of Glycine as a Function of Concentration[103]

	Concentration of glycine (μg/spot)			
	2	5	10	20
Range (% R)	84.3–86.0	77.3–77.9	63.6–65.1	55.3–57.1
Mean (% R)	85.1	77.5	64.4	56.3
S.D.* (% R)	0.92	0.27	0.66	0.76
Equivalent change in measured concentration of glycine (μg)	0.18	0.20	0.50	1.40
Probable % relative error	9.0	4.0	5.0	7.0

*See footnote to Table 7.01.

(From Frodyma, M. M. and Frei, R. W., *J. Chromatogr.*, 17, 131 (1965). With permission of Elsevier Pub. Co.)

TABLE 7.13

Reproducibility of Scanning for the Glycine Ninhydrin Complex (0.6 μg Glycine Per Spot)[124]

	Reflectance at 510 nm		Fluorescence quenching	
Chromatogram	Peak area (mm²) mean value n = 6	S.D.* (%)	Peak area (mm²) mean value n = 6	S.D. (%)
1	340	1.8	2,025	4.1
2	340	2.6	1,800	3.8
3	360	2.8	1,890	5.8
4	360	5.3	1,600	4.1
5	330	2.7	1,705	3.1
Mean value (n = 5)	350		1,805	
S.D. (%)	4.0		9.1	

*See footnote to Table 7.01.

(From Zürcher, H., Pataki, G., Borko, J., and Frei, R. W., *J. Chromatogr.*, 43, 457 (1969). With permission of Elsevier Pub. Co.)

thin-layer chromatograms working with a Joyce, Loebl Chromoscan in the reflectance mode. Heathcote et al.[105] have also applied the method of Heathcote and Haworth[104] to the quantitative determination of amino acids in urine. In applying the methodology to urine samples, it was found that oligopeptides and inorganic salts present in the samples caused distortions in the chromatograms. As available desalting techniques were found to remove only the inorganic salts, a new method that also removed the basic oligopeptides had to be devised. It was found that the use of a column containing an ion-retardation resin could accomplish this under standardized conditions.

A column (1.5 x 12.5 cm) was prepared from an aqueous suspension of Bio-Rad AG11A8 ion-retardation resin (50-100 mesh) and was washed with 100 ml of distilled water to remove im-

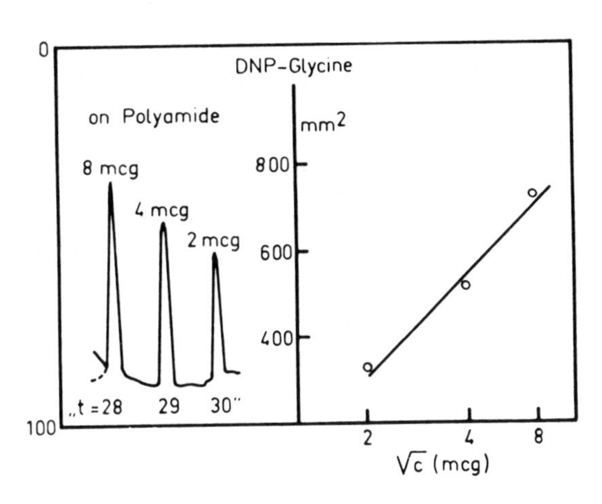

FIGURE 7.27 Peaks and calibration curve of glycine (after ninhydrin spray) on silica gel, measured with the Zeiss Chromatogram spectrophotometer at 510 nm.[21] (From Pataki, G., *Chromatographia,* 1, 492 (1968). With permission of Pergamon Press.)

purities. A 2-ml sample of urine was then placed on the inside wall of the column with a pipette without disturbing the surface of the resin. After addition of 1 ml of water to wash the sample onto the column, followed by running the water through the column so that the surface of the resin was free of water, the amino acids were washed from the column with 19 ml of water at a flow rate of 2 ml/min. The first 13 ml of eluate were discarded. The remaining 6 ml, which contained the amino acids, was collected. The column was cleaned for further use by washing it with 100 ml of distilled water at a flow rate of 10 ml/min.

From 20-40 μl samples were spotted on a cellulose layer (layer thickness 400 μm) with a 5-μl capillary pipette, taking care to maintain a small, uniform spot size at the origin. Two-dimensional development was used, with drying between developments. The solvent system for the first dimension was 2-propanol-butanone-1*N* HCl (60:15:25), while 2-methylbutanol-2-butanone-propanone-methanol-water-(0.88) ammonia (50:20:10:5:15:5) was the solvent system for the second dimension. After chromatography, the dried plates were sprayed with ninhydrin-cadmium acetate chromogenic reagent until translucent, heated in a convection oven at 60° for 15 min, then stored in a dark, ammonia-free atmosphere for 4 hr prior to measurement. The peak areas of the separated amino acids were then measured using a Chromoscan in the reflectance mode (double-beam), and the calculated peak areas were compared with standard calibration curves described in a preceding paper.[104]

Urine samples from four normal and four pathological subjects were analyzed by the TLC method, and results were compared with those obtained using the Technicon® automatic ion-exchange technique. The desalting technique was found to have the disadvantage that it removed the amino acids of taurine and urea. However, these are contained in the first 13 ml of eluate from the column. The results obtained by the in situ reflectance-TLC method were in good agreement with those obtained by the standard column methods.

As many as 50 samples could be passed through the column before the resin needed to be removed, washed, and repacked. The authors suggest that the method provides a rapid method for qualitative screening that is easily adapted to quantitative results.

Nester/Faust[106] have published a method for amino acid analysis based on the same principles. After the plate is sprayed with a ninhydrin solution, the amino acids are quantitatively analyzed by scanning with their Model 900 Chromatogram Scanner. An accuracy of ±2.8% is claimed for valine after one-dimensional chromatography. Rather unexpectedly for this determination, a calibration curve of peak area vs. concentration, which is both perfectly linear and passes through the origin, is presented.

One of the first evaluations of ninhydrin complexes of amino acids on paper chromatograms by spectral reflectance dates back to 1953 when Goodban and Stark[107] determined the amino acid content of sugar-beet juices. Because they used a rather crude measuring device, they only obtained semiquantitative results. Measurements by the transmission mode were doubtless superior in those days.

c. Sugars

The feasibility of this approach was also demonstrated by the successful determination of mixtures of sugars.[108,109] For aldopentoses the following procedure is recommended:

1. Chromatographic resolution of sugar mixtures on cellulose thin-layers by the procedure set forth by Vomhof and Tucker;[110]
2. Chromogenesis by dipping the chroma-

togram in ethyl acetate containing 2% each of aniline and trichloroacetic acid;

3. Air-drying the plates for 30 min, followed by heating at 110°C for 6 min;

4. Removal of the spots and preparation of samples and standards for reflectance measurements at 500 nm, as described earlier.

The dipping technique provides a more uniform distribution of the spray reagent over the entire chromatoplate. The reproducibility obtained under these conditions for arabinose is depicted in Table 7.14. An accuracy of about 5% is reported for this method; the sensitivity is 5 μg/spot. The sugar-aniline compounds are stable for at least one week.

A naphtharesorcinol/phosphoric acid mixture described by McCready and McComb[111] (0.2% NR and 85% H_3PO_4 9:1, v/v) was found useful for hexoses as well as for pentoses.[109] Using the dipping technique, stable red spots were obtained for fructose, and blue-green and blue spots for rhamnose, arabinose, and glucose.

Bevenue and Williams[112] used similar visualization techniques before the quantitative evaluation of raffinose or melibiose on paper chromatograms by diffuse reflectance techniques. Plotting of reflectance density (log 1/reflectance) vs. concentration gave linear relationships only in very limited concentration ranges.

TABLE 7.14

Reproducibility of Reflectance Readings Obtained for Aniline Compounds of Equal Concentrations of Arabinose Adsorbed on Cellulose, Using a Reference Cell Containing Background Material Scraped from the Chromatoplate of Sample 1*[108]

Sample (25 μg)	Percent reflectance	Deviation from mean
1	91.2	+1.2
2	88.0	−1.4
3	89.7	+0.3
4	90.6	+1.2
5	88.9	−0.5
6	88.0	−1.4
7	88.8	−0.6
8	88.3	−1.1
9	91.0	+1.6
	Mean 89.4	S.D.† +1.3

*Readings taken at 500 nm.
†See footnote to Table 7.01.

The same workers[113] described a similar method for the quantization of reducing sugars, such as fructose, glucose, or arabinose. After the chromatograms were developed by the descending technique with a solvent mixture of ethyl acetate:pyridine:water (8:2:1), they were air-dried for 60 min and dipped in a reagent solution containing 1 g 4,5-dinitroveratiole/100 ml of acetone. After the sheets were air-dried for 5 to 10 min, they were dipped in a 1 *M* KOH solution in 95% EtOH, air-dried again, and heated at 60°C in a humid atmosphere for 10 min. Spectral reflectance was measured immediately with a Photovolt reflectance unit Model 501-A (no longer available) equipped with a filter for 515 nm. Calibration curves reflection density (Reflectance) vs. log concentration were constructed for analysis; the useful linear range was between 2 and 20 μg/spot.

McCready and McComb[111] successfully used a spray reagent described earlier in this section for the determination of several hexoses and pentoses (see Figure 7.28). The Photovolt instrument was used for quantization by reflectance spectroscopy. Small amounts of galactose were determined by Owens et al.,[114] who used the same procedure. The overall accuracy of all these procedures was

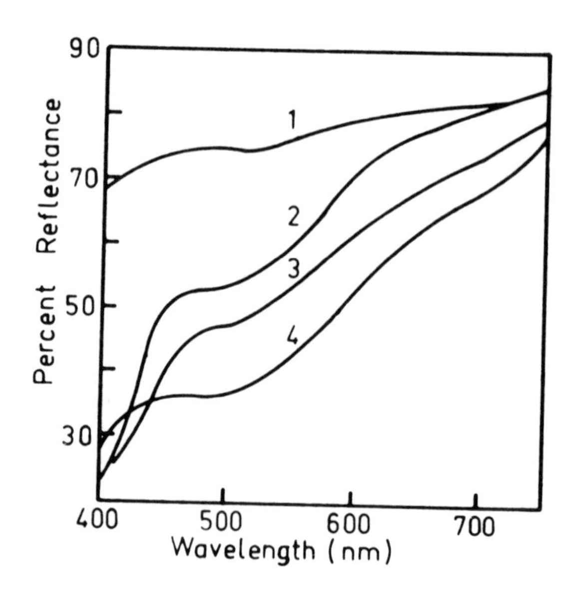

FIGURE 7.28 Spectrophotometric reflectance curves and colors produced on paper chromatograms from 1, background; 2, galacturonic acid; 3, galactose; and 4, arabinose. With aniline-trichloroacetic acid indicator. (From McCready, R. N. and McComb, E. A., *Anal. Chem.*, 26, 1645 (1954). With permission of Am. Chem. Soc.)

reported to be ±5% (after one-dimensional chromatography).

d. Antioxidants

Nester/Faust[106] have proposed a visible reflectance technique for the determination of antioxidants in rubber. As an example, antioxidants such as phenyl-β-naphthaline, 2,4,6-tri-*tert*-butylphenol, 6-ethoxy-2,2,4-trimethyl-1,2-dihydroquinoline, butylated, styrenated cresol, and hexylphenyl-*p*-phenylenediamine were separated on silica-gel layers according to a method proposed by Davies.[115] Visualization was carried out with a phosphomolybdic acid solution (10% w/v methanol) and after heating at 105°C for 10 min, the blue spots on yellow background were recorded in the reflectance mode with the Uniscan 900.

e. Miscellaneous

Weiss has recently described a method for the determination of estriol in the urine of pregnant women.[116] The estriol was hydrolyzed, and the sample was then subjected to an extraction procedure, following which the hydrolyzed estriol was coupled with Fast Black Salt K. Standard samples and extracts were spotted on the chromatogram and, following development, the chromatograms were scanned with a Zeiss Chromatogram spectrophotometer at 550 nm. Quantitative determinations may be made to as low as 0.005 μg, and automatic data collection and evaluation may be used. The method has the advantage of relative simplicity combined with sensitivity and specificity. Weiss believes it could be used for routine clinical use, and that the possibilities for automation and the handling of large numbers of samples should make it quite attractive.[116]

Massa et al.[117] have reported an in situ reflectance method for the determination of pilocarpine on thin-layer chromatograms. The chromatographic procedure is used to separate pilocarpine from isopilocarpine and pilocarpic acid. The spots are made visible by spraying with a modified Dragendorff reagent, following which quantitative analysis is carried out by reflectance. The method has been applied to pharmaceutical preparations.[117] Massa et al. have also reviewed methods for the determination of such compounds as codeine, morphine, papaverine, noscarpine, and narceine and discussed the use of reflectance in such analyses.[118]

A Zeiss ERI-10 densitometer has recently been used to study quantitatively the formation of ring-closure products following a ring-closure.[119] The reaction mixture is spotted on a paper chromatogram, the chromatogram is developed, and then it is sprayed with an ethanolic solution of 1.0 *M* H_2SO_4 containing 3% *p*-dimethylaminobenzaldehyde. The chromatogram is then heated at 80° for 5 min to obtain the colored salt. In situ reflectance measurements are used to quantitate the ring-closure reaction. A linear working range of from 10-50 μg was found for 2,4-dioxo-6-methyltetrahydro-*s*-triazine.

In another recent study, in situ reflectance spectroscopy on chromatographic thin-layers was used to study structural relationships and their influence in pi-complex formation.[120] Reflectance spectra of the complexes were measured on a Farrand UV-VIS Chromatogram Analyzer by plotting reflectance of the sample with reference to background at 10 nm intervals between 420 and 650 nm in the single-beam mode of operation. For these measurements, the monochromator was removed from the analyzer leg and replaced with an auxiliary filter No. 3-73, which provides a sharp-cut at about 400 nm. The monochromator in the exciter leg of the instrument provided monochromatic light. Slit aperture reducers appropriate to the response characteristics of the 1P28 photomultiplier tube were used for the measurements. In the range where the tube has optimum response (approximately 450 to 600 nm), a 0.031-in. slit aperture reducer was used. Outside this range, larger aperture reducers (0.062, 0.125 in.) were used. The 3/16-11/32 slit set was used for all measurements.

The indoles studied contained the following substituents in the 5-position: -CN, -Cl, -H, -OH, $-NH_2$. The spectra of the complexes these indoles, which act as electron-donors, form with the electron-acceptor reagent 2,4,7-trinitro-9-fluorenone (TNF) were quite revealing. The absorption band having the narrowest bandwidth and lowest intensity was that of the 5-cyanoindole complex. In this compound, the -CN group was electron-withdrawing and thus reduced the donor ability of the indole ring. The absorption maximum for this complex was at about 445 nm. Changing the substituent to increase the donor ability of the indole ring resulted in broader, more intense absorption bands and also produced a bathochromic shift in the absorption maxima.

The possibility of performing a quantitative analysis of the complexed indoles was also investigated.[120] Measurements were performed using the Farrand instrument in the double-beam mode on the TNF-complexes of 3-indoleacetic acid and 5-hydroxyindole acetic acid at the wavelength of maximum absorption of these complexes. A linear relationship between concentration and peak area was found in the range 0.5 to 10 μg.

4. Ultraviolet Reflectance Spectroscopy of Other Organic Systems

As early as 1958 and 1959 Zeitlin and Niimoto[121,122] reported evidence for the use of diffuse reflectance spectroscopy in the UV region of the spectrum. They measured UV reflectance spectra of a number of ketones on filter paper.

Korte and Weitkamp[123] used a modified form of the Kubelka-Munk equation in carrying out the determination of 2,3,6-trimethylfluorenone on paper chromatograms. The deviations from the Kubelka-Munk theory observed by these workers were partially attributable to the fact that they used only one layer of paper and did not fulfill conditions for infinite layer thickness. It was not until 1965 that the first application of this technique to TLC was published.[2]

a. Aspirin and Salicylic Acid

Because of the large number of substances having characteristic UV spectra that may be employed in their analysis, it was felt that the application of UV reflectance spectroscopy to TLC would prove to be invaluable, especially for those in the pharmaceutical and biochemical sciences. The feasibility of this approach was ascertained by using it to determine the composition of mixtures of aspirin (acetylsalicylic acid) and salicylic acid that had been separated on silica-gel plates.[2] This particular system was selected for study not only because other spectra of the two compounds were suitable, but also because it presented no difficulties in locating the compounds after resolution. Both appear as yellowish-brown spots when the plates have dried.

Aspirin and salicylic acid, of Merck U.S.P. and Chase U.S.P. purity, respectively, were dried over sulfuric acid for 24 hr before use. Stock solutions of the compounds in chloroform, 0.10 *M*, were used to prepare the dilution series used in this study. Development was carried out by the ascending technique with a mixture of hexane: glacial acetic acid:chloroform (85:15:10). R_f values observed for salicylic acid and aspirin were 0.35 and 0.2, respectively. Although in preparative work the spots may be removed from the chromatoplates immediately after having been developed, in the analysis of the mixtures the plates were dried in an oven at 90°C for 2 hr. Under these conditions, a quantitative conversion of aspirin to salicylic acid took place. This can be seen from Figure 7.29, which shows the gradual conversion at various drying conditions. A shift in absorption maximum from 278 nm (for aspirin) to 302 nm (for salicylic acid) is observed. The preparation of samples and the cell used for obtaining the reflectance spectra have already been described. The aspirin is determined as salicylic acid at 302 nm; precisions of 0.37% R and 0.47% R standard deviation for salicylic acid and aspirin, respectively, have been reported. Linear calibration curves (absorbance vs. $\sqrt{\text{conc.}}$) were obtained over a concentration region of 0.2 to 1.7 μmole/spot.

Aspirin has also been used as a model system in an extensive error analysis of UV reflectance spectroscopy.[91]

b. Amino Acid Derivatives

Dinitrophenylation of amino acids leads to amino acid derivatives (DNP-amino acids), which

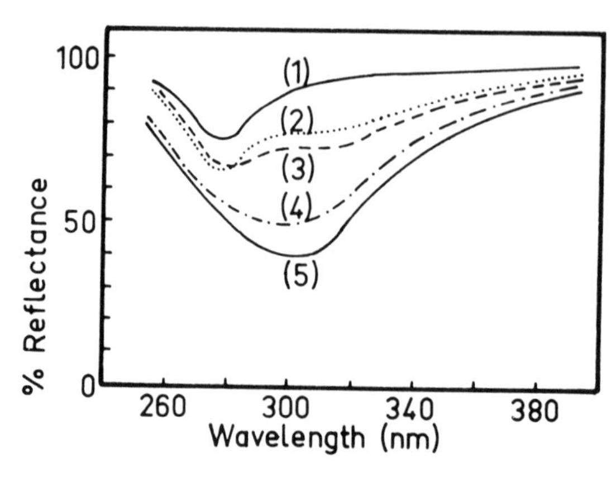

FIGURE 7.29 Reflectance spectra obtained for aspirin adsorbed on silica gel G at indicated intervals after spotting. (1) After 15 min at room temperature; (2) after 2 more hr at room temperature; (3) after an additional 5 min at 90°C; (4) after an additional 10 min at 90°C; (5) after still another 10 min at 90°C. (From Frodyma, M. M., Lieu, V. T., and Frei, R. W., *J. Chromatogr.*, 18, 520 (1965). With permission of Elsevier Pub. Co.)

are UV active, and thus suitable for analysis by the fluorescence quenching or reflectance mode. An additional advantage is the separability of DNP-amino acids from interfering salts.

Similar arguments can be presented for the reaction of amino acids to form 3-phenyl-2-thiohydantoins (PTH-amino acids) when treated with phenyl isothiocyanate. Since most PTH-amino acids are soluble in organic solvents, this method can be used to separate the amino acids from interfering impurities; again they show UV activity.

Pataki[96] discussed in detail the preparation and chromatographic separation of these derivatives. Pataki[21] and Zürcher et al.[124] discussed the qualitative and quantitative evaluation of DNP- and PTH-amino acids by diffuse reflectance spectroscopy, after separation on thin-layer chromatograms. Measurements were carried out with the Zeiss Chromatogram spectrophotometer.

UV reflectance spectra for PTH-histidine are shown in Figure 7.30. Whereas the λ_{max} values (absorption maxima) of DNP derivatives (e.g., glycine or adenine) do not show any shifts after treatment of the spots with HCl or NaOH, significant changes are observed in the case of PTH-histidine (Figure 7.30) and other PTH-amino acids.[21]

Shifting of maxima as a result of chemical or temperature treatment can, provided the shifts are controllable and characteristic of a compound, serve as an additional means of identification and characterization.[26]

Reflectance spectroscopy can also yield quantitative results. Peaks and corresponding calibration curve for DNP-glycine, recorded with the Zeiss Chromatogram scanner, are shown in Figure 7.31. The DNP-amino acids were separated on silica gel with a chloroform:benzyl alcohol: acetic acid mixture (70:30:3, v/v) or on polyamide layers with a benzene:acetic acid mixture (4:1, v/v). Analysis for glycine was made at 350 nm. A comparative investigation on the reproducibility of diffuse reflectance spectroscopic and fluorimetric data was carried out with amino acids and nucleo derivatives.[124] Table 7.15 shows the results obtained for DNP- and PTH-amino acids after one-dimensional separation on silica-gel layers with starch used as a binder. The data for both derivatives are in fairly good agreement (5 to 6% relative standard deviation, if measured on different chromatograms). The comparative reproducibilities are depicted in Table 7.16 for the three major in situ techniques.

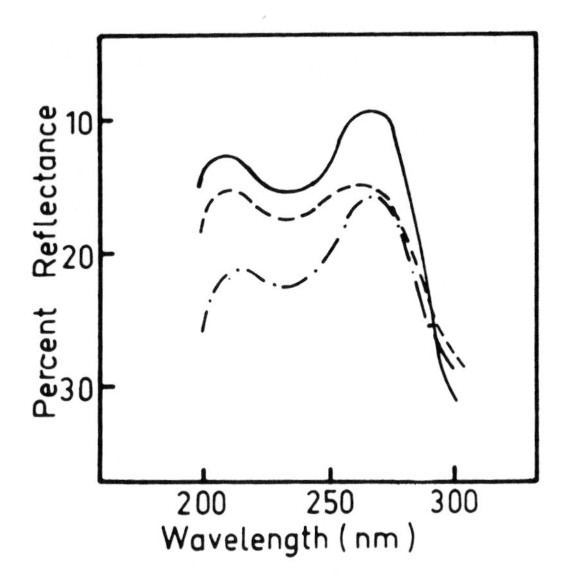

FIGURE 7.30 Reflectance spectrum of PTH-histidine (5 μg), measured with the Zeiss Chromatogram spectrophotometer, before and after spraying the layer with 1 *N* HCl and 1 *N* NaOH (two different plates) on silica gel with fluorescence indicator. ——— unsprayed, λ_{max} = 268 nm; —·—·— sprayed with 1 *N* HCl, λ_{max} = 267 nm; — — — sprayed with 1 *N* NaOH, λ_{max} = 258 nm. (From Pataki, G., *Chromatographia,* 1, 492 (1968). With permission of Pergamon Press.)

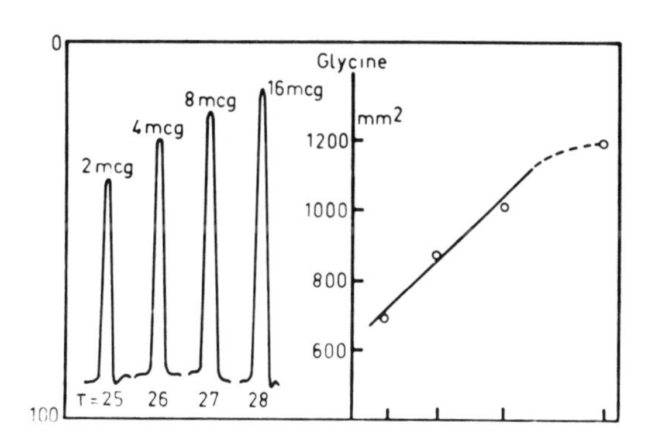

FIGURE 7.31 Peaks and calibration curve of DNP-glycine on polyamide layer, measured with the Zeiss Chromatogram scanner at 350 nm. ("t" values indicate time interval between drying and measurement.) (From Pataki, G., *Chromatographia,* 1, 492 (1968). With permission of Pergamon Press.)

Results obtained by reflectance and fluorescence quenching are compared on the one hand and fluorescence measurements carried out with two different instruments on the other. It can be

TABLE 7.15

Reproducibility of Scanning of DNP- and PTH-Amino Acids[124]

	DNP-glycine (2 μg), measured at 350 nm		PTH-glycine (2 μg), measured at 268 nm	
Chromatogram	Peak area (mm^2) mean value (n = 6)	S.D.* (%)	Peak area (mm^2) mean value (n = 6)	S.D. (%)
1	535	1.8	570	4.2
2	560	1.0	630	4.3
3	540	2.6	570	4.2
4	480	4.0	590	4.0
5	550	3.8	630	3.9
Mean value (n = 5)	530		600	
S.D. (%)	6.0		5.0	

*See footnote to Table 7.01.

(From Zürcher, H., Pataki, G., Borko, J., and Frei, R. W., *J. Chromatogr.*, 43, 457 (1969). With permission of Elsevier Pub. Co.)

TABLE 7.16

Summary of the Reproducibilities of Scanning Under Rigidly Standardized Conditions[124]

	S.D.* (%)	
Substance/technique	On the same chromatogram	On different chromatograms
Amino acids		
Reflectance	2–5.3	4
Fluorescence quenching	3–6	9.1
DNP-amino acids		
Reflectance	1–4	6
Fluorescence quenching	5–7.4	8.6–9
PTH-amino acids		
Reflectance	3.9–4.2	5
Fluorescence quenching	5–7	12
Nucleic-acid derivatives		
Reflectance	–	6–12.5†
Fluroescence quenching	–	13.5–15.2†
DANS-amino acids		
Fluorescence‡	1.9–3.6	7.5
Fluorescence §	2–5	9.1

*See footnote to Table 7.01.
†Two-dimensional chromatograms.
‡Measurements with the Zeiss Chromatogram spectrophotometer.
§Measurements with the Camag Turner Scanner.

(From Zürcher, H., Pataki, G., Borko, J., and Frei, R. W., *J. Chromatogr.*, 43, 457 (1969). With permission of Elsevier Pub. Co.)

seen that reflectance spectroscopy is superior to the quenching technique. In the case of fluorescence measurements, only small differences from one instrument to the other can be observed. It must be pointed out that the reproducibilities discussed[124] can only be reached if the technique of layer preparation, application of substances, chromatography, drying, and scanning, as well as some other influencing factors, are rigidly standardized.

c. Nucleo Derivatives

Lieu et al.[3] described a method for the analysis of nucleotides by in situ spectral reflectance based on the spot-removal technique.

The nucleotides are employed in the form of aqueous stock solutions. A 2.0 mg/ml solution of each nucleotide is prepared by dissolving appropriate amounts of adenosine-3′(2′)-phosphoric acid monohydrate (3′- + 2′-AMP), disodium guanosine-3′(2′)-phosphate monohydrate (3′- + 2′-GMP), uridine-3′(2′)-phosphoric acid (3′- + 2′-UMP), cytidine-2′(3′)-phosphate (2′- + 3′-CMP), and adenosine-5′-diphosphate (5′-ADP) in distilled water. Where necessary a few drops of concentrated ammonium hydroxide are added to achieve solution. The first three nucleotides, which are mixtures of the 3′ and 2′ isomers, were obtained from California Corporation for Biochemical Research, grade-A quality. The other two were obtained from Nutritional Biochemical Corporation of Cleveland, Ohio.

After spotting, the cellulose plates are dried at 85°C for 15 min and then developed by the ascending technique with the use of a mixture of saturated ammonium sulfate, 1 *M* sodium acetate, isopropyl alcohol (80:18:2), according to a procedure devised by Randerath.[125] The developed plates are dried at 85°C for 60 min. Since the resolved compounds appear as light- to dark-blue spots under UV light (max at 254 nm), no difficulty is experienced in locating them on the plates. For qualitative analysis, each spot is excised and then centered in a windowless reflectance cell on top of a uniformly thick layer of 40 mg of cellulose powder that has been packed in the cell by means of a fitted tamp made of an aluminum planchet affixed to a cork stopper (see Instrumentation, Chapter 4, Section 3). The same cell can be used for quantitative work. Samples are prepared in the usual manner.

Except for AMP and DMP, the reflectance spectra of the various nucleotides adsorbed on cellulose are such as to permit the ready identification of the adsorbed species. It is not possible to distinguish between the isomers of AMP and the isomers of GMP solely on the basis of spectral data, as identical spectra are obtained for the members of each isomer pair. By making use of both spectral data and R_f values, however, all the resolved nucleotides can be identified unequivocally. It is noteworthy that in most cases the absorption maximum of the reflectance coincides closely with the absorption maximum of the corresponding transmittance spectrum obtained with an aqueous solution of the nucleotide. This would seem to indicate that these data (Table 7.17) can be used interchangeably for most purposes.

With regard to the precision of the method, standard deviations ranging from 0.39 to 0.78 reflectance unit have been reported by Lieu et al.[3] The accuracy is 2 to 6%. Calibration curves % R vs. log C were used.

Pataki[21] and Frei et al.[126-128] investigated complex systems of nucleo derivatives with the Zeiss Chromatogram spectrophotometer. The spectra of a number of nucleo derivatives investigated are presented in Figure 7.32; the reflectance maxima of 16 compounds are listed in Table 7.18. For a reproducibility study of the λ_{max} values, six individual spectra of hypoxanthine were recorded independently and from separate chromatoplates over a period of six days. The data fluctuated no more than 2 nm. Shifts of λ_{max} values with varying concentrations were studied for ranges of 0.5 to 5.0 μg/spot; the maximum shift was ±2 nm. The possibility of using the spectra for identification purposes is therefore obvious.

The stability of these nucleo derivatives was excellent over prolonged periods of time. The λ_{max} values did not fluctuate more than ±0.82% over a period of 24 hr; therefore, a quantitative study of these systems was carried out.[127,128] For the first phase of this project a group of five compounds was chosen for investigation and the method was applied to the determination of nucleo derivatives in biological material.

Chromatographically pure hypoxanthine, uridine, uracil, inosine, and thymine were used for the preparation of stock solutions in 0.04% (v/w) NaOH. In many cases adenine was added as internal standard. The plates were coated with Cellulose MN-300 (Macherey, Nagel & Co., Düren,

TABLE 7.17

R_f Values and Absorption Maxima of Nucleotides Adsorbed on MN-Cellulose and Absorption Maxima of Nucleotides in Aqueous Solution[3]

		R_f value		Absorption maximum (nm)	
Nucleotide	Isomer	MN-300	MN-300G	Reflectance	Transmittance
AMP	2′	0.35	0.35	260	260
AMP	3′	0.21	0.28	260	260
ADP	5′	0.45	0.48	259	259
CMP	2′ + 3′	0.73	0.73	281	280
GMP	2′	0.51	0.58	256	252
GMP	3′	0.39	0.49	256	252
UMP	3′ + 2′	0.68	0.73	264	262

(From Lieu, V. T., Frodyma, M. M., Higashi, L. S., and Kunimoto, L. H., *Anal. Biochem.*, 19, 454 (1967). With permission of Academic Press.)

TABLE 7.18

Reflectance Maxima of Nucleo Derivatives and Related Compounds Adsorbed on Cellulose[126]

Compound	λ_{max}(nm)	Compound	λ_{max}(nm)
Xanthine	270–271	Nicotinamide	265
Guanine	248/277–278	Cytidine	280–281
Uridine	265	Guanosine	255
Inosine	250	AMP-3′	262
Uric acid	288	AMP-2′	261–262
Nicotinic acid	263	AMP-5′	261–262
Hypoxanthine	251	CMP-3′	279
Uracil	261	GMP-3′	255–256

(From Frei, R. W., Zürcher, H., and Pataki, G., *J. Chromatogr.*, 43, 551 (1969). With permission of Elsevier Pub. Co.)

Germany). All solvents used were of reagent grade.

Samples and standards were applied with 5-μl capillaries (Microcaps, Drummond Scientific Co. Ltd., Broomall, Pennsylvania). Two-dimensional chromatography was carried out according to a previously described method.[129] The spots were viewed under an UV lamp (Camag Ltd., Muttenz, Switzerland) at 254 nm and marked on the backside of the plates with a grease pencil. Reflectance measurements were made with the Zeiss Chromatogram spectrophotometer. The chromatographic peaks were evaluated by planimetry with a planimeter manufactured by Ott AG., Kempten, Bavaria, Germany. A few calibration plots (peak areas vs. square root of concentration) are given in Figure 7.33. The curves were computed statistically and correlation coefficients ranging between 0.980 and 0.994 were obtained.

Klaus[130] has introduced the use of an internal standard in one-dimensional TLC of bands to compensate for fluctuations inherent in the chromatographic procedure. Such an internal standard would be even more desirable in two-dimensional chromatography, and the advantage of using an internal standard for the problem at hand was clearly demonstrated with hypoxanthine (Table 7.19). Three micrograms of uracil and adenine were added as internal standards to 3 μg of hypoxanthine. Chromatography was carried out in the usual way, once with a set of 6 and once with a set of 12 samples, which were then analyzed on the Zeiss instrument. Improvements in reproducibility between 3 and 4% relative

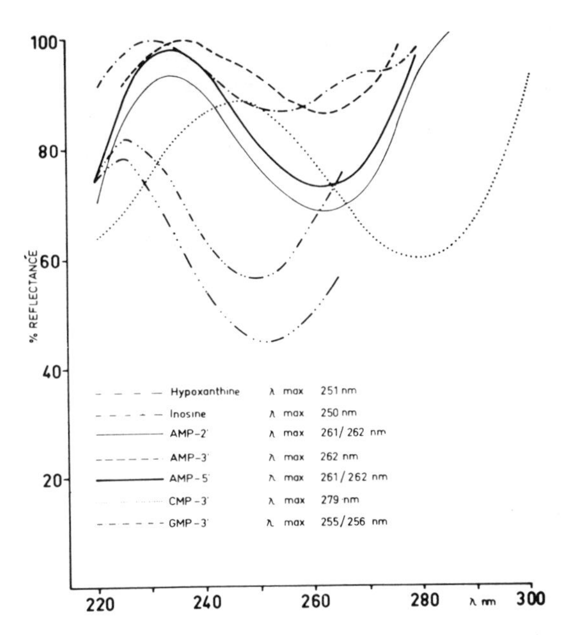

FIGURE 7.32 Reflectance spectra of some nucleo derivatives and related compounds adsorbed on cellulose (5 μg/spot). —•••—•••— hypoxanthine, λ_{max} = 251 nm; —••—••— Inosine, λ_{max} = 250 nm; ———— AMP-2', λ_{max} = 261/262 nm; — — — — — AMP-3', λ_{max} = 262 nm; ———— AMP-5', λ_{max} = 261/262 nm; ••••••• CMP-3', λ_{max} = 279 nm; —•—•—•—•— GMP-3', λ_{max} = 255/256 nm. (From Frei, R. W., Zürcher, H., and Pataki, G., *J. Chromatogr.*, 43, 551 (1969). With permission of Elsevier Pub. Co.)

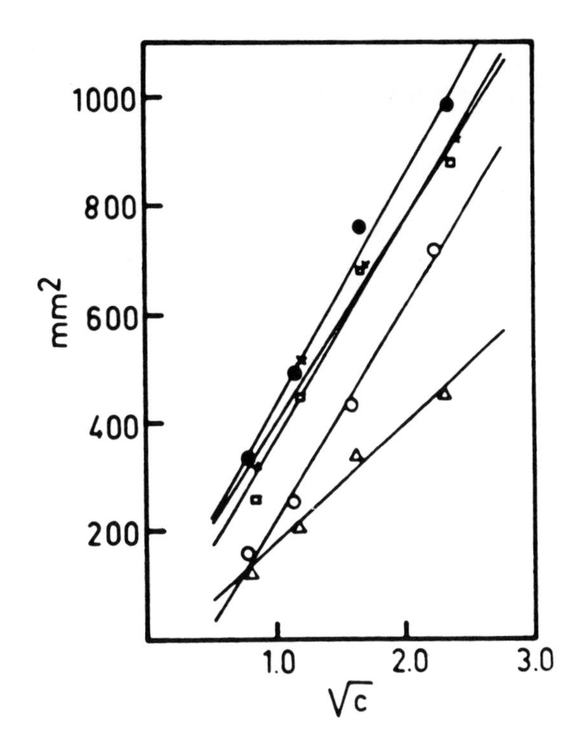

FIGURE 7.33 Calibration curves of some nucleo derivatives determined by in situ reflectance spectroscopy on cellulose layers. •, hypoxanthine; x, thymine; □, xanthine; Δ nicotinamide; O, AMP-3'. (From Frei, R. W., Zürcher, H., and Pataki, G., *J. Chromatogr.*, 45, 284 (1969). With permission of Elsevier Pub. Co.)

standard deviation were observed with the use of both uracil and adenine as internal standards. Since uracil is itself a component of the biological samples of interest, it was decided to use adenine, which is not present and does not interfere with any of the naturally occurring nucleo derivatives on the chromatogram. The resulting chromatogram, as well as the corresponding chromatographic peaks, is shown in Figure 7.34. For quantitative work, the peak evaluation method (also shown in Figure 7.34) proved to be the best of a number of techniques tested, particularly for not completely resolved double peaks.

Calibration curves of ratios (substance/adenine) plotted vs. concentration showed the usual bent shape (Figure 7.35). Reasonably linear curves were obtained, with the origin generally differing somewhat from zero, if the ratios were plotted vs. the square root of concentration (Figure 7.35). Both calibration curves were used for evaluation purposes. The linear plots generally gave somewhat better results. The results of this analysis are presented in Table 7.20. The total time per analysis is two days, which includes chromatographic separation on the first day and evaluation of data by the internal standard method on the second. Actual working time for one technician is 6 to 8 hr. Nine standards of three different concentrations and three samples were chromatographed on 12 plates in the same chromatographic tank. The number of plates and the separation time remain the same no matter how many components are being determined in the mixture, which makes the method even more attractive timewise.

The use of one set of calibration curves for the determination of mixtures analyzed on different days was also investigated (Table 7.21). An average error of about 16% was found, as compared to a 4 to 5% error with standards chromatographed simultaneously with every determination (Table 7.20). The timesaving factor is so enormous, however, that the latter method may have merit for certain applications.

TABLE 7.19

A Comparison of the Reproducibility of Data with and without the Use of Internal Standards[127]

	Hypoxanthine	Uracil	Adenine	Hypoxanthine/ uracil	Hypoxanthine/ adenine
Mean and standard deviation* (n = 6)	752 ± 69	790 ± 65	780 ± 75	0.96 ± 0.069	0.96 ± 0.05
S.D. (%)	9.1	8.3	9.6	6.3	5.1
Mean and standard deviation (n = 12)	700 ± 70	760 ± 81	740 ± 60	0.924 ± 0.066	0.948 ± 0.065
S.D. (%)	10.0	10.7	8.1	7.1	6.9

*See Footnote to Table 7.01.

(From Frei, R. W., Zürcher, H., and Pataki, G., *J. Chromatogr.*, 45, 284 (1969). With permission of Elsevier Pub. Co.)

TABLE 7.20

Analysis of a Known Artificial Mixture of Nucleo Derivatives[127]

		Amount found*		Percent deviation	
Compound	Amount present (μg/spot)	Conc.	$\sqrt{C}$	Conc.	$\sqrt{C}$
Hypoxanthine	4.12	4.56	4.26	10.7	3.4
Inosine	3.90	4.01	4.04	2.8	3.6
Thymine	4.35	4.06	4.11	6.7	5.5
Uridine	3.83	3.80	3.56	0.7	7.0
Uracil	3.60	3.43	3.65	4.7	1.4
		Average deviation		5.12%	4.02%

*Average of three independent analyses.

(From Zürcher, H., Pataki, G., Borko, J., and Frei, R. W., *J. Chromatogr.*, 43, 457 (1969). With permission of Elsevier Pub. Co.)

Cartilage red bone-marrow extract (Rumalon, Robapharm Ltd., Basle, Switzerland) was examined for these nucleo derivatives. The same chromatographic procedure was used;[129] a complete spotchart and data on a preliminary investigation of this system are given in Figure 7.36. After freeze-drying the various batches, 5:1 dilutions were made and analyzed similarly to the artificial mixture. The reproducibility of hypoxanthine in Rumalon was checked. A relative standard deviation of 605 mm^2±37 or 6.1% was found. This is in the order of magnitude of reproducibilities observed with the artificial mixture.

The previously discussed work was recently extended to bases (AMP-3′, AMP-2′, AMP-5′, GMP-3′, and CMP-3′).[128] The chromatographic procedure remains the same, and adenine is again used as an internal standard. Figure 7.37 shows the five nucleotides, together with internal standard and previously investigated nucleo derivatives after chromatographic separation. The corresponding chromatographic peaks are depicted in the same drawing. The quantitative results are presented in Table 7.22. The average deviation ∿ 7.0% is 2 to 3% higher than for the previous mixture (Table 7.22), but the results encourage further investigation of this system.

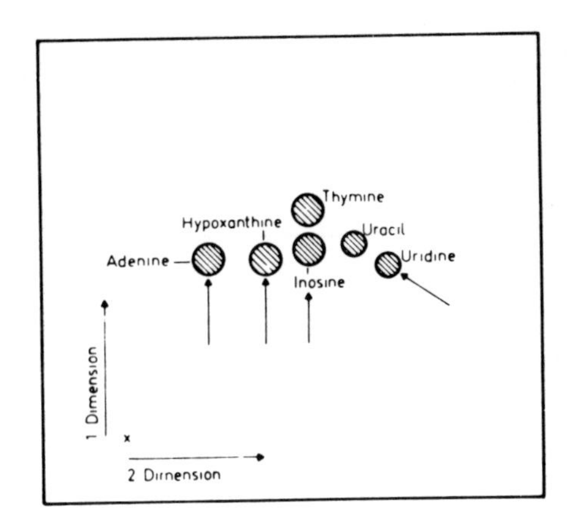

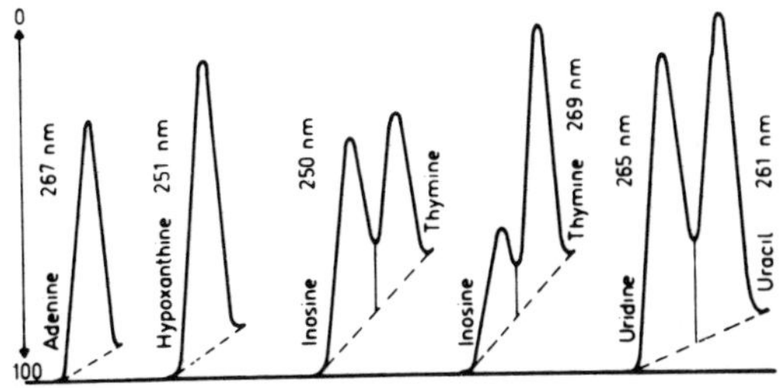

FIGURE 7.34 Thin-layer chromatogram and corresponding chromatographic peaks for nucleo derivatives separated on purified cellulose. Scanning speed, 7.5 cm/min; recorder speed, 8 cm/min. Arrows indicate the scan direction. (From Frei, R. W., Zürcher, H., and Pataki, G., *J. Chromatogr.*, 45, 284 (1969). With permission of Elsevier Pub. Co.)

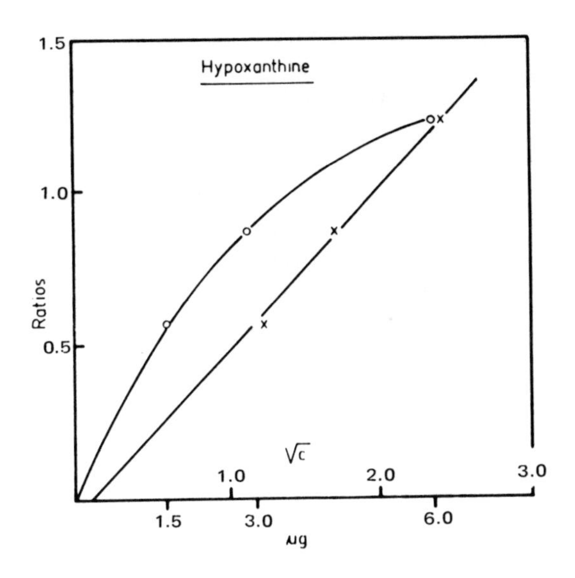

FIGURE 7.35 Calibration curves of peak area ratios hypoxanthine/adenine plotted vs. concentration (O–O) and vs. square root of concentration (x–x) of hypoxanthine. (From Frei, R. W., Zürcher, H., and Pataki, G., *J. Chromatogr.*, 45, 284 (1969). With permission of Elsevier Pub. Co.)

In conclusion, one can say that diffuse reflectance spectroscopy can be used also as a means of quantitative determination of nucleo derivatives in biological systems. In the case of Rumalon the method has now been adopted as a semiroutine analysis for periodic checks of the nucleo-derivative content of the various extraction batches before they enter the pharmaceutical processing stage.

In the case of nicotinamide and nicotinic acid an in situ fluorescence method was preferred for analysis.[131] In this project, the different shapes of calibration curves obtained for fluorimetric, reflectance spectroscopic, and fluorescence quenching measurements were discussed on the basis of nicotinamide. Calibration curves for nicotinamide were recorded with the Zeiss instrument, using reflectance spectroscopy (Figure 7.37). For UV reflectance and quenching the same shape of calibration curve was obtained, thus supporting the assumption that both methods work on the same principle, mainly the UV absorption of the compound of interest. Both curves can be rendered linear over a useful concentration range by plotting (peak area)^2, which is essentially reflectance $(R)^2$ vs. concentration. This in turn would be a rough approximation of the Kubelka-Munk function.[132] In the case of reflectance spectroscopy, this UV absorption is measured directly. In the quenching method, the compound acts as a UV filter, and as a result, the degree of excitation of the fluorescent background is lowered, and a dark spot appears where the substance is adsorbed. This darkening effect (called quenching of fluorescence even though it is not a quenching phenomenon in the true sense) is again an indirect measure of the concentration of this UV filter, or in our case, the nicotinamide. The somewhat lower sensitivity of the quenching method can be attributed to the fact that with the available instrumental setup it is more difficult to choose optimum conditions. Therefore, fluorescence quenching offers no advantage over reflectance spectroscopy. It suffers from poorer reproducibility of data, owing to higher background fluctuations, and it is less specific. If, however, one possesses only the Turner instrument, which does not permit reflectance measurements, then the quenching method generally offers a reasonable alternative for the analysis of UV-active substances, in particular those that do not fluoresce or

TABLE 7.21

The Use of One Set of Calibration Curves for the Analysis of Mixtures on Various Days[127]

Compound	Amount present (μg)	Mixture 1		3		1		2	
		Calibration curves set 2		3		3		3	
		Amount found (μg)							
		$\sqrt{C}$	Conc.	$\sqrt{C}$	Conc.	$\sqrt{C}$	Conc.	$\sqrt{C}$	Conc.
Hypoxanthine	4.12	5.48	5.3	4.97	4.75	5.06	4.8	4.24	4.50
Percent error		33.0	28.6	20.6	15.3	22.8	16.5	2.9	9.2
Inosine	3.9	–	–	4.45	4.6	–	–	3.03	3.00
Percent error		–	–	14.1	18.0	–	–	22.3	23.1
Thymine	4.35	5.15	5.25	5.15	5.25	3.8	3.63	3.28	3.18
Percent error		18.4	20.7	18.4	20.7	12.6	16.5	24.6	26.9
Uridine	3.83	–	–	4.00	4.15	–	–	2.82	2.85
Percent error		–	–	4.4	8.3	–	–	26.4	25.6
Uracil	3.6	3.6	3.45	2.72	2.68	3.76	3.63	3.53	3.45
Percent error		0.2	4.2	24.4	25.6	4.4	0.8	1.9	4.2
Average percent error for one analysis		17.1	17.8	16.4	17.6	13.3	11.3	15.6	17.8
Total average percent error								15.6	16.2

(From Zürcher, H., Pataki, G., Borko, J., and Frei, R. W., *J. Chromatogr.*, 43, 457 (1969). With permission of Elsevier Pub. Co.)

TABLE 7.22

Analysis of a Known Artificial Mixture of Nucleotides[128]

Compound	Amount present μg/spot	Amount found (μg)*		Percent deviation	
		Conc.	$\sqrt{C}$	Conc.	$\sqrt{C}$
AMP-3′	2.00	1.90	1.84	5	8
AMP-2′	2.00	1.94	1.90	3	5
AMP-5′	2.00	1.81	1.82	9.5	9
GMP-3′	2.00	1.83	1.83	8.5	8.5
CMP-3′	2.00	1.83	1.89	8.5	5.5
Average deviation				6.9	7.2

*Average of five analyses.

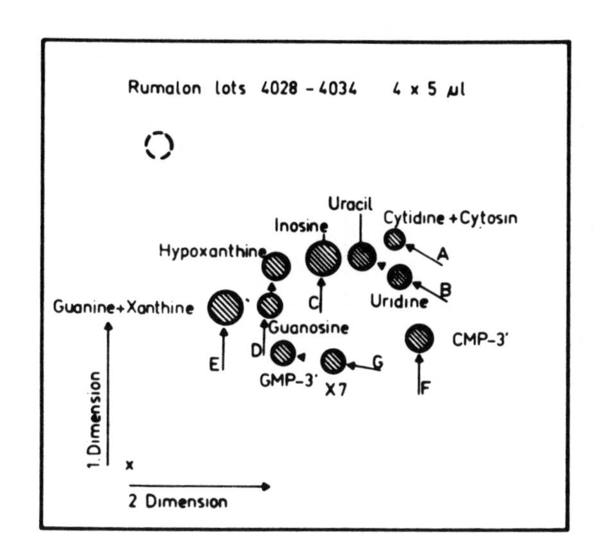

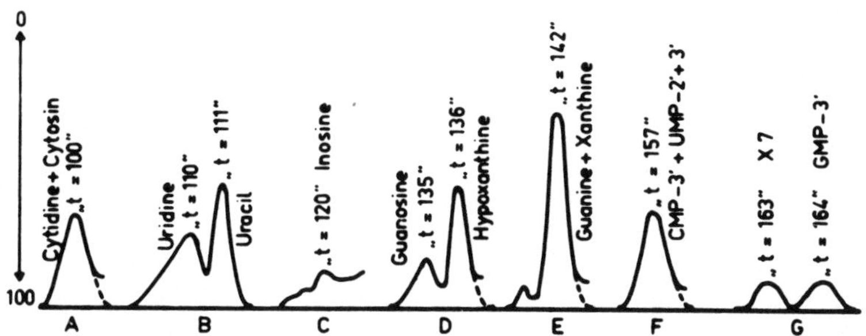

FIGURE 7.36 Thin-layer chromatogram and corresponding chromatographic peaks for nucleo derivatives in cartilage extract (Rumalon, Robapharm), separated on purified cellulose. Arrows indicate the scan direction. (From Pataki, G., *Chromatographia,* 1, 492 (1968). With permission of Pergamon Press.)

cannot be rendered fluorescent very easily. True fluorescence spectroscopy in turn is based on the emission of secondary radiation by the compound of interest itself, upon being excited with a suitable energy source. The method's major advantage is its high sensitivity, which is usually from 10 to 100 times greater than that of reflectance spectroscopy (see also Figure 7.38). For compounds that exhibit a stable and natural fluorescence, without necessitating a spraying procedure, the reproducibility is comparable with reflectance spectroscopy. The calibration plot, fluorescence (plotted as peak area) vs. concentration, is linear at lower concentration, i.e., up to 0.3 µg/spot (Figure 7.38), and is so smoothly bent that it can easily be used for analytical purposes. The deviation from linearity at higher concentrations can be attributed to the substrate, which partially absorbs the emitted fluorescence.

Koof and Noronha have reported the analysis of adenine, adenosine, and adenosine monophosphate by in situ reflectance on silica-gel thin-layers following chromatography.[133] Adenosine monophosphate alone could be analyzed by chromatography using as solvent system chloroform-methanol-water-ammonia (25%) (55:55:15:6). After drying, the plates are scanned with a Zeiss Chromatogram spectrophotometer at 261 nm, the absorption maximum of the compound. In the presence of adenine and adenosine, the solvent system used was chloroform-acetone-methanol-water (80:30:20:2). The R_f values of adenine and adenosine are 0.48 and 0.36, respectively, while adenosine monophosphate remains at the origin. The quantitative determinations for the three compounds are again carried out at 261 nm. The plate is developed a second time after the determination of adenine and adenosine with chloroform-methanol-water-ammonia (25%) (55:55:15:6). The adenosine monophosphate has an R_f of 0.40 in this system and may be quantitatively determined by

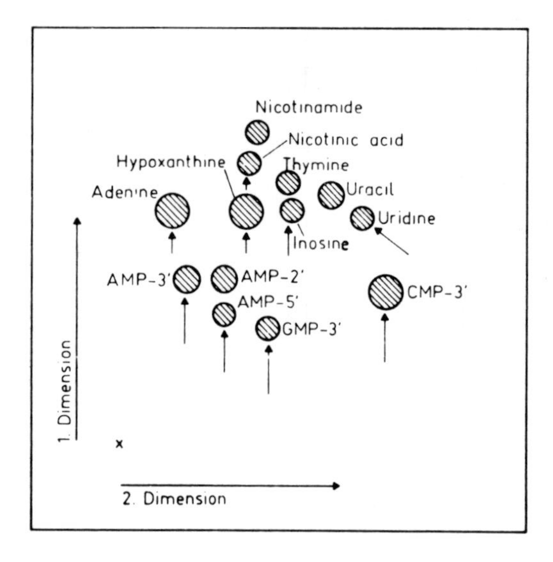

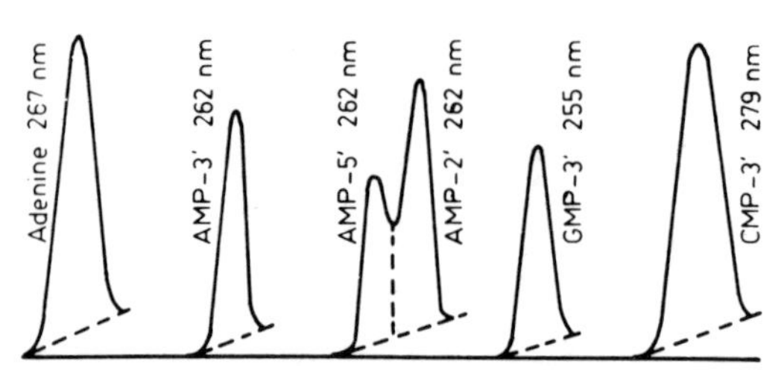

FIGURE 7.37 Thin-layer chromatogram of an artificial mixture of nucleo derivatives separated on purified cellulose, and corresponding chromatographic peaks for five bases and an internal standard (adenine). Arrows mark the scan direction.[128]

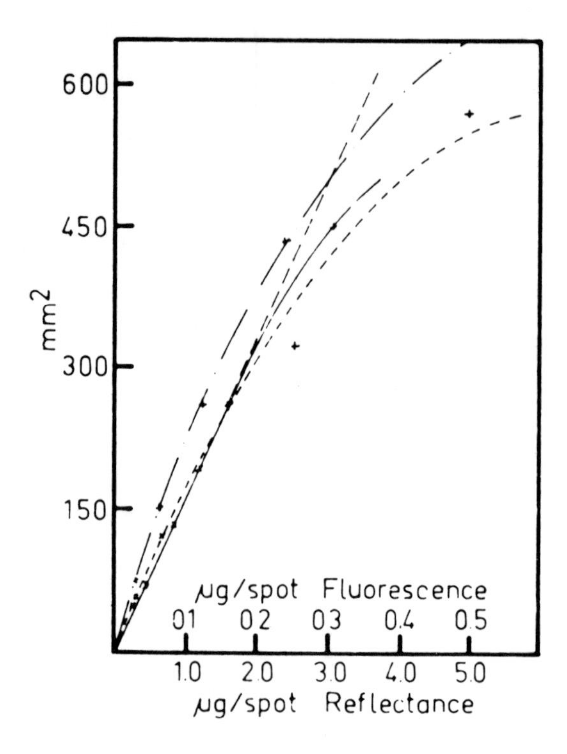

FIGURE 7.38 Calibration curves for nicotinamide obtained by fluorescence spectroscopy (theoretical) (— — — —), fluorescence spectroscopy (experimental) (————————), reflectance spectroscopy (•—•—•—•—), and fluorescence quenching (- - - - - - - -). (From Frei, R. W., Kunz, A., Pataki, G., Prims, T., and Zürcher, H., *Anal. Chim. Acta*, 45, 284 (1969). With permission of Elsevier Pub. Co.)

scanning. The other two compounds run near the solvent front in the second development. A relative standard deviation of 2.4% is reported for the determination of adenosine monophosphate by this method.

d. Vitamins

In this study, the UV reflectance technique was applied to the nondestructive analysis of five B-group vitamins – thiamine hydrochloride, pyridoxine hydrochloride, nicotinic acid, nicotinamide, and *p*-aminobenzoic acid.[4] Because these vitamins are colorless, it is first necessary to locate them once they have been resolved on thin-layer plates. Two methods of achieving this were investigated; one involved the addition of a luminous pigment to the silica gel G adsorbent and examination of the chromatoplates under UV illumination, and the other employed the direct scanning technique described under General Experimental Procedure, Chapter 7, Section 1d.

The vitamins of Nutritional Biochemical purity were employed in the form of aqueous stock solutions, 0.10 *M* except for *p*-aminobenzoic acid which was 0.020 *M*. The use of two adsorbent systems was investigated, Merck silica gel G and Merck silica gel G to which 2% of a luminous pigment had been added. The latter mixture was prepared by weighing out silica gel G and the luminous pigment "KS Super" (Ridel-Dehaën AG., Seelze-Hannover, Germany) in a 98:2 ratio and by shaking them together in a closed flask.

Development was carried out with the use of a mixture of glacial acetic acid:acetone:methoanol: benzene (5:5:20:70) according to a procedure devised by Gänshirt and Malzacher.[134] The developed plates were dried at 45°C for 20 min. Reflectance spectra recorded directly from the chromatoplates are presented in Figure 7.39. As can be seen from this figure, each of the vitamins, with the exception of nicotinic acid and nicotinamide, has a unique reflectance spectrum. Spectra

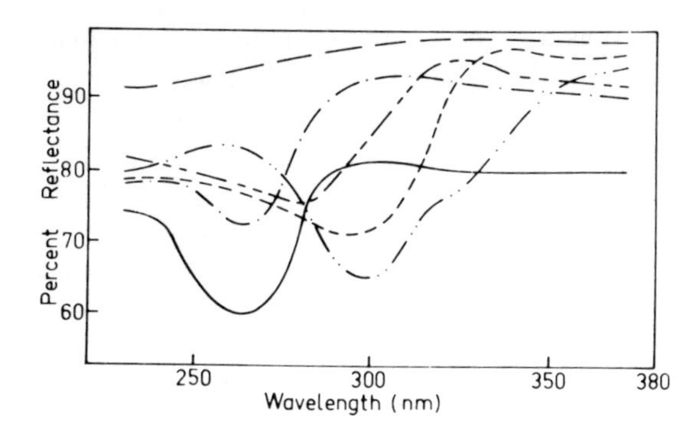

FIGURE 7.39 Reflectance spectra of A, thiamine hydrochloride; B, pyridoxine hydrochloride; C, nicotinic acid; D, nicotinamide; and E, *p*-aminobenzoic acid adsorbed on silica gel; F, silica gel G containing 2% luminous pigment, using silica gel G as a reference standard. (From Frodyma, M. M. and Lieu, V. T., *Anal. Chem.*, 39, 814 (1967). With permission of Am. Chem. Soc.)

and R_f values are essentially identical on silica gel with and without luminous pigment (Table 7.23).

Of the two methods, the one using only silica gel G as an adsorbent provides the greater sensitivity. However, the other is somewhat less time-consuming in that it does not require the use of a spectrophotometer to locate the spots on the developed plates. Both methods are rapid and nondestructive and have been used successfully in the analysis of multivitamin preparations. For quantitative analysis, samples are prepared and packed in the windowless cells as described before. The calibration curves (% R vs. log C, μg/spot) are not linear, and the use of the Kubelka-Munk function may be advisable. The reproducibility of reflectance measurements of nicotinic acid and nicotinamide at 264 nm was ±0.33% R to ±0.39% R mean standard deviation for each of the 4 samples (conc. 2.0 μmole/80 mg). An accuracy of 2.3% was found.

Other work on reflectance spectroscopy of nicotinic acid and nicotinamide has been reported by Pataki[21] and Frei et al.[131] The determination of microgram quantities of vitamin K_1 in infant formula products by reflectance spectroscopy in situ on thin-layer chromatograms has also recently been reported.[135] A liquid-liquid extraction was used to separate the lipids from the product. The fat-soluble vitamins were then separated from neutral and polar lipids on a neutral alumina column, and thin-layer chromatography was used to resolve the vitamin K_1 from other fat-soluble vitamins and quinones. Quantitative analysis was then performed by reflectance spectroscopy.

e. Hormones

Struck et al.[14] described a procedure for the determination of Δ^4-androstene-3,17-dione and testosterone. It is one of the earlier studies that were carried out with the Zeiss Chromatogram spectrophotometer.

It is reported that 0.2 to 0.5 μg of the studied

TABLE 7.23

Absorption Maxima, R_f Values, and Sensitivities of Vitamins Adsorbed on Silica Gel G and on Silica Gel G-Luminous Pigment Mixture[4]

Vitamin	Absorption maximum (nm)		R_f value		Sensitivity (μmole)	
	Silica gel G-lum.pig. mixture	Silica gel G	Silica gel G-lum.pig. mixture	Silica gel G	Silica gel G-lum.pig. mixture	Silica gel G
Thiamine hydrochloride	278	278	0.00	0.00	0.01	0.01
Pyridoxine hydrochloride	298	298	0.17	0.18	0.05	0.01
Nicotinamide	263–268	264	0.51	0.51	0.02	0.01
Nicotinic acid	262–268	264	0.66	0.65	0.02	0.01
p-Aminobenzoic acid	295	295	0.85	0.86	0.01	0.005

(From Frodyma, M. M. and Lieu, V. T., *Anal. Chem.*, 39, 814 (1967). With permission of the American Chemical Society.)

compounds can be estimated with an accuracy of ±10% standard deviation on different two-dimensional chromatograms. Two-dimensional chromatography was carried out on silica gel GF (with fluroescence indicator). The solvents used were (1) chloroform:acetone (98:2) and (2) chloroform:acetone (80:20). Total separation time was 65 min. Under UV light as little as 1 μg of the hormones could be detected as dark spots on a fluorescent background. The separation step was followed by in situ spectral reflectance measurements. The spectra of the two compounds in solution and adsorbed on silica gel respectively are shown in Figure 7.40. Absorption maxima are shifted bathochromically by at least 10 nm in the case of reflectance spectra.

f. Pharmaceutical Compounds and Drugs

Stahl and co-workers have investigated a number of systems using the Zeiss Chromatogram spectrophotometer. Reflectance spectra of caffeine,[136] phenacetine,[137] and the strychnous alkaloids strychnine and brucine,[12] were obtained directly from the chromatoplates. The spectra of five opium alkaloids were obtained in an investigation of drug samples and, using appropriate absorption maxima, their quantitative determination was carried out.[138] Band positions of individual phenyl ethers and phenyl propanes were studied[139,140] with the intention of characterizing chemical classes of drugs. It was possible to separate chromatographically mixtures of oils obtained from certain plants and to identify the zones selectively by spectral reflectance. Similar investigations served to identify phthalides in essential oils.[141] Other applications have been the

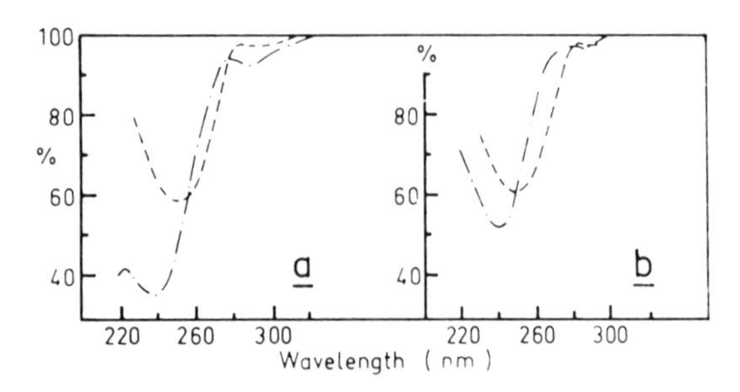

FIGURE 7.40 Transmittance (—•—•—) and reflectance spectra of a, testosterone and b, Δ^4-androstene-3,17-dione. (From Struck, H., Karg, H., and Jork, H., *J. Chromatogr.*, 36, 74 (1968). With permission of Elsevier Pub. Co.)

investigation of aromatic hydroxy, aldehydo derivatives[137] (e.g., vanillin, anisaldehyde, and asarylaldehyde) and a study of the pyrethrines in pyrethrum and peony flowers.[142] Kraus et al.[143] reported the measurement of Scylla glycosides by spectral reflectance with the Zeiss instrument after separation on silica-gel thin-layer plates. Schunack and co-workers[144] discussed similar procedures for the quantization of some xanthine derivatives in blood. The determination of theophylline, 2-hydroxypropyl-theophylline, and 2,3-dihydroxypropyltheophylline was outlined in detail. Accuracies ranging from ±4 to 18% were reported for the various compounds.

Reflectance has also been used by Schunack and co-workers[145] for the analysis of chemically unrelated drugs in several combinations. One preparation contained codeine phosphate (220 nm), phenobarbital (220 nm), caffeine (275 nm), acetylsalicyclic acid (300 nm), and phenacetin (249 nm). The absorption maxima of the compounds are noted in brackets. The second preparation contained amitriptyline (242 nm) and chlorodiazoepoxide (242 nm). Various solvent systems were used to separate the compounds on silica-gel thin-layers. Linear working ranges and instrumental detection limits were established in the microgram range. Reproducibilities were excellent, with the medium deviation of results calculated from 3 single measurings on the same thin-layer plate showing a maximum of 2.7%. The method has the advantage of combining thin-layer chromatography (a well-established method of separation) with a direct method of quantitation (in situ reflectance), thus eliminating the need for elution methods. More recently, Röder et al. have used diffuse reflectance for the determination of opium alkaloids separated on thin-layer chromatograms.[146]

A method for the direct determination of physostigmine by reflectance spectroscopy following thin-layer chromatography has also been reported recently.[147] Physostigmine is a naturally occurring alkaloid found in Calabar beans. It is a cholinesterase inhibitor and is used in the treatment of some diseases of the eye. The drug is usually administered as an eyewash in aqueous solution containing up to 1% physostigmine sulfate. Physostigmine is hydrolyzed in water to a colorless phenolic compound, eseroline, which subsequently oxidizes to rubreserine and other colored oxidation products.

Physostigmine was separated from eseroline and rubreserine on alumina thin-layers using a chloroform-acetone (5:4) solvent system. To obtain well-defined, elliptical spots, the plates were activated at 110°C for one hour and then stored over a saturated solution of sodium bromide for three days prior to use. Solutions were applied to the chromatogram with a 1-μl syringe, and the chromatograms were developed to a height of 10 cm. Spectra of the compounds were recorded from 225 to 350 nm, and the spectrum of rubreserine was also recorded from 360 to 600 nm. Instrumental detection limits for the three compounds on the Zeiss Chromatogram spectrophotometer were 0.05 μg.

To study the reproducibility of the determination, 5 spots (1 μl/spot) of a 0.25% solution of physostigmine sulfate were applied to each of 5 plates. Relative standard deviation for the 5 plates was 5.81%. A single spot was also scanned 25 times to test the reproducibility of a single measurement raising the scanning head and then removing and replacing the plate between measurements. The relative standard deviation was 1.25%. Plotting the Kubelka-Munk function vs. the concentration of the drug revealed a linear working range from 0.05 to 3 μg. At higher concentrations, the plot curved. The effect of changing the slit-widths and slit-lengths was also studied. The results show that the method may be used for the determination of the amount of degradation products (eseroline and rubreserine) present in a sample of physostigmine. The method is quicker than elution techniques, and spraying is not necessary to produce colored spots suitable for densitometry.

Schlemmer has used a Zeiss spectrophotometer equipped with a Camag-Z Scanner to analyze drug mixtures by reflectance spectrometry.[148] As many drugs are sold in mixed formulations, an analytical method that achieves a rapid separation of the constituents followed by quantitation appears attractive.

In the first application,[99] nicotinic acid and etofylline were separated on thin-layers and scanned (nicotinic acid at 261.5 nm; etofylline at 273 nm). While these compounds may be analyzed easily in solution by spectrophotometric measurements when they are present in a mixture in pure form, interferences were found in the analysis of samples from coated tablets. Peak area vs. concentration produced a linear plot from 0.25 to 4 μg for nicotinic acid and from 0.5 to 8 μg for etofylline. The analyses at 2 μg showed a coefficient of variation of the order of 2.5.

Quantitative TLC was also tried for the analysis of cyclobarbital (calcium salt) and pentobarbital (as free acid) in a mixed formulation with acetylsalicyclic acid and phenacetin.[148] Again, the two compounds of interest could be separated on thin-layer plates, but development for only 8.5 cm was recommended due to the tendency of the compounds to tail at longer migration distances. Treating the developed chromatograms with ammonia vapor produced a bathochromic shift in the spectra of the barbiturates, which made them readily detectable in the 5 to 25 μg range at 234 nm. Without treatment, the absorbance maxima are at 220 nm, and the wider slits needed at this wavelength lower resolution. Less background interference is also obtained at the longer wavelength. Using lower concentrations of the preparation would permit the analysis of acetylsalicyclic acid and phenacetin, but this analysis may be done as readily in solution.[148]

Sodium oestrone sulfate and sodium equilin sulfate were separated on silica-gel thin-layers inpregnated with silver nitrate.[148] As the silver nitrate is sensitive to light, scanning in the UV-region could not be carried out quantitatively. Therefore, the developed chromatograms were exposed to sulfuryl chloride vapors for one minute, then to a steaming water bath for one minute. Drying the plates at 100° for 10 min produced uniformly distributed hydrochloric and sulfuric acids on the plate, resulting in charring of the spots. The chromatograms were then scanned at 412 nm.

Schlemmer[148] concluded that in situ reflectance measurements on thin-layer chromatograms provide a useful method for the analysis of pharmaceutical products. The method is sufficiently accurate and reproducible, yet is not particularly time-consuming. The authors share this opinion. Obviously, the range of potential applications is tremendous, and the present success of in situ reflectance techniques will result in a significant increase in such work.

5. Reflectance Spectroscopy of Inorganic Systems

Clark[149] published a comprehensive review of the application of diffuse reflectance spectroscopy to inorganic chemistry, giving 32 literature references. The application of the technique to the structural analysis of metal complexes was emphasized particularly by this author, who has published several papers in this field. The first inorganic application of spectral reflectance to the problem of evaluating paper chromatograms was made by Vaeck,[150,151] who, in determining divalent nickel in microgram quantities on paper, found a linear relationship between the Kubelka-Munk function and nickel concentrations from 30 to 100 mg/l. A critical comparison of the reflectance and transmission techniques for the analysis of copper spots on paper chromatograms was carried out by Ingle and Minshall,[152] who found that the paper appears optically more uniform in reflected light. A precision of ±0.43 in the percent reflectance measurements was reported for 4 replicate analyses of copper rubeanate on paper.

In view of these results it was decided to investigate the applicability of this technique to inorganic compounds (trace metals in particular) separated on thin-layer plates.

a. Rubeanic Acid Complexes

As the first system to be investigated, the author chose rubeanic acid, a widely used chelating agent for nickel, cobalt, and copper-trace analysis in solution as well as on column or paper chromatograms. The behavior of such complexes adsorbed on media commonly used in chromatography, under varying temperature, pH, and humidity conditions, was also investigated.[153]

The reagents used in this investigation, solvents, rubeanic acid (0.1% in ethanol), and metal chloride stock solutions (1 g of Cu^{2+}, Ni^{2+}, Co^{2+}, Zn^{2+}, Mn^{2+}, Fe^{3+}, Cr^{3+} per liter) were reagent grade; doubly distilled water was used throughout. TLC-grade alumina G and silica gel G from Merck Darmstadt with a particle size of 10 to 40 μm and 15% plaster of Paris binder were used. The cellulose layers were prepared from MN cellulose 300 of 10-μm maximum particle size (Macherey, Nagel & Co., Düren, Germany) without binder.

After the 20 x 5-cm plates were coated, they were heated at 110°C (1 hr for silica gel and alumina, 15 min for cellulose) and then stored in a dry cabinet over silica-gel adsorbent. The spots were applied with a 10-μl Hamilton microsyringe. Then the samples were dried for 5 min at 110°C, cooled to room temperature, and chromatographed in a preconditioned (12 hr) chromatography chamber; the solvent systems used for the chromatographic separations were a mixture of 0.5% concentrated hydrochloric acid in acetone for separation on silica-gel layers, a mixture of methyl ethyl ketone:conc. HCl:H_2O (15:3:2) for cellulose layers, and a mixture of 25% 12 N HCl in n-butanol for alumina plates. After separation, the plates were dried in air or at 110°C for 5 min, and the spots were made visible by spraying or dipping with the rubeanic-acid reagent. The cells and the technique used in preparing the separated material for examination have been described earlier; although the prepared barium sulfate plates were used occasionally, the plate adsorbent usually served as the reference standard. The 20-30-mg samples removed from the plates for analysis of the spots were weighed to ±0.1 mg and ground in a small agate mortar for two periods of 1 min each to ensure homogeneity and a more uniform particle size. Samples that were not analyzed immediately were stored in a desiccator over silica gel.

The reflectance spectra of the rubeanic-acid chelates adsorbed on silica gel G and recorded with a Spectronic 505 spectrophotometer are shown in Figure 7.41; to the eye the spots appear blue-purple for nickel, orange-yellow for cobalt, and olive-green for copper. The reflectance spectra have absorption peaks somewhat broader than those of the corresponding transmission spectra in aqueous solution and show a shift in absorption maxima to higher wavelength.

An investigation of the influence of pH on

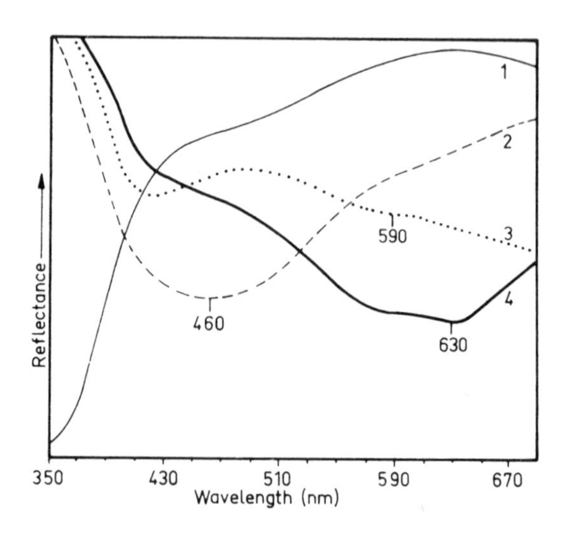

FIGURE 7.41 Reflectance spectra of 1, rubeanic acid and the complexes of 2, cobalt, 3, copper, and 4, nickel adsorbed on silica gel G. (From Frei, R. W. and Ryan, D. E., *Anal. Chim. Acta*, 37, 187 (1967). With permission of Elsevier Pub. Co.)

complex formation revealed that for quantitative work the best results are obtained at pH 7 to 8; this was achieved by subjecting the plates to ammonia vapor for 10 min before the complexing step to neutralize the small amount of hydrochloric acid left behind after the separation of the metals. Better results for cellulose layers were obtained by first spraying or dipping the plates, after separation, in an ethanolic rubeanic-acid solution, drying them at 110°C for 5 min, and then dipping them in a 5% ammonia solution and redrying at 110°C for 5 to 10 min.

To determine the best treatment for the chromatoplates after applying the complexing agent, the plates were heated for different periods at temperatures commonly used in chromatographic work; results are shown in Table 7.24. The cobalt, nickel, and copper complexes are very stable; the rubeanic-acid background, however, disappears almost completely after heating at 110°C for 5 min.

On all adsorbents the intensity of the complex is enhanced by heating. This was greater for alumina and silica gel than for cellulose; the increase was also greater for cobalt and nickel chelates. It is probable that the surface water of the adsorbent is removed on heating, leaving adsorption sites free for the metal complex and resulting in stronger complex-adsorbent interaction. The initial increase in intensity after heating for 30 min at 60°C or for 5 min at 110°C suggests that most of the water has been removed; a further small increase in intensity occurs after heating at 100°C for 30 min and at 110°C for 2 hr; after that further heating produces little change. The process is reversible: a decrease in the intensity of heated samples was observed after storing the samples in a humid atmosphere.

The experiments show that heating of the chromatoplates at 110°C for 5 min is satisfactory: analysis time and adsorption intensity are most favorable and the reagent background is sufficiently reduced to avoid any interference. Samples were stored in a desiccator over silica gel to avoid fluctuations; reflectance readings of individual samples kept in a desiccator varied by only 1 to 2% R units over several weeks.

The sensitivity with the spectrophotometer, using 20-mg samples from the chromatoplates, was 0.05 μg for nickel and copper, and 0.1 μg for cobalt. The visual sensitivity was 0.03 μg/spot for nickel and copper, and 0.05 μg for cobalt if the initial spot size was small (diam. 0.5 cm) and no excessive tailing occurred.

Reflectance-concentration plots give relatively smooth curves over the concentration range 0.05 to 20 μg/spot, but at higher concentrations the

TABLE 7.24

A Study of Temperature and Time Effect on the Reflectance Maxima of Rubeanic-acid Complexes Adsorbed on Various Adsorbents[153]

			% R				
			60°C		110°C		
Adsorbent	Complex of	Air-dry	5 min	30 min	5 min	30 min	2 hr
Cellulose	Co	86.7	86.4	85.7	85.9	85.0	–
	Ni	84.8	84.9	84.2	84.2	83.6	–
	Cu	87.0	87.0	86.7	86.5	86.2	–
Alumina	Co	89.8	88.6	84.5	84.2	83.2	82.5
	Ni	83.0	82.9	80.2	80.0	79.6	79.1
	Cu	87.8	87.6	86.8	87.0	86.6	86.9
Silica gel	Co	87.4	86.4	83.7	83.9	83.1	82.3
	Ni	83.8	82.9	81.6	81.2	80.4	79.9
	Cu	85.2	84.9	84.0	83.8	83.4	83.7

(From Frei, R. W. and Ryan, D. E., *Anal. Chim. Acta*, 37, 187 (1967). With permission of Elsevier Pub. Co.)

curves flatten to such a degree that accuracy is very poor.

To study the adherence of the systems to the Kubelka-Munk law, the Kubelka-Munk functions $[(1\text{-}R)^2/2R]$ were plotted against concentration. A linear curve was obtained for up to 10 μg of cobalt and nickel and for the entire concentration range for copper. The reproducibility of the method ranged from 0.55 to 0.77% R standard deviation for the spraying technique.

For cellulose layers, which are mechanically very stable, dipping rather than spraying gives better control and is more convenient. The plates are dipped in a 0.1% rubeanic-acid-in-ethanol solution contained in a 20 x 25-cm photographic tray.

Reproducibility data for nickel, cobalt, and copper, chromatographed on cellulose and treated by dipping technique, are presented in Table 7.25. The improvement in reproducibility is particularly significant for cobalt; data obtained on cellulose layers (using the spraying technique) are similar to results for silica gel and alumina. The reproducibility for cobalt separated on cellulose on different days and treated by the dipping technique (four trials) was ±0.70% R average standard deviation: this is a decided improvement on similar trials using the spraying technique (standard deviation 1.37% R). Although the best reproducibility and accuracy are achieved by processing standards and samples simultaneously, a single calibration curve can be used with satisfactory results. It is to be expected that the accuracy varies with concentration, despite the good reproducibility of reflectance readings.[3] For nickel (Table 7.26) the accuracy decreases at low concentrations to 20% relative error for 0.1 μg/spot; below 0.5 μg the sensitivity limit is approached (0.05 μg/spot with a probable 50% relative error). The best accuracy was observed for 10 μg/spot; the accuracy decreases with a further increase in concentration, since the calibration curve flattens off at concentrations greater than 20 μg. Similar error patterns were found in a fundamental error analysis for in situ reflectance work.[3]

b. Pyridine-2-aldehyde-2-quinolylhydrazone (PAQH) Complexes

The new chelating reagent PAQH, developed by Heit and Ryan,[154] was investigated extensively in connection with reflectance spectroscopy and TLC.[155,157] A method was devised for trace analysis of nickel, copper, and cobalt, as well as iron. The procedure is almost identical to that described for rubeanic acid. The PAQH reagent solution is 0.03% (w/v) in ethanol (0.01*N* in HCl for stabilization).[155] When using this chelating agent, control of pH was found to be even more important. After an extensive study of pH conditions, it was found that rendering the spray reagent 0.01 *M* in sodium hydroxide shortly before use resulted in highest spot intensities for cobalt.

With nickel, PAQH reacts as a tridentate

TABLE 7.25

Probable Relative Error in the Measurement of the Concentration of Nickel, Cobalt, and Copper Separated on Cellulose and Developed by the Dipping Technique[153]

	Nickel	Cobalt	Copper
Range (% R)*	78.0–79.0	80.1–81.0	84.8–85.6
Mean (% R)	78.4	85.1	85.1
S.D.† (% R)	0.44	0.43	0.39
Equivalent change in measured concentration (μg)	0.06	0.06	0.07
Probable % relative error	3.7	3.7	4.4

*For 12 replicate samples of 1.6 μg cation/spot.
†See footnote to Table 7.01.

(From Frei, R. W. and Ryan, D. E., *Anal. Chim. Acta*, 37, 187 (1967). With permission of Elsevier Pub. Co.)

TABLE 7.26

Probable Relative Error in the Measurement of the Concentration of Nickel as a Function of Concentration[153]

	Concentration of Ni (μg/spot)							
	0.1	0.5	1.0	2.0	4.0	6.0	10	20
S.D.*	0.50	0.50	0.50	0.50	0.50	0.50	0.50	0.50
Equivalent change in measured concentration of nickel (μg/spot)	0.02	0.04	0.05	0.07	0.10	0.14	0.22	1.0
Probable % error	20	8.0	5.0	3.5	2.5	2.3	2.2	5.0

*See footnote to Table 7.01.

(From Frei, R. W. and Ryan, D. E., *Anal. Chim. Acta*, 37, 187 (1967). With permission of Elsevier Pub. Co.)

reagent to form either square planar (coordination number four) or octahedral (coordination number six) complexes with metals; both 1:1 and 2:1 (reagent to metal) complexes would be expected for nickel. Figure 7.42 shows the reflectance spectra for nickel on silica gel after spraying with solutions varying from 0.025 to 1.0*N* in sodium hydroxide. A red complex with a maximum at 520 nm is predominant at low basicity, but conversion to a brown species (maximum 480 to 490 nm) occurs with increasing sodium hydroxide concentration; maximum sensitivity is obtained with 1.0*N* sodium hydroxide spray solution when nearly quantitative formation of the brown species has occurred. The brown complex is the 2:1 species normally encountered in solution. The red species observed at low basicity on silica gel is presumably the 1:1 planar complex; the 1:1 species is readily observed on silica gel but not in solution, and the pronounced influence of adsorbent on equilibrium conditions is evident. The formation of the two species, depending on pH, was also observed for copper. For both nickel and copper, spraying with a 1.0*N* sodium hydroxide solution is recommended.

The results of the temperature study with the nickel chelate are illustrated in Figure 7.43, which shows spectra obtained with various treatments after spraying. Immediate heating of the plate results in a considerable loss in sensitivity (curve 1). If the plate is allowed to stand for 1 hr there is a sensitivity increase of at least 5% (curve 2) since the reaction is not quantitative immediately after spraying. Curve 2 also shows that a significant portion of the nickel seems to be present in the red complex form after standing for one hour. However, if the sample is now heated at 110°C for 5 min the intensity increases and the red complex disappears to a large extent (curve 3). The increase in peak height at 480 nm occurs at the expense of the red modification, which is converted to the brown species. Further heating results only in a minor increase of the peak, and the red complex disappears completely (curve 4); a small batho-

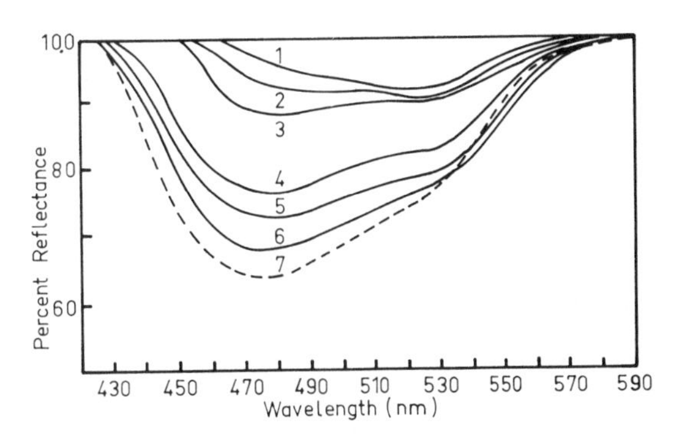

FIGURE 7.42 Reflectance spectra of nickel complex adsorbed on silica gel and sprayed with NaOH solutions. 1, 0.025 *N* NaOH; 2, 0.05 *N* NaOH; 3, 0.10 *N* NaOH; 4, 0.25 *N* NaOH; 5, 0.50 *N* NaOH; 6, 0.75 *N* NaOH; 7, 1.0 *N* NaOH. (From Frei, R. W., Liiva, R., and Ryan, D. E., *Can. J. Chem.*, 46, 167 (1968). With permission of Nat. Research Council of Canada.)

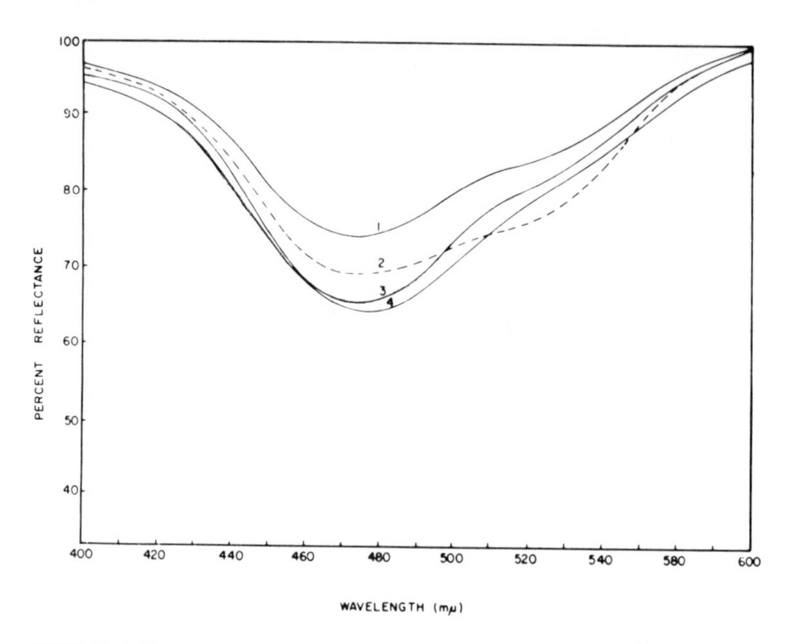

FIGURE 7.43 Reflectance spectra of nickel complex on silica gel. 1, heated at 110°C for 5 min immediately after spraying; 2, air-dried for 1 hr; 3, as 2 and then heated for 5 min at 110°C; 4, as 2 and then heated for 30 min at 110°C. (From Frei, R. W., Liiva, R., and Ryan, D. E., *Can. J. Chem.*, 46, 167 (1968). With permission of Nat. Research Council of Canada.)

chromic shift is also observed upon prolonged heating.

The samples are air-dried on cellulose for one hour. Heating of the cellulose plates results in a dark-yellow background coloration that is probably due to attack of the cellulose by sodium hydroxide.

As a result of temperature studies, heating at 110°C for 5 min after 1 hr of air-drying is recommended for nickel and copper. For cobalt, immediate heating at 110°C for 5 min is suggested.

Both the pH and temperature studies are good examples of the use of reflectance spectroscopy in the investigation of adsorption properties of systems on chromatographic adsorbents.

The sensitivity of this method was found to be about 0.01 μg/spot for all three cations at a 50% accuracy level. Linear calibration curves [$T(R_\infty)$ vs. concentration] for concentrations up to 1.5 μg/spot were obtained. Absorption maxima used for analytical purposes are listed in Table 7.27.

The reproducibility of the method is similar to that of the rubeanic-acid method, with a slight improvement on cellulose plates in connection with the dipping technique.

Iron, zinc, manganese, cadmium, and lead ions react with PAQH to form colored products but do not interfere in the determination of cobalt, nickel, and copper. Iron in commercial silica gel can be removed by chromatographing the plates with the solvent mixture; iron impurities moving with the solvent front are then cut off, and the plate is used for analysis. Cadmium, lead, and iron in the sample move with the solvent front. Zinc and cadmium are chromatographically separated from nickel, cobalt, and copper and, in addition, even tenfold excesses of these metal complexes fade almost completely upon heating at 100°C on silica gel.

A combination of the two proposed methods has been used for the determination of trace metals in cereal.[156] Nickel and copper were determined as the rubeanic-acid complex and cobalt as the PAQH complex. A standard ashing procedure was used. Results for the analysis of some oat samples are presented in Tables 7.28 and 7.29.

In another study, PAQH was used as a semi-selective scavenger, and the metal chelates extracted in chloroform were chromatographically separated. This was followed by a reflectance spectroscopic investigation.[157] This approach to

TABLE 7.27

Absorption Maxima (nm) of PAQH Complexes[155]

	Aqueous solution	Silica gel	Cellulose
Cobalt	500 (pH 7–8)	508(0.01 *N* NaOH)	536(0.01 *N* NaOH)
Nickel	480 (pH 9)	490(1.0 *N* NaOH)	520(1.0 *N* NaOH)
Copper	475 (pH 9)	485(1.0 *N* NaOH)	505(1.0 *N* NaOH)

(From Frei, R. W., Liiva, R., and Ryan, D. E., *Can. J. Chem.*, 46, 167 (1968). With permission of the National Research Council of Canada.)

TABLE 7.28

Analysis of Oats*[156]

	Amount of metal found (ppm) Sample No.				Average and standard deviation† (ppm)	Rel. S.D. (%)
	1	2	3	4		
Cobalt	0.03	0.02	0.02	0.03	$0.02_5 \pm 0.00_5$	± 20
Copper	3.45	3.68	3.50	3.27	3.48 ± 0.17	± 4.9
Nickel	0.96	1.05	0.90	0.99	$0.98 \pm 0.06_3$	± 6.4

*The value for each sample is the mean obtained from four spots developed simultaneously.

†See footnote to Table 7.01.

(From Frei, R. W., *J. Chromatogr.*, 34, 563 (1968). With permission of Elsevier Pub. Co.)

TABLE 7.29

Recovery of Metals from a Sample of Oats*[156]

	Cobalt		Copper		Nickel	
Amount present in original samples (ppm)	0.03	0.03	3.48	3.48	0.96	0.96
Amount added (ppm)	0.05	0.10	3.0	6.0	1.0	3.0
Total amount found (ppm)	0.07	0.15	6.69	9.98	2.08	4.12

*Mean of four results.

(From Frei, R. W., *J. Chromatogr.*, 34, 563 (1968). With permission of Elsevier Pub. Co.)

metal trace analysis has several advantages over the methods previously proposed. The extracted chelates can be applied directly to the chromatogram without prior sample preparation; the sensitivity is increased through evaporation of the organic solution of extracted chelates to the small volume required for chromatographic separation; and adsorbent impurities do not interfere. In addition, faster and clearer separation of the metal chelates in comparison to separation of free metal ions can usually be achieved, because less-polar solvents can be used for the separation process on highly polar adsorbents such as alumina and silica gel. Since the chelates themselves have an intense color, no spraying reagent is needed for the detection of the metals, and variables such as background coloration and irregular color development can be eliminated. Extraction procedures have been described by Singhal and Ryan.[158] The complexes are formed in a solution adjusted by an ammonia-ammonium chloride buffer to about pH 10 and are extracted with chloroform. To ensure complete recovery of cobalt, a second extraction with amyl alcohol is carried out.

The separation of iron, nickel, copper, and cobalt is of major interest since earlier work has shown that their PAQH complexes are most suitable for analytical purposes. Best separation is achieved on alumina layers with chloroform used as the chromatographic solvent.

The R_f values for the Fe^{3+}, Fe^{2+}, Ni^{2+}, Cu^{2+}, and Co^{2+} complexes were determined from six separate runs (Table 7.30). The other metal complexes had R_f values in the range of 0.03 to 0.06 on alumina plates and slightly higher on alumina sheets. Their separation from the copper, the iron(III), and sometimes the cobalt complex was therefore possible only by resorting to two-dimensional chromatography, using a chloroform:2% methanol mixture for the second dimension.

The same order of separation was observed on alumina plates and alumina sheets, but the actual R_f values, particularly for iron(III) and copper, differ considerably (Table 7.30). The separation on alumina sheets was actually found to be somewhat more efficient.

A light yellow spot resulting from an excess of reagent extracted along with the complexes usually appeared between the copper and nickel spots, completely separated from the other components. The development time for the complete separation of the iron(III), nickel, copper, cobalt, and reagent spot ranged between 20 and 40 min.

The stability of these chelates at higher temperatures was also investigated by spectral reflectance. Complete decomposition of the adsorbed metal chelates was observed at the temperatures and times indicated in Table 7.31. The decomposition appeared as a progressive fading of the spots until they could no longer be distinguished from the background. The color did not reappear after cooling.

Interesting relationships were observed among R_f values, decomposition temperatures, and position of the absorption maxima.[157]

In the analytical procedure, up to tenfold excesses of interfering metal can be tolerated. The detection limits are 0.01 μg/spot for iron, nickel,

TABLE 7.30

R_f Values for PAQH Complexes Separated on Alumina Plates and Sheets[157]

Metal	Alumina plates			Alumina sheets (Eastman Kodak)		
	Range	Average	Rel. S.D.* (%)	Range	Average	Rel. S.D. (%)
Fe^{3+}	0.76–0.86	0.82	3.1	0.52–0.60	0.56	4.3
Ni^{2+}	0.31–0.38	0.35	5.5	0.36–0.39	0.37	3.5
Cu^{2+}	0.06–0.08	0.07	7.1	0.26–0.28	0.27	5.7
Fe^{2+}	0.05–0.06	0.05	7.0	0.08–0.10	0.09	5.2
Co^{2+}	0.00	0.00	0	0.00	0.00	0

*See footnote to Table 7.01.

(From Frei, R. W., Ryan, D. E., and Stockton, C. A., *Anal. Chim. Acta,* 42, 159 (1968). With permission of Elsevier Pub. Co.)

TABLE 7.31

Decomposition Study of the Chelates Adsorbed on Alumina[157]

Metal	Decomposition temperature (°C)	Time (hr)
Fe^{3+}	180	1.5
Ni^{2+}	180	2
Cu^{2+}	180	2
Co^{2+}	200	4

(From Frei, R. W., Ryan, D. E., and Stockton, C. A., *Anal. Chim. Acta,* 42, 159 (1968). With permission of Elsevier Pub. Co.)

and copper, and 0.007 μg/spot for cobalt. Drying of the chromatoplates at 110°C for five min immediately after development is recommended.

For quantitative work, the reproducibility is considerably improved because of the absence of spraying procedures.

The total iron content of a solution can be determined by oxidizing all the iron(II) to the iron(III) state with H_2O_2 in acetic acid medium (pH 5). An elegant method was also possible for the simultaneous determination of Fe^{2+} and Fe^{3+},[157] since both form stable and intensely colored chelates. Their R_f values are indeed so different (0.05 and 0.82) that, if no interfering ions are present, they can be separated within a few minutes. Identification of the metals can be made on the basis of color and R_f values. The same group of metal chelates was successfully investigated by UV reflectance spectroscopy.[23] A further spectral reflectance study of these complexes was carried out on various chromatographic adsorbents[159] in an attempt to enhance the use of the analytical techniques discussed above.

c. Dithizone and Oxine Complexes

Another investigation was carried out on a group of dithizone and oxine complexes[160] for the purpose of developing a qualitative analysis scheme, based on a combination of spectra and R_f values, analogous to the one worked out for amino acids.[20]

After spotting of the cations (nitrates in aqueous solution), the plates were dried with compressed air and then developed in one dimension by the ascending technique with the use of 25% (v/v) hydrochloric acid (12 *N*) in 1-butanol. The developed plates were dried at 75°C for 30 min and then sprayed with a mixture consisting of equal volumes of dithizone (0.05% w/v in carbon tetrachloride) and 8-hydroxyquinoline (1% w/v in carbon tetrachloride) applied by an atomizer. The plates were air-dried for 10 min. The reflectance spectra of the samples were recorded with a Beckman DK-2 spectrophotometer. After the first set of spectra was obtained, the reference and sample cells were exposed to ammonia fumes for at least 2 min in a chamber containing 15 *M* ammonia, and a second set of reflectance data was recorded. Some reflectance spectra obtained before and after treatment with ammonia are shown in Figure 7.44.

Table 7.32 lists R_f values and spectral data for all the cations investigated. It clearly demonstrates the possible use of this method as a fingerprinting device for many metals.

d. Complexes of 1-(2-pyridylazo)-2-naphthol

Galik and Vincourova[161] described an approach similar to the one discussed earlier[157] with PAQH complexes. PAN complexes of cobalt, copper, nickel, and iron(III) were extracted into

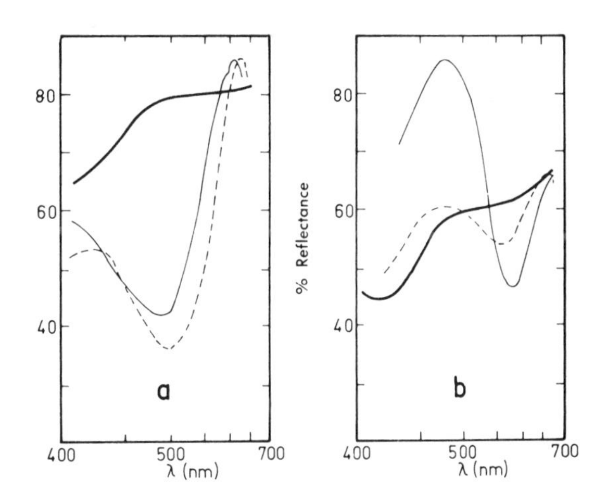

FIGURE 7.44 a. Reflectance spectra obtained for 10 μg of chromium (————), lead (– – – – – –), and silver (————) cations adsorbed on cellulose after chromatoplates had been sprayed with dithizone-oxine reagent. b. Reflectance spectra obtained for 10 μg of chromium (————), lead (————), and silver (– – – – – –) cations adsorbed on cellulose after chromatoplates had been sprayed with dithizone-oxine reagent and subsequently exposed to ammonia fumes. (From Zaye, D. F., Frei, R. W., and Frodyma, M. M., *Anal. Chim. Acta,* 39, 13 (1967). With permission of Elsevier Pub. Co.)

TABLE 7.32

R_f Values and Spectral Data for Cations Absorbed on MN-Cellulose[160]

	$R_f$value		After spraying with dithizone-oxine reagent		After spraying with dithizone-oxine reagent followed by ammonia treatment	
Cation	MN-300	Paper	Absorption maximum (nm)	Sensitivity (μg)	Absorption maximum (nm)	Sensitivity (μg)
Aluminum	0.04	0.04	–	–	380	2.3
Bismuth	0.56	0.60	505	2.2	515	2.2
Cadmium	0.90	0.83	498	1.6	519	1.6
Chromium	0.07	–	–	–	414	6.5
Cobalt	0.41	0.39	512	0.5	564	0.5
Copper	0.51	0.51	502	0.6	398	0.6
Iron	0.93	0.92	380	2.7	597	0.9
Lead	0.30	–	498	0.5	556	2.2
Manganese	0.16	–	510	2.2	507	2.6
Mercury	0.88	0.86	492	0.5	–	–
Nickel	0.09	0.04	503	1.2	514	0.6
Silver	0.00	0.00	492	2.0	574	0.6
Tin	0.80	–	510	0.9	–	–
Zinc	0.87	0.79	513	0.7	526	0.7

(From Zaye, D. F., Frei, R. W., and Frodyma, M. M., *Anal. Chim. Acta*, 39, 13 (1967). With permission of Elsevier Pub. Co.)

chloroform and separated on Silulfol (silica gel) thin-layer plates. The most effective separation was achieved with a 20% (v/v) ethanol-benzene mixture, followed (after intermediate drying in cold air) by elution with 8% (v/v) acetone in dichloroethane. After drying, the chromatograms were scanned in the reflectance mode by a Zeiss, Jena (DDR) recording reflectance filter-photometer ERI-10 with filter No. 2 and without the use of the built-in integrator. The chromatogram and corresponding peaks obtained with the previously mentioned solvent systems are shown in Figure 7.45. The peak areas were integrated manually and calibration plots peak area vs. concentration (μg/spot) were employed for a quantitative analysis.

The instrumental detection limits were reported for 10 mm^2 peaks as 0.016 μg for Co and Cu, 0.007 μg for Ni, and 0.012 μg for Fe. Interfering metal chelates from metals such as vanadium, gallium, or indium can be eliminated in the separation process. The stability of the four metal chelates is such that the instrument response does not change significantly over a period of five days. The method was checked as to its suitability for routine analysis of traces of cobalt, copper, nickel, and iron in organic salts (see Table 7.33).

As an example, six determinations of iron in NH_4Cl by this method gave an iron concentration of $4.8 \times 10^{-4}\% \pm 0.64 \times 10^{-4}$, which corresponds to a relative standard deviation of about 13%. This compares to standard transmission spectroscopic methods for iron with relative standard deviations

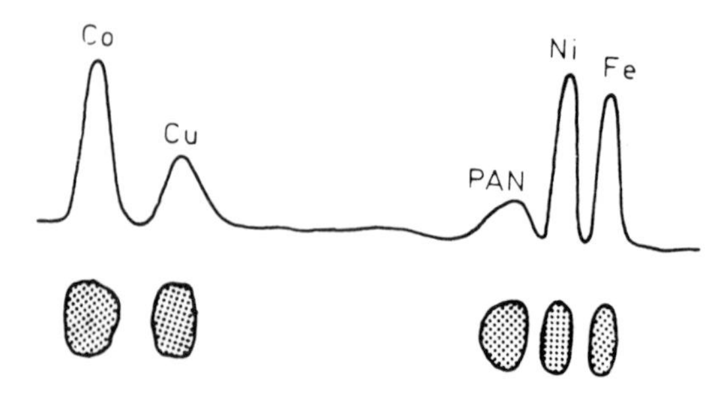

FIGURE 7.45 Typical chromatogram obtained by development with 20% (v/v) ethanol in benzene and subsequent elution with 8% (v/v) acetone in dichloroethane. (From Galik, A. and Vincourova, A., *Anal. Chim. Acta*, 46, 113 (1969). With permission of Elsevier Pub. Co.)

TABLE 7.33

Accuracy of the Method Proposed by Galik and Vincourova[161]

	Cobalt		Copper		Nickel		Iron	
Sample	μg added	% Recovery	μg added	% Recovery	μg added	% Recovery	μg added	% Recovery
$CaCl_2$	3.0	92	2.86	117	2.0	95	2.1	78
$CaCl_2$	2.0	105	1.91	94	2.0	98	2.1	89
$CaCl_2$	2.0	125	1.91	106	1.0	94	1.0	101
$CaCl_2$	4.0	136	0.95	126	2.0	105	2.1	106
$Ba(NO_3)_2$	2.0	105	1.91	105	2.0	94	2.1	99
$Sr(NO_3)_2$	2.0	110	1.91	126	1.0	95	1.0	107
NH_4Cl	1.0	120	1.91	99	2.0	85	1.0	101
NH_4HCO_3	3.0	92	1.91	98	3.0	129	3.1	98

(From Galik, A. and Vincourova, A., *Anal. Chim. Acta,* 46, 113 (1969). With permission of Elsevier Pub. Co.)

of about 4 and 8%. Agreement with results obtained by conventional methods was good.

Other investigators[162,163] have used PAN as a chromogenic spray reagent for a number of metal ions separated on cellulose layers impregnated with liquid ion exchanger Primene JM-T hydrochloride. In one phase of this work, parameters involved in the direct reflectance densitometric determination of zinc as the PAN complex were investigated.[162]

The Joyce, Loebl Chromoscan with TLC attachment was used for these measurements and, as in previous studies concerned with spraying procedures and in situ determination of trace metals,[153,155] fluctuations in the spraying step and the difficulty of distributing the chromogenic material uniformly were found to give the most serious errors (see Table 7.34). The stability of the zinc spot was found to be satisfactory for quantitative work, and, in spite of the spraying error, an accuracy of 5% was deemed possible for this method applied at the 1-μg level.

In a later publication, a similar method was discussed for Co, Cu, Zn, Cd, Pb, Bi(III), and UO_2^{2+}. Again, a PAN solution (0.1% in ethanol) was used for the visualization step, followed by exposure of the plates to ammonia vapor. Separation of the ions was carried out according to a method described also by Graham et al.,[162,164] using aqueous hydrochloric acid as the mobile phase and a sandwich-type chamber.

After densitometric (reflectance mode) evaluation of the chromatograms, peak areas were plotted vs. concentration in μg/spot (see Figure 7.46) and the metal ions were evaluated quantitatively. Results for the analysis of Zn, Cd, and Co are given in Table 7.35.

Calibration plots as depicted in Figure 7.46 are linear up to 5 μg/spot. For higher concentration, plots of peak area vs. $\sqrt{C}$ might be used. The method was compared to a spot-removal technique, followed by absorptiometric analysis. Both methods were found to be comparable in accuracy (∿±4%), but the in situ reflectance method is faster and permits a larger range of metals to be investigated.

PAN was also used as a detecting agent in an extensive study on the separation of a large number of heavy metal ions on various adsorbent layers.[165,166] Diffuse reflectance studies carried out on these systems suggested that similar quantitative evaluation methods could be used after separation and visualization of the metal ions.[23]

TABLE 7.34

Fluctuations Due to Spraying Procedure[162]

Plate No.	S.D.*† (%)
1	4.3
2	3.2
3	5.2
4	3.9

*See footnote to Table 7.01.
†Determined from eight spots per plate.

(From Graham, R. J. T., Bark, L. S., and Tinsley, D. A., *J. Chromatogr.,* 39, 211 (1969). With permission of Elsevier Pub. Co.)

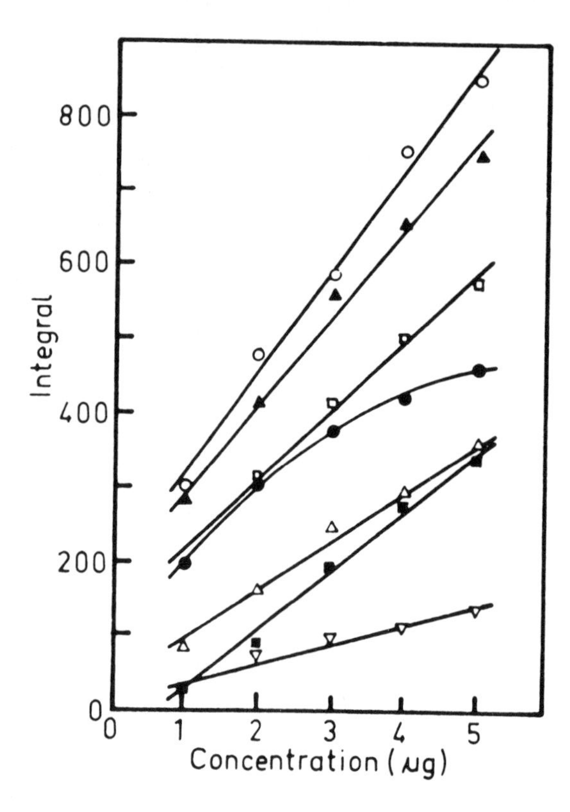

FIGURE 7.46 Densitometric calibration curves for metal ions in the range 1.0 to 5.0 μg of metal ion. ▲, Co(II); O, Cu(II); •, Zn(II); ▽, Pb(II); □, Cd(II); Δ, Bi(III); ■, UO_2^{2+}. (From Graham, R. J. T., Bark, L. S., and Tinsley, D. A., *J. Chromatogr.*, 39, 218 (1969). With permission of Elsevier Pub. Co.)

e. Complexes of 4-(2-Thiazolylazo)-resorcinol

The chelating reagent 4-(2-thiazolylazo)-resorcinol (TAR) was chosen for study as a reagent for the in situ reflectance analysis of metal ions separated on thin-layer chromatograms as it was believed it would offer advantages in stability, sensitivity, and ease of scanning over other reagents described in preceding sections of this chapter.[167] The metals chosen for study were again cobalt, nickel, and copper. All chromatographic solvents were redistilled and only double distilled water was used in the studies. Separations were carried out on precoated plastic-backed cellulose chromatogram sheets (Eastman Chromagram 6064). Chromatographic impurities were removed by an ascending development of the sheets with methanol-acetone-6N HCl (5:5:1) prior to use. The spray reagent consisted of 0.1% or 0.05% TAR in 95% ethanol and was applied to the chromatograms using an aerosol atomizer. The chromatograms were developed using one of the following two solvent systems:

(a) acetone – 6 *N* HCl (9:1), or
(b) acetone – 6 *N* HCl – acetic acid – water (8:1:1:1).

To measure the spectra of the TAR complexes,[167] 1-μg spots were cut out of the chromatogram sheets, backed with a blank sheet to approximate an infinite layer thickness, and mounted directly on the sample port of a standard reflectance attachment of a Spectronic 505 spectroreflectometer. The spectra were measured in the double-beam mode using as reference a reagent background cut from the same plate as the chelate. The spectrum of the chelate was measured against a barium sulfate standard. When time studies were in progress, the sample was left in position over the sample port, while during temperature and pH studies, great care was taken to remount the sample in the same position. This is particularly critical, as a minor change in position can result in a dramatic change in the intensity of the spectrum obtained. The sample and reflectance sphere were covered with a black cloth during all measurements to exclude stray light.

TAR forms purple complexes with cobalt, copper, and nickel in weakly alkaline media, while TAR is itself yellow-orange under these conditions. After spraying, the chromatograms were dried for 30 min in a chamber containing ammonia vapor, then dried in air at room temperature for 10 min to remove excess adsorbed ammonia. After a further 20 min, a time study of the spectra was begun. The absorption peaks were observed to broaden with time and, as the ammonia was desorbed, the wavelength maximum shifted to shorter wavelength. Maximum absorption was attained after 48 hr, indicating the rather slow kinetics of the reaction. Drying the chromatograms with an air gun at 40° for 10 min was found to produce the same effect as drying at room temperature for 48 hr. The bathochromic shift from 560 nm to 550 nm was again observed. As the reagent background decreases substantially on drying, this is also of analytical importance. Further studies revealed that similar results could be achieved by drying the chromatogram in an oven at 60° for 1 hr. In all cases, the resulting chelates showed no significant change from the maximum value for several hours. Similar results

TABLE 7.35

The Direct Reflectometric Determination of Zn(II), Cd(II), and Co(II)[163]

	Standard solutions				Test solution (added)	Test solution (found)	Percent error
Zn(II) (μg)	0.50	0.75	1.00	1.25	1.00	0.97	-3.0
Mean integral	420	608	743	901		724	
Cd(II) (μg)	0.5	0.75	1.00	1.25	1.00	1.03	+3.0
Mean integral	117	168	208	248		213	
Co(II) (μg)	0.75	1.00	1.25	1.50	1.00	0.96	-4.0
Mean integral	102	136	169	201		130	

(From Graham, R. J. T., Bark, L. S., and Tinsley, D. A., *J. Chromatogr.*, 39, 218 (1969). With permission of Elsevier Pub. Co.)

TABLE 7.36

Absorption Maxima, Stabilities, and Detection Limits for Cobalt, Copper, and Nickel Chelates of TAR Adsorbed on Basic Cellulose and Silica Gel[167]

	On basic cellulose		On basic silica gel		
Chelates	λ max., nm	Stability (in hr)	λ max., nm	Stability (in hr)	Detection limits (μg/spot) for both adsorbents
Co-TAR	550	2–18	560	2–20	0.001
Cu-TAR	555	2–20	560	1–5*	0.002
Ni-TAR	550–580	3–5*	550–580	1–5*	0.001

*Maximum time of investigation.

(From Frei, R. W. and Miketukova, V., *Mikrochim. Acta*, 29 (1971). With permission of Springer-Verlag.)

were obtained for the copper and nickel complexes. While the copper complexes had somewhat narrower absorption bands than cobalt, the nickel complex had much broader peaks that tended to broaden with time and shift to higher wavelengths. All complexes could be determined by scanning the chromatogram at a wavelength setting of 550 to 560 nm.

The metal-TAR complexes were also studied on silica-gel thin-layers.[167] Similar results were obtained, although the more active silica-gel surface was observed to produce a bathochromic shift. Experimental results are summarized in Table 7.36. Reflectance spectra of the cobalt-TAR complex on cellulose are shown in Figure 7.47. The effect of basicity on the chromatogram was studied, and it was found that failure to include the ammonia treatment produced a two-colored precipitate that was unsuitable for analytical work. Again, time studies following treatment with ammonia revealed a hypsochromic shift in the wavelength of the absorption maximum as the ammonia was desorbed.

Heating the silica-gel chromatograms at 70° with an air gun for 10 min produced a large bathochromic shift (2 to 30 nm), which is greater than could normally be explained by adsorbent-adsorbate interactions and suggests the possible formation of a new chelate species. The shift proved to be reversible on prolonged cooling.[167]

An investigation of the quantitative analytical

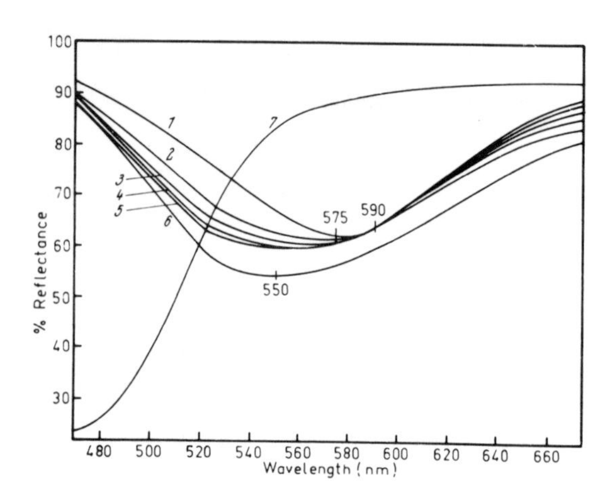

FIGURE 7.47 Reflectance spectra of the cobalt-TAR complex on cellulose as a function of time and basicity. Curves *1-6:* Co-TAR complex; *1*, after 10 min drying with air at room temperature, total time 20 min; *2*, 65 min; *3*, 2 hr; *4*, 4 hr., *5*, 24 hr; *6*, 48 hr; *7*, reagent background after 24 to 48 hr. (From Frei, R. W. and Miketukova, V., *Mikrochim. Acta,* 29 (1971). With permission of Springer-Verlag.)

application of the TAR-chelates was carried out using a Farrand UV-VIS Chromatogram Analyzer.[167] In these tests, cobalt was chromatographed using solvent system (b) and on silica-gel sheets (Eastman Chromagram 6061), and two clean sheets were used as backing to provide an infinite layer thickness. The sheets were scanned 2 to 3 hr after the treatment with ammonia vapor for 30 min and after drying at about 40° for 45 min. Peak areas were measured with a Gelman planimeter and also using the instrument's integrator, with little difference in the results. Results of the reproducibility studies are shown in Table 7.37. Better results were achieved in the double-beam mode, but reproducibilities become quite poor as the detection limit of 0.01 μg/spot is approached due to an unfavorable signal-to-noise ratio. Plotting the square of peak areas or integrator counts against the concentration (an approximation to the Kubelka-Munk function) produced linear calibration plots extending from the detection limit up to one microgram.

A number of metallic ions could cause interferences due to the relative nonspecificity of TAR.[167] On cellulose sheets using solvent system (a), vanadium (V) would interfere in the determination of nickel, while uranium (IV) and lead (II) could interfere in the determination of copper due to poor chromatographic resolution. Problems could also be caused by hundredfold excesses of Cd(II), Bi(III), Zn(II), and Mn(II). Fe(III) runs with the organic solvent front, so causes no appreciable interferences even when present in large amounts.

If the silica-gel sheets and solvent system (b) are used, a tenfold excess of Mn(II) will interfere in the analysis of cobalt, and vanadium (V) and lead (II) may cause interferences due to tailing. Excesses of An(II) and U(IV) will interfere in the copper analysis. Thus, the interferences may generally be eliminated by choosing the proper chromatographic system or else by a selective

TABLE 7.37

Reproducibility as a Function of Concentration for Cobalt Developed on Silica-gel Layers and Recorded with the Farrand Chromatogram Scanner at 580 nm[167]

Conc. μg Co/spot	Number of determinations (n)	Single-beam average reading and S.D.* (in.²)	Coefficient of variation (% S.D.)	Double-beam coefficient of variation (% S.D.)
0.05	9	0.33 ± 0.034	10.27	
0.2	9	1.07 ± 0.059	5.60	
0.6	6		3.43	1.48

*See footnote to Table 7.01.

(From Frei, R. W. and Miketukova, V., *Mikrochim. Acta,* 29 (1971). With permission of Springer-Verlag.)

extraction procedure prior to chromatographic analysis.[167]

f. Miscellaneous

On the basis of previously discussed principles,[153,155] a method was developed for the quantitative determination of copper, nickel, and zinc.[168] Copper and nickel were determined in the presence of 11 other cations without interference by employing neocuproine and dimethylglyoxime, respectively, as chromogenic reagents. In the case of zinc, 3,3-dimethylnaphthidine was used with interferences occurring from tin, cadmium, and iron. Accuracies of 2.1% and 2.8% were reported for nickel and copper, and 5.6% for zinc.

Diffuse reflectance spectroscopy has also been used for the identification of various chromium(III) complexes on TLC sheets.[169] The application of this technique to the analysis of the photolysis products of $[Cr(en_2)ox]^+$ has been described.

Webb et al.[170] discussed the determination of iron, manganese, zinc, and copper in plant material by paper chromatography and in situ reflectance spectroscopy. They discussed the ashing, sampling, and chromatographic procedures in detail. After chromatographic separation on Whatman No. 1 sheets by the descending technique, the manganese spot was detected with formaldoxime reagent, analyzed at 450 nm. Copper was made visible with a sodium diethyldithiocarbamate solution (wavelength of analysis 435 nm). Potassium thiocyanate reagent was used for iron and measured at 480 nm. Zincon reagent served to detect and analyze zinc at 620 nm. Results for various materials are presented in Table 7.38. A comparison of the technique to several other methods shows good agreement. For metal concentrations > 2 μg/spot, the investigators[170] recommend elution and analysis by transmission spectroscopy.

TABLE 7.38

Iron, Manganese, Copper, and Zinc Contents of Samples of Plant Material Analyzed in Other Laboratories by Different Methods*[170]

	Amounts in dry matter, ppm			
Found	Kale leaf	Tomato leaf	Cabbage leaf	Oat grain
Iron				
By six other methods (Range of mean values)	140–144	180–210	95–116	91–102
By reflectance method[116]	144	194	112	73
Manganese				
By four other methods (Range of mean values)	37–41	466–532	47–55	65–70
By reflectance method[116]	39	468	49	61
Copper				
By three other methods (Range of mean values)	3.1–4.2	–	3.7–5.1	4.7–6.3
By reflectance method[116]	3.9	–	5.5	4.6
Zinc				
By spectrographic method (One value only reported)	33	95	60	43
By reflectance method[116]	35	105	55	48

*Values by method discussed are means of four separate determinations. Other values are means of two or more determinations.

(From Webb, R. A., Hallas, D. G., and Stevens, H. M., *Analyst*, 94, 794 (1969). With permission of the Society for Analytical Chemistry.)

In summary, the importance of the in situ reflectance technique as a tool for both inorganic and inorganic-analytical chemists to use in the study of complex metal chelate systems will become more and more apparent. Much information can now be obtained on mechanisms of complex formation that are difficult or impossible to obtain from solution systems. The increased knowledge of adsorption phenomena gained through this technique will enable chromatographers to begin moving away from the present empirical approach.

REFERENCES

1. **Frei, R. W. and Frodyma, M. M.,** *Anal. Biochem.*, 9, 310 (1964).
2. **Frodyma, M. M., Lieu, V. T., and Frei, R. W.,** *J. Chromatogr.*, 18, 520 (1965).
3. **Lieu, V. T., Frodyma, M. M., Higashi, L. S., and Kunimoto, L. H.,** *Anal. Biochem.*, 19, 454 (1967).
4. **Frodyma, M. M. and Lieu, V. T.,** *Anal. Chem.*, 39, 814 (1967).
5. **Frei, R. W., Nomura, N. S., and Frodyma, M. M.,** *Mikrochim. Acta,* 1099 (1967).
6. **Frei, R. W. and Nomura, N. S.,** *Mikrochim. Acta,* 565 (1968).
7. **Frei, R. W., Zürcher, H., and Pataki, G.,** *J. Chromatogr.*, 43, 551 (1969).
8. **Frei, R. W., Zürcher, H., and Pataki, G.,** *J. Chromatogr.*, 45, 284 (1969).
9. **Gänshirt, H.,** in *Dünnschicht-Chromatographie–Ein Laboratoriums-handbuch,* 2nd ed., Stahl, E., Ed., Springer-Verlag, Berlin, 1967, 142.
10. Zeiss Information No. 50-657/K-a, Carl Zeiss Inc., Oberkochen, Germany, 1968.
11. **Jork, H.,** *Z. Anal. Chem.*, 236, 310 (1968).
12. **Jork, H.,** *Cosmo Pharma,* 1, 3 (1967).
13. **Stahl, E. and Jork, H.,** *Zeiss Inform. (No. 68),* 16, 52 (1968).
14. **Struck, H., Karg, H., and Jork, H.,** *J. Chromatogr.*, 36, 74 (1968).
15. **Jork, H.,** *J. Chromatogr.*, 33, 297 (1968).
16. **Jork, H.,** *Cosmo Pharma,* 4, 12 (1968).
17. **Klaus, R.,** *J. Chromatogr.*, 16, 311 (1964).
18. **Klaus, R.,** *Pharm. Ztg.*, 112, 480 (1967).
19. **De Galan, L., van Leeuwen, J., and Camstra, K.,** *Anal. Chim. Acta,* 35, 395 (1966).
20. **Frei, R. W., Fukui, I. T., Lieu, V. T., and Frodyma, M. M.,** *Chimia,* 20, 23 (1966).
21. **Pataki, G.,** *Chromatographia,* 1, 492 (1968).
22. **Frodyma, M. M., Frei, R. W., and Williams, D. J.,** *J. Chromatogr.*, 13, 61 (1964).
23. **Frei, R. W.,** unpublished data.
24. **Braun, W. and Kortüm, G.,** *Zeiss Inform.*, 16, 27 (1968).
25. **Braun, W. and Kortüm, G.,** *Zeiss Mitteilungen,* 4, 379 (1968).
26. **Lieu, V. T., Frei, R. W., Frodyma, M. M., and Fukui, I. T.,** *Anal. Chim. Acta,* 33, 639 (1965).
27. **Klaus, R.,** *J. Chromatogr.*, 34, 539 (1968).
28. **Huber, W.,** *J. Chromatogr.*, 33, 378 (1968).
29. **Koopmans, H. J. and Bouwmeester, P. C.,** *Chromatographia,* 1, 83 (1971).
30. **Touchstone, J. C., Levin, S. S., and Murawec, T.,** *Anal. Chem.*, 43, 858 (1971).
31. **Goldman, J. and Goodall, R. R.,** *J. Chromatogr.*, 32, 24 (1968).
32. **Goldman, J. and Goodall, R. R.,** *J. Chromatogr.*, 40, 345 (1969).
33. **Goldman, J. and Goodall, R. R.,** *J. Chromatogr.*, 47, 386 (1970).
34. **Novacek, V.,** *Am. Lab.*, 27 (1969).
35. **Frei, R. W.,** *J. Chromatogr.*, 64, 285 (1972).
36. **Tausch, W.,** *Sonderdruck aus messtechnik,* 80, 38 (1971).
37. **Seiler, N. and Möller, H.,** *Chromatographia,* 2, 319 (1969).
38. **Getz, M. E.,** in *Methods in Residue Analysis. Pesticide Chemistry,* Vol. 4, Tahori, A. S., Ed., Gordon and Breach Science Publishers, New York, 1971.
39. **Pollack, V. and Boulton, A. A.,** *J. Chromatogr.*, 63, 87 (1971).
40. **Treiber, L. R., Nordberg, R., Lindstedt, S., and Stollenberger, P.,** *J. Chromatogr.*, 63, 211 (1971).
41. **Treiber, L. R.,** *J. Chromatogr.*, 69, 399 (1972).
42. **Frei, R. W. and Miketukova, V.,** *Mikrochim. Acta,* 290 (1971).
43. **Boulton, A. A. and Pollack, V.,** *J. Chromatogr.*, 45, 189 (1969).
44. **Pollack, V. and Boulton, A. A.,** *J. Chromatogr.*, 45, 200 (1969).
45. **Pollack, V. and Boulton, A. A.,** *J. Chromatogr.*, 46, 247 (1970).

46. **Pollack, V. and Boulton, A. A.,** *J. Chromatogr.,* 50, 39 (1970).
47. **Pollack, V. and Boulton, A. A.,** *J. Chromatogr.,* 50, 30 (1970).
48. *Quantitative Paper and Thin-Layer Chromatography,* Shellard, E. J., Ed., Academic Press, New York, 1967.
49. **Jork, H.,** *J. Chromatogr.,* 48, 372 (1970).
50. **Waksmundzki, A. and Rozylo, J. K.,** *Chem. Anal.* (Warsaw), 15, 1079 (1970).
51. **Waksmundzki, A. and Rozylo, J. K.,** *Chem. Anal.* (Warsaw), 16, 1161 (1971).
52. **Kortüm, G., Braun, W., and Herzog, G.,** *Angew. Chem.,* 2, 333 (1963).
53. **Kortüm, G.,** *Reflexionsspektroskopie,* Springer-Verlag, Berlin, 1968, 309.
54. **Abbott, D. C. and Thompson, J.,** in *Residue Rev.,* Gunther, F. A., Ed., 11, 1 (1965).
55. **Frei, R. W. and Freeman, C. D.,** *Residue Rev.,* 1214 (1968).
56. **Abbott, D. C., Blake, K. W., Tarrant, K. R., and Thompson, J.,** *J. Chromatogr.,* 30, 136 (1967).
57. **Beroza, M., Hill, K. R., and Norris, K. H.,** *Anal. Chem.,* 40, 1611 (1968).
58. **Watts, R. R.,** *J. Assoc. Agr. Chem.,* 48, 1161 (1965).
59. **Getz, M. E.,** *J. Assoc. Agr. Chem.,* 45, 393 (1962).
60. **Petrowitz, H.-J. and Wagner, S.,** *Chem.-Ztg.,* 95, 331 (1971).
61. **Getz, M. E. and Hill, K. R.,** Paper No. 54, American Chemical Society Division of Pesticides Chemistry, 163rd ACS National Meeting, Boston, April (1972).
62. **Kynast, G.,** *Z. Anal. Chem.,* 250, 105 (1970).
63. **Kynast, G.,** *Z. Anal. Chem.,* 251, 161 (1970).
64. **Kynast, G.,** *Chromatographia,* 3, 425 (1970).
65. **Kynast, G.,** *Z. Anal. Chem.,* 256, 20 (1971).
66. **Kossman, K.,** in *Methods in Residue Analysis. Pesticide Chemistry,* Vol. IV, Tahori, A. S., Ed., Gordon and Breach Science Publishers, New York, 1971.
67. **Hutzinger, O., Jamieson, W. D., MacNeil, J. D., and Frei, R. W.,** *J. Assoc. Anal. Chem.,* 54, 1100 (1971).
68. **MacNeil, J. D., Frei, R. W., and Hutzinger, O.,** *Int. J. Environ. Anal. Chem.,* 1, 205 (1972).
69. **Frei, R. W., MacNeil, J. D., and Hutzinger, O.,** *Int. J. Environ. Anal. Chem.,* 2, 1 (1972).
70. **MacNeil, J. D., Frei, R. W., Safe, S., and Hutzinger, O.,** *J. Assoc. Anal. Chem.,* 55, 1270 (1972).
71. **MacNeil, J. D., Frei, R. W., and Hutzinger, O.,** *Can. J. Chem.,* in press.
72. **MacNeil, J. D., Frei, R. W., and Hutzinger, O.,** to be published.
73. **MacNeil, J. D., Frei, R. W., and Hutzinger, O.,** presented at Trace Analysis Symp., Halifax, to be published.
74. **Foster, R.,** *Organic Charge Transfer Complexes,* Academic Press, New York, 1969.
75. Hais, I. M. and Macek, K., Eds., *Handbuch der Papierchromatographie,* Vol. 1, Fisher, Jena, 1958.
76. **Block, R. J., Durrum, E. L., and Zweig, G.,** *A Manual of Paper Chromatography and Paper Electrophoresis,* 2nd ed., Academic Press, New York, 1958.
77. **Shellard, E. J.,** in *Quantitative Paper and Thin-layer Chromatography,* Shellard, E. J., Ed., Academic Press, London and New York, 1968, 51.
78. **Frei, R. W., Zeitlin, H., and Frodyma, M. M.,** *Chem. Rundschau* (Solothurn), 19, 411 (1966).
79. **Frei, R. W.,** in *Recent Progress in Thin-layer Chromatography and Related Methods,* Vol. II, Niederwieser, A., and Pataki, G., Eds., Ann Arbor Science Publishers, Ann Arbor, Mich., 1970, Chapt. I.
80. **Butler, C. G., Linley, P. A., and Rowson, J. M.,** Scientiae Pharmaceuticae - II. Proceedings of the 25th Congress of Pharmaceutical Sciences, Prague, August, 1965.
81. **Latner, A. L. and Park, D. C.,** *Clin. Chim. Acta,* 11, 538 (1965).
82. **Kremers, B., Briere, R. O., and Batsakis, J. G.,** *Am. J. Med. Technol.,* 33, No. 1, 1967.
83. **Yamaguchi, K., Fujii, S., Tobata, T., and Kato, S.,** *J. Pharm. Soc. Jap.,* 74, 1322 (1954).
84. **Yamaguchi, K., Fukushima, S., and Ito, M.,** *J. Pharm. Soc. Jap.,* 75, 556 (1955).
85. **Kortüm, G. and Vogel, J.,** *Angew. Chem.,* 71, 451 (1959).
86. **Mottier, M.,** *Mitt. Geb. Lebensmittelunters. Hyg.,* 47, 372 (1956).
87. **Frodyma, M. M. and Frei, R. W.,** *J. Chem. Educ.,* 46, 522 (1969).
88. **Frei, R. W. and Zeitlin, H.,** *Anal. Chim. Acta,* 32, 32 (1965).
89. **Zeitlin, H., Frei, R. W., and McCarter, M.,** *J. Catal.,* 4, 77 (1965).
90. **Frei, R. W. and Frodyma, M. M.,** *Anal. Chim. Acta,* 32, 501 (1965).
91. **Lieu, V. T. and Frodyma, M. M.,** *Talanta,* 13, 1319 (1966).
92. **Frei, R. W., Ryan, D. E., and Lieu, V. T.,** *Can. J. Chem.,* 44, 1945 (1966).
93. **Yamaguchi, K., Fukushima, S., and Ito, M.,** *J. Pharm. Soc. Jap.,* 76, 339 (1956).
94. **Garside, C. and Riley, J. P.,** *Anal. Chim. Acta,* 46, 179 (1969).
95. **Riley, J. P. and Wilson, T. R. S.,** *J. Mar. Biol. Assoc. U. K.,* 45, 583 (1965).
96. **Pataki, G.,** *Techniques of Thin-layer Chromatography in Amino Acid and Peptide Chemistry,* Ann Arbor Science Publishers, Ann Arbor, Mich., 1968.
97. **Frei, R. W. and Frodyma, M. M.,** *Chem. Rundschau* (Solothurn), 19, 26 (1966).
98. **Moffat, E. D. and Little, R. I.,** *Anal. Chem.,* 31, 926 (1959).
99. **Brenner, M. and Niederwieser, A.,** *Experientia,* 16, 378 (1960).
100. **Bull, H. B., Hahn, J. W., and Baptist, V. R.,** *J. Am. Chem. Soc.,* 71, 550 (1949).

101. **Frodyma, M. M. and Frei, R. W.,** *J. Chromatogr.*, 15, 501 (1964).
102. **El Khadem, H. S., El-Shafei, Z. M., and Abdel Rahman, M. M. A.,** *Anal. Chem.*, 35, 1766 (1963).
103. **Frodyma, M. M. and Frei, R. W.,** *J. Chromatogr.*, 17, 131 (1965).
104. **Heathcote, J. G. and Haworth, C.,** *J. Chromatogr.*, 43, 84 (1969).
105. **Heathcote, J. G., Davies, D. M., Haworth, C., and Oliver, R. W. A.,** *J. Chromatogr.*, 55, 377 (1971).
106. Nester/Faust Applications Bulletin. (Nester/Faust MFG Corp., 2401 Ogletown Road, Newark, Del. 19711.)
107. **Goodban, A. E., Stark, J. B., and Owens, H. S.,** *J. Agric. Food Chem.*, 1, 261 (1953).
108. **Thaller, F. J.,** The Determination of Sugars by Spectral Reflectance, M.Sc. thesis, University of Hawaii, Hawaii, 1965.
109. **Frei, R. W., Thaller, F. J., and Frodyma, M. M.,** unpublished data.
110. **Vomhof, D. W. and Tucker, T. C.,** *J. Chromatogr.*, 17, 150 (1965).
111. **McCready, R. N. and McComb, E. A.,** *Anal. Chem.*, 26, 1645 (1954).
112. **Bevenue, A. and Williams, K. T.,** *Arch. Biochem. Biophys.*, 73, 291 (1958).
113. **Bevenue, A. and Williams, K. T.,** *J. Chromatogr.*, 2, 199 (1959).
114. **Owens, H. S., McComb, E. A., and Deming, G. W.,** *Proc. Am. Soc. Sugar Beet Technol.*, 1955.
115. **Davies, J. R.,** *J. Chromatogr.*, 28, 451 (1967).
116. **Weiss, P. A. M.,** *Geburtshilfe Frauenheilkd.*, 30, 634 (1970).
117. **Massa, V., Gal, F., Susplugas, P., and Maestre, G.,** *Trav. Soc. Pharm. Montp.*, 30, 367 (1970).
118. **Massa, V., Gal, F., Susplugas, P., and Maestre, G.,** *Trav. Soc. Pharm. Montp.*, 30, 273 (1970).
119. **Safta, M. and Ostrogovich, G.,** *J. Chromatogr.*, 69, 219 (1972).
120. **Hutzinger, O., Heacock, R. A., MacNeil, J. D., and Frei, R. W.,** *J. Chromatogr.*, 68, 173 (1972).
121. **Zeitlin, H. and Niimoto, A.,** *Nature*, 181, 1616 (1958).
122. **Zeitlin, H. and Niimoto, A.,** *Anal. Chem.*, 31, 1167 (1959).
123. **Korte, F. and Weitkamp, H.,** *Angew. Chem.*, 70, 434 (1958).
124. **Zürcher, H., Pataki, G., Borko, J., and Frei, R. W.,** *J. Chromatogr.*, 43, 457 (1969).
125. **Randerath, K.,** *Thin-layer Chromatography*, 2nd ed., Academic Press, New York, 1966, 223.
126. **Frei, R. W., Zürcher, H., and Pataki, G.,** *J. Chromatogr.*, 43, 551 (1969).
127. **Frei, R. W., Zürcher, H., and Pataki, G.,** *J. Chromatogr.*, 45, 284 (1969).
128. **Frei, R. W.,** unpublished data.
129. **Pataki, G.,** *J. Chromatogr.*, 29, 126 (1967).
130. **Klaus, R.,** *J. Chromatogr.*, 40, 235 (1969).
131. **Frei, R. W., Kunz, A., Pataki, G., Prims, T., and Zürcher, H.,** *Anal. Chim. Acta*, 49, 527 (1970).
132. **Stahl, E. and Jork, H.,** *Zeiss Inform.*, 16, 52 (1968).
133. **Koof, H. P. and Noronha, R. V.,** *Anal. Chem.*, 250, 124 (1970).
134. **Gänshirt, H. and Malzacher, A.,** *Naturwissenschaften*, 47, 279 (1960).
135. **Manes, J. D., Jr., Fluckiger, H. B., Spaeth, D. G., and Schneider, D. L.,** *Fed. Amer. Soc. Exper. Biol.*, 31, 714A (1972).
136. **Jork, H.,** Proc. of the 3rd Chromatography Symposium, Brussels, 1966, 296.
137. **Jork, H.,** Paper presented at the 4th International Symposium of Chromatography and Electrophoresis, Brussels, 1966.
138. **Stahl, E. and Jork, H.,** *Arzneim. Forsch.*, 18, 1231 (1968).
139. **Jork, H.,** *Z. Anal. Chem.*, 221, 17 (1966).
140. **Stahl, E. and Jork, H.,** *Arch. Pharm.*, 299, 670 (1966).
141. **Stahl, E. and Bohrman, H.,** unpublished data.
142. **Pfeifle, J.,** Ph.D. thesis, University of the Saarland, Saarbrücken, Germany, 1966.
143. **Kraus, M., Mutschler, E., and Rochelmeyer, H.,** *J. Chromatogr.*, 40, 244 (1969).
144. **Schunack, W., Eich, E., Mutschler, E., and Rochelmeyer, H.,** *Arzneim. Forsch.*, 19, 1754 (1969).
145. **Eich, E., Geissler, H., Mutschler, E., and Schunack, W.,** *Arzneim. Forsch.*, 19, 1895 (1969).
146. **Röder, K., Eich, E., and Mutschler, E.,** *Arch. Pharm.*, 304, 297 (1971).
147. **Smith, G.,** *Proc. Soc. Anal. Chem.*, 66 (1971).
148. **Schlemmer, W.,** *J. Chromatogr.*, 63, 121 (1971).
149. **Clark, R. T. H.,** *J. Chem. Educ.*, 41, 488 (1964).
150. **Vaeck, S. V.,** *Nature*, 172, 213 (1953).
151. **Vaeck, S. V.,** *Anal. Chim. Acta*, 10, 48 (1954).
152. **Ingle, R. B. and Minshall, E.,** *J. Chromatogr.*, 8, 369 (1962).
153. **Frei, R. W. and Ryan, D. E.,** *Anal. Chim. Acta*, 37, 187 (1967).
154. **Heit, M. L. and Ryan, D. E.,** *Anal. Chim. Acta*, 32, 448 (1965).
155. **Frei, R. W., Liiva, R., and Ryan, D. E.,** *Can. J. Chem.*, 46, 167 (1968).
156. **Frei, R. W.,** *J. Chromatogr.*, 34, 563 (1968).
157. **Frei, R. W., Ryan, D. E., and Stockton, C. A.,** *Anal. Chim. Acta*, 42, 159 (1968).
158. **Singhal, S. P. and Ryan, D. E.,** *Anal. Chim. Acta*, 37, 91 (1967).
159. **Frei, R. W. and Zeitlin, H.,** *Can. J. Chem.*, 47, 3902 (1969).

160. Zaye, D. F., Frei, R. W., and Frodyma, M. M., *Anal. Chim. Acta,* 39, 13 (1967).
161. Galik, A. and Vincourova, A., *Anal. Chim. Acta,* 46, 113 (1969).
162. Graham, R. J. T., Bark, L. S., and Tinsley, D. A., *J. Chromatogr.,* 39, 11 (1969).
163. Graham, R. J. T., Bark, L. S., and Tinsley, D. A., *J. Chromatogr.,* 39, 218 (1969).
164. Graham, R. J. T., Bark, L. S., and Tinsley, D. A., *J. Chromatogr.,* 35, 416 (1968).
165. Miketukova, V. and Frei, R. W., *J. Chromatogr.,* 47, 427 (1970).
166. Miketukova, V. and Frei, R. W., *J. Chromatogr.,* 47, 441 (1970).
167. Frei, R. W. and Miketukova, V., *Mikrochim. Acta,* 29 (1971).
168. Frodyma, M. M., Zaye, D. F., and Lieu, V. T., *Anal. Chim. Acta,* 40, 451 (1968).
169. Kirk, A. D., Moss, K. C., and Valentin, J. G., *J. Chromatogr.,* 36, 332 (1968).
170. Webb, R. A., Hallas, D. G., and Stevens, H. M., *Analyst,* 94, 794 (1969).

Derivation of the Kubelka-Munk Function

When a plane parallel layer of thickness d is irradiated diffusely and monochromatically with a beam of intensity I_o, the radiation flow in the positive x-direction can be represented by I, and the radiation flow in the negative x-direction (caused by scattering) can be represented by J. An infinitesimally thin layer – dx – parallel to the surface is penetrated by the radiation in all possible directions with respect to the normal. The average path of the radiation $d\phi$, therefore, is not dx but

$$d\phi_I = dx \int_0^{\pi/2} \frac{\delta I \; d\varphi}{I\delta\varphi\cos\varphi} \equiv udx \text{ or } d\phi_J = dx \int_0^{\pi/2} \frac{\delta J \; d\varphi}{J\delta\varphi\cos\varphi} \equiv vdx \tag{a}$$

where $\frac{\delta J}{\delta\varphi}$ and $\frac{\delta I}{\delta\varphi}$ stand for the angular distribution of the radiation.

Assuming conditions for ideal diffuse radiation,

$$\frac{\delta I}{\delta\varphi} = I \sin 2\,\varphi \text{ and } \frac{\delta J}{\delta\varphi} = J \sin 2\,\varphi \tag{b}$$

where $u = v = 2$. This factor is included in s (scattering coefficient) and k (absorption coefficient). The component $kIdx$ of I is adsorbed in the layer dx while the component $sIdx$ is scattered backward. The radiation J in the negative x-direction contributes radiation $sJdx$ by scattering in the positive x-direction. The change in intensity of I in the layer dx, therefore, is composed of the following elements:

$$dI = -(k + s)\,Idx + sJdx \tag{c}$$

By analogy, the decrease in intensity of J is

$$dJ = +(k + s)\,Jdx - sIdx \tag{d}$$

These are the basic differential equations that describe the absorption and scattering processes. The indefinite integrals are:

$$I = A(1 - \beta)e^{\sigma x} + B(1 + \beta)e^{-\sigma x} \tag{e}$$

$$J = A(1 + \beta)e^{\sigma x} + B(1 - \beta)e^{-\sigma x} \tag{f}$$

with

$$\sigma \equiv \sqrt{K(K + 2S)} \tag{g}$$

and

$$\beta \equiv \frac{\sigma}{K + 2S} = \sqrt{K/(K + 2S)}. \tag{h}$$

The constants A and B are determined by the limiting conditions. If one integrates for the entire thickness, d, of the layer the conditions

for

$$X = O{:}I = I_o$$

and for

$$X = d{:}I = I_{(x = d)};\ J = O$$

are valid, and one obtains

$$A = -\frac{(1 - \beta)e^{-\sigma d}}{(1 + \beta)^2 e^{\sigma d} - (1 - \beta)^2 e^{-\sigma d}} I_o \tag{i}$$

$$B = \frac{(1 + \beta)e^{\sigma d}}{(1 + \beta)^2 e^{\sigma d} - (1 - \beta)^2 e^{-\sigma d}} I_o \tag{j}$$

The transmission of the layer is therefore given by

$$T = \frac{I(x = d)}{I_o} = \frac{4\beta}{(1 + \beta)^2 e^{\sigma d} - (1 - \beta)^2 e^{-\sigma d}}$$

$$= \frac{2\beta}{(1 + \beta)^2 \sin h\sigma d + 2\beta \cos h\sigma d}, \tag{k}$$

and the diffuse reflectance by

$$R = \frac{I(x = o)}{I_o} = \frac{(1-\beta)^2\,(e^{\sigma d} - e^{-\sigma d})}{(1 + \beta)^2 e^{\sigma d} - (1 - \beta)^2 e^{-\sigma d}}$$

$$= \frac{(1 - \beta)^2 \sin h\sigma d}{(1 + \beta)^2 \sin h\sigma d + 2\beta \cos h\sigma d}. \tag{l}$$

For $s = 0$ (nonscattering layer) and $\beta = 1$, Equation j becomes Bouguer-Lambert's Law $T = e^{-kd}$, and R' becomes zero.

For infinite layer thickness, d approaches 0 and one obtains R'_∞

$$R'_\infty = \frac{1 - \beta}{1 + \beta} = \frac{S + K - \sqrt{K(K + 2S)}}{S}. \tag{m}$$

These conditions are achieved experimentally with 1-mm layers of fine powders and R'_∞ can therefore be measured. Equation m can be transformed to a more convenient form

$$\frac{(1 - R'_\infty)^2}{2R'_\infty} = \frac{k}{s}. \tag{n}$$

APPENDIX 2

Kubelka-Munk Values for Reflectance Measurements Between 0.0 and 99.9%. (Courtesy of Carl Zeiss, Oberkochen.)

$$\frac{(1-R)^2}{2\ R} = \frac{k}{s}$$

R [%]	.0	.1	.2	.3	.4	.5	.6	.7	.8	.9
0	∞	499.00	249.00	165.66	124.00	99.002	82.336	70.432	61.504	54.560
1	49.005	44.460	40.672	37.468	34.721	32.340	30.257	28.420	26.786	25.325
2	24.009	22.820	21.738	20.750	19.845	19.012	18.243	17.532	16.871	16.255
3	15.681	15.144	14.641	14.168	13.722	13.303	12.906	12.532	12.176	11.840
4	11.520	11.215	10.925	10.649	10.385	10.133	9.8925	9.6617	9.4406	9.2285
5	9.0249	8.8294	8.6413	8.4604	8.2862	8.1184	7.9565	7.8004	7.6496	7.5040
6	7.3633	7.2272	7.0955	6.9680	6.8444	6.7248	6.6087	6.4961	6.3869	6.2808
7	6.1778	6.0777	5.9804	5.8858	5.7937	5.7041	5.6169	5.5320	5.4492	5.3686
8	5.2899	5.2133	5.1385	5.0655	4.9943	4.9248	4.8569	4.7906	4.7258	4.6624
9	4.6005	4.5400	4.4807	4.4228	4.3661	4.3106	4.2563	4.2031	4.1510	4.1000
10	4.0500	4.0009	3.9529	3.9058	3.8596	3.8144	3.7699	3.7263	3.6836	3.6416
11	3.6004	3.5600	3.5202	3.4812	3.4429	3.4053	3.3683	3.3320	3.2962	3.2611
12	3.2266	3.1927	3.1593	3.1265	3.0942	3.0625	3.0312	3.0005	2.9702	2.9404
13	2.9111	2.8822	2.8538	2.8258	2.7983	2.7712	2.7444	2.7181	2.6921	2.6666
14	2.6414	2.6165	2.5921	2.5680	2.5442	2.5207	2.4976	2.4748	2.4523	2.4302
15	2.4083	2.3867	2.3654	2.3444	2.3237	2.3033	2.2831	2.2632	2.2435	2.2241
16	2.2049	2.1860	2.1674	2.1489	2.1307	2.1128	2.0950	2.0775	2.0601	2.0430
17	2.0261	2.0094	1.9929	1.9766	1.9605	1.9446	1.9289	1.9133	1.8979	1.8827
18	1.8677	1.8529	1.8382	1.8237	1.8093	1.7952	1.7811	1.7672	1.7535	1.7400
19	1.7265	1.7133	1.7001	1.6871	1.6743	1.6616	1.6490	1.6365	1.6242	1.6120
20	1.5999	1.5880	1.5762	1.5645	1.5529	1.5415	1.5301	1.5189	1.5078	1.4968
21	1.4859	1.4751	1.4644	1.4539	1.4434	1.4330	1.4228	1.4126	1.4025	1.3926
22	1.3827	1.3729	1.3632	1.3536	1.3441	1.3347	1.3253	1.3161	1.3069	1.2979
23	1.2889	1.2800	1.2711	1.2624	1.2537	1.2451	1.2366	1.2282	1.2198	1.2115
24	1.2033	1.1951	1.1871	1.1791	1.1711	1.1633	1.1555	1.1477	1.1401	1.1325
25	1.1250	1.1175	1.1101	1.1027	1.0955	1.0882	1.0811	1.0740	1.0669	1.0600
26	1.0530	1.0462	1.0393	1.0326	1.0259	1.0192	1.0126	1.0061	0.9997	0.9932
27	0.9869	0.9805	0.9742	0.9680	0.9618	0.9557	0.9496	0.9436	0.9376	0.9316
28	0.9257	0.9199	0.9140	0.9083	0.9026	0.8969	0.8913	0.8857	0.8801	0.8746
29	0.8691	0.8637	0.8583	0.8530	0.8477	0.8424	0.8372	0.8320	0.8269	0.8217
30	0.8167	0.8116	0.8066	0.8017	0.7967	0.7918	0.7870	0.7822	0.7774	0.7726
31	0.7679	0.7632	0.7586	0.7539	0.7494	0.7448	0.7403	0.7358	0.7313	0.7269
32	0.7225	0.7181	0.7138	0.7095	0.7052	0.7010	0.6967	0.6926	0.6884	0.6843
33	0.6802	0.6761	0.6720	0.6680	0.6640	0.6600	0.6561	0.6522	0.6483	0.6444
34	0.6406	0.6368	0.6330	0.6292	0.6255	0.6218	0.6181	0.6144	0.6108	0.6072
35	0.6036	0.6000	0.5965	0.5929	0.5894	0.5860	0.5825	0.5791	0.5756	0.5723
36	0.5689	0.5655	0.5622	0.5589	0.5556	0.5524	0.5491	0.5459	0.5427	0.5395
37	0.5364	0.5332	0.5301	0.5270	0.5239	0.5208	0.5178	0.5148	0.5118	0.5088
38	0.5058	0.5028	0.4999	0.4970	0.4941	0.4912	0.4883	0.4855	0.4827	0.4798
39	0.4771	0.4743	0.4715	0.4688	0.4660	0.4633	0.4606	0.4579	0.4553	0.4526
40	0.4500	0.4474	0.4448	0.4422	0.4396	0.4371	0.4345	0.4320	0.4295	0.4270
41	0.4245	0.4220	0.4196	0.4172	0.4147	0.4123	0.4099	0.4075	0.4052	0.4028
42	0.4005	0.3981	0.3958	0.3935	0.3912	0.3890	0.3867	0.3845	0.3822	0.3800
43	0.3778	0.3756	0.3734	0.3712	0.3691	0.3669	0.3648	0.3627	0.3606	0.3585
44	0.3564	0.3543	0.3522	0.3502	0.3481	0.3461	0.3441	0.3421	0.3401	0.3381
45	0.3361	0.3341	0.3322	0.3303	0.3283	0.3264	0.3245	0.3226	0.3207	0.3188

APPENDIX 2 (Continued)

$$\frac{(1-R)^2}{2R} = \frac{k}{s}$$

R [%]	.0	.1	.2	.3	.4	.5	.6	.7	.8	.9
46	0.3170	0.3151	0.3133	0.3114	0.3096	0.3078	0.3060	0.3042	0.3024	0.3006
47	0.2988	0.2971	0.2953	0.2936	0.2919	0.2901	0.2884	0.2867	0.2850	0.2833
48	0.2817	0.2800	0.2783	0.2767	0.2751	0.2734	0.2718	0.2702	0.2686	0.2670
49	0.2654	0.2638	0.2623	0.2607	0.2591	0.2576	0.2561	0.2545	0.2530	0.2515
50	0.2500	0.2485	0.2470	0.2455	0.2441	0.2426	0.2411	0.2397	0.2383	0.2368
51	0.2354	0.2340	0.2326	0.2312	0.2298	0.2284	0.2270	0.2256	0.2243	0.2229
52	0.2215	0.2202	0.2189	0.2175	0.2162	0.2149	0.2136	0.2123	0.2110	0.2097
53	0.2084	0.2071	0.2058	0.2046	0.2033	0.2021	0.2008	0.1996	0.1984	0.1971
54	0.1959	0.1947	0.1935	0.1923	0.1911	0.1899	0.1888	0.1876	0.1864	0.1852
55	0.1841	0.1829	0.1818	0.1807	0.1795	0.1784	0.1773	0.1762	0.1751	0.1740
56	0.1729	0.1718	0.1707	0.1696	0.1685	0.1675	0.1664	0.1653	0.1643	0.1632
57	0.1622	0.1612	0.1601	0.1591	0.1581	0.1571	0.1561	0.1551	0.1541	0.1531
58	0.1521	0.1511	0.1501	0.1491	0.1482	0.1472	0.1462	0.1453	0.1443	0.1434
59	0.1425	0.1415	0.1406	0.1397	0.1388	0.1378	0.1369	0.1360	0.1351	0.1342
60	0.1333	0.1324	0.1316	0.1307	0.1298	0.1289	0.1281	0.1272	0.1264	0.1255
61	0.1247	0.1238	0.1230	0.1222	0.1213	0.1205	0.1197	0.1189	0.1181	0.1173
62	0.1165	0.1157	0.1149	0.1141	0.1133	0.1125	0.1117	0.1109	0.1102	0.1094
63	0.1087	0.1079	0.1071	0.1064	0.1056	0.1049	0.1042	0.1034	0.1027	0.1020
64	0.1012	0.1005	0.0998	0.0991	0.0984	0.0977	0.0970	0.0963	0.0956	0.0949
65	0.0942	0.0935	0.0929	0.0922	0.0915	0.0909	0.0902	0.0895	0.0889	0.0882
66	0.0876	0.0869	0.0863	0.0856	0.0850	0.0844	0.0838	0.0831	0.0825	0.0819
67	0.0813	0.0807	0.0800	0.0794	0.0788	0.0782	0.0776	0.0771	0.0765	0.0759
68	0.0753	0.0747	0.0741	0.0736	0.0730	0.0724	0.0719	0.0713	0.0707	0.0702
69	0.0696	0.0691	0.0685	0.0680	0.0675	0.0669	0.0664	0.0659	0.0653	0.0648
70	0.0643	0.0638	0.0633	0.0627	0.0622	0.0617	0.0612	0.0607	0.0602	0.0597
71	0.0592	0.0587	0.0582	0.0578	0.0573	0.0568	0.0563	0.0559	0.0554	0.0549
72	0.0544	0.0540	0.0535	0.0531	0.0526	0.0522	0.0517	0.0513	0.0508	0.0504
73	0.0499	0.0495	0.0491	0.0486	0.0482	0.0478	0.0473	0.0469	0.0465	0.0461
74	0.0457	0.0453	0.0449	0.0444	0.0440	0.0436	0.0432	0.0428	0.0424	0.0421
75	0.0417	0.0413	0.0409	0.0405	0.0401	0.0398	0.0394	0.0390	0.0386	0.0383
76	0.0379	0.0375	0.0372	0.0368	0.0365	0.0361	0.0357	0.0354	0.0350	0.0347
77	0.0344	0.0340	0.0337	0.0333	0.0330	0.0327	0.0323	0.0320	0.0317	0.0313
78	0.0310	0.0307	0.0304	0.0301	0.0298	0.0294	0.0291	0.0288	0.0285	0.0282
79	0.0279	0.0276	0.0273	0.0270	0.0267	0.0264	0.0261	0.0259	0.0256	0.0253
80	0.0250	0.0247	0.0244	0.0242	0.0239	0.0236	0.0233	0.0231	0.0228	0.0225
81	0.0223	0.0220	0.0218	0.0215	0.0213	0.0210	0.0207	0.0205	0.0202	0.0200
82	0.0198	0.0195	0.0193	0.0190	0.0188	0.0186	0.0183	0.0181	0.0179	0.0176
83	0.0174	0.0172	0.0170	0.0167	0.0165	0.0163	0.0161	0.0159	0.0157	0.0154
84	0.0152	0.0150	0.0148	0.0146	0.0144	0.0142	0.0140	0.0138	0.0136	0.0134
85	0.0132	0.0130	0.0129	0.0127	0.0125	0.0123	0.0121	0.0119	0.0118	0.0116
86	0.0114	0.0112	0.0110	0.0109	0.0107	0.0105	0.0104	0.0102	0.0100	0.0099
87	0.0097	0.0096	0.0094	0.0092	0.0091	0.0089	0.0088	0.0086	0.0085	0.0083
88	0.0082	0.0080	0.0079	0.0078	0.0076	0.0075	0.0073	0.0072	0.0071	0.0069
89	0.0068	0.0067	0.0065	0.0064	0.0063	0.0062	0.0060	0.0059	0.0058	0.0057
90	0.0056	0.0054	0.0053	0.0052	0.0051	0.0050	0.0049	0.0048	0.0047	0.0046

$$\frac{(1-R)^2}{2R} = \frac{k}{s}$$

R [%]	.0	.1	.2	.3	.4	.5	.6	.7	.8	.9
91	0.0045	0.0043	0.0042	0.0041	0.0040	0.0039	0.0039	0.0038	0.0037	0.0036
92	0.0035	0.0034	0.0033	0.0032	0.0031	0.0030	0.0030	0.0029	0.0028	0.0027
93	0.0026	0.0026	0.0025	0.0024	0.0023	0.0023	0.0022	0.0021	0.0020	0.0020
94	0.0019	0.0018	0.0018	0.0017	0.0017	0.0016	0.0015	0.0015	0.0014	0.0014
95	0.0013	0.0013	0.0012	0.0012	0.0011	0.0011	0.0010	0.0010	0.0009	0.0009
96	0.0008	0.0008	0.0008	0.0007	0.0007	0.0006	0.0006	0.0006	0.0005	0.0005
97	0.0005	0.0004	0.0004	0.0004	0.0003	0.0003	0.0003	0.0003	0.0002	0.0002
98	0.0002	0.0002	0.0002	0.0001	0.0001	0.0001	0.0001	0.0001	0.0001	0.0001
99	0.0001	0.0000	0.0000	0.0000	0.0000	0.0000	0.0000	0.0000	0.0000	0.0000

AUTHOR INDEX

I

J

K

L

M

N

O

SUBJECT INDEX

A

B

C

D

E

F

G

H

I

K

L

M

N

O

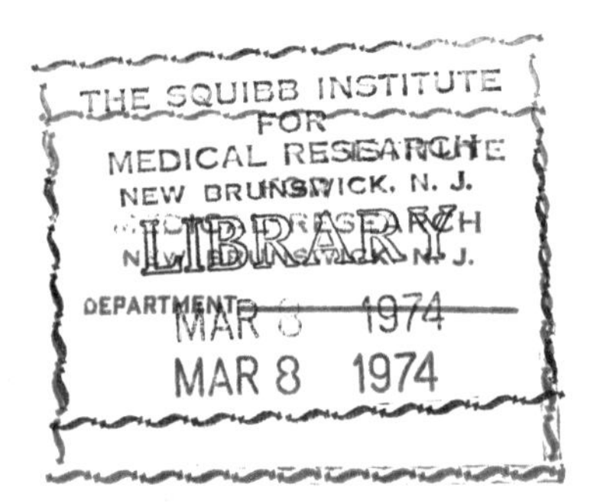